Lecture Notes in Networks and Systems 835

The series "Lecture Notes in Networks and Systems" publishes the latest developments in Networks and Systems—quickly, informally and with high quality. Original research reported in proceedings and post-proceedings represents the core of LNNS.

Volumes published in LNNS embrace all aspects and subfields of, as well as new challenges in, Networks and Systems.

The series contains proceedings and edited volumes in systems and networks, spanning the areas of Cyber-Physical Systems, Autonomous Systems, Sensor Networks, Control Systems, Energy Systems, Automotive Systems, Biological Systems, Vehicular Networking and Connected Vehicles, Aerospace Systems, Automation, Manufacturing, Smart Grids, Nonlinear Systems, Power Systems, Robotics, Social Systems, Economic Systems and other. Of particular value to both the contributors and the readership are the short publication timeframe and the world-wide distribution and exposure which enable both a wide and rapid dissemination of research output.

The series covers the theory, applications, and perspectives on the state of the art and future developments relevant to systems and networks, decision making, control, complex processes and related areas, as embedded in the fields of interdisciplinary and applied sciences, engineering, computer science, physics, economics, social, and life sciences, as well as the paradigms and methodologies behind them.

Indexed by SCOPUS, INSPEC, WTI Frankfurt eG, zbMATH, SCImago.

All books published in the series are submitted for consideration in Web of Science.

For proposals from Asia please contact Aninda Bose (aninda.bose@springer.com).

José Bravo · Gabriel Urzáiz
Editors

Proceedings of the 15th International Conference on Ubiquitous Computing & Ambient Intelligence (UCAmI 2023)

Volume 1

 Springer

Editors
José Bravo
MAmI Research Lab
Castilla-La Mancha University
Ciudad Real, Spain

Gabriel Urzáiz
Anahuac University
Mérida, Yucatán, Mexico

ISSN 2367-3370 ISSN 2367-3389 (electronic)
Lecture Notes in Networks and Systems
ISBN 978-3-031-48305-9 ISBN 978-3-031-48306-6 (eBook)
https://doi.org/10.1007/978-3-031-48306-6

This Springer imprint is published by the registered company Springer Nature Switzerland AG
The registered company address is: Gewerbestrasse 11, 6330 Cham, Switzerland

Paper in this product is recyclable.

Preface

Ubiquitous computing (UC) is a paradigm that allows software applications to obtain and process environmental information in order to make users feel that changes in the environment do not affect the functionality provided to them, thus aiming at making the technology invisible. UC is made possible by the confluence of computing, communication and control technologies, and it involves measuring and considering context variables (e.g., time and location of the users) to appropriately adapt the functionality of software applications to user needs.

On the other hand, ambient intelligence (AmI) represents functionality embedded into the environment that allows systematic and unattended sensing, as well as proactive acting, to provide smart services. Thus, AmI solutions frequently appear in domains of "being helped," e.g., social, psychological, healthcare and instrumental scenarios.

The UCAmI Conference presents advances in both AmI application scenarios and their corresponding technical support through UC. In doing this, this forum covers a broad spectrum of contributions from the conception of technical solutions to the assessment of the benefits in particular populations. The aim is for users of these smart environments to be unaware of the underlying technology, while reaping the benefits of the services it provides. Devices embedded within the environment are aware of the people's presence and subsequently react to their behaviors, gestures, actions and context.

During the last years, the interest in ubiquitous computing and ambient intelligence has grown considerably, due to new challenges posed by society, demanding highly innovative services for several application domains, such as vehicular ad hoc networks, ambient-assisted living, e-health, remote sensing, home automation and personal security. The COVID-19 pandemic has made us not only more aware of the ubiquity of information technologies in our daily life but also of their need to seamlessly support our everyday activities.

We are concerned with the sound development of UC and AmI as the only way to properly satisfy the expectations around this exciting intersection of information, communications and control technologies. Therefore, this UCAmI edition involves research work in five tracks: AmI for health and A3L, Internet of everything and sensors, smart environments, human–computer interaction and data science.

We received 107 submissions for this 15th edition of UCAmI authored by 207 researchers from 18 countries. A total of 297 reviews were performed, reaching the high average of 2.19 reviews per submission. We would like to thank all the authors who submitted their work for consideration, as well as the reviewers who provided their detailed and constructive reviews. Many thanks also to the track chairs for the great commitment shown in organization and execution of the papers reviewing process.

Finally, we are happy to return to the same place of the 2011 edition of the conference: Riviera Maya (Mexico).

November 2023   José Bravo
                                                                                          Gabriel Urzáiz

Organization

Our Staff

José Bravo (General Chair) Castilla-La Mancha University, Spain
Gabriel Urzaiz (Local Chair) Anahuac University, Mexico

Steering Committee

José Bravo, Spain
Pino Caballero, Spain
Macarena Espinilla, Spain
Jesús Favela, Mexico
Diego López-De-Ipiña, Spain
Chris Nugent, UK
Sergio F. Ochoa, Chile
Ramón Hervás, Spain
Gabriel Urzaiz, Mexico
Vladimir Villareal, Panama
Jesús Fontecha, Spain
Iván González, Spain

Organization Committee

Cosmin Dobrescu, Spain
David Carneros, Spain
Laura Villa, Spain
Luis Cabañero, Spain
Tania Mondéjar, Spain
Esperanza Johnson, Spain
Alejandro Pérez, Spain
Brigitte Nielsen, Panama
Paloma Bravo, Spain

Track Chairs

AmI for Health & (A3L) (Ambient, Active & Assisted Living)

Jesús Fontecha, Spain
Ian Cleland, UK

Smart Environment

Macarena Espinilla, Spain
Kåre Synnes, Sweden
Chris Nugent, UK

Internet of Everything (IoT + People + Processes) and Sensors

Joaquín Ballesteros, Spain
Cristina Santos, Portugal

Data Science

Marcela Rodríguez, Mexico
Alberto Morá, Mexico

Human–Computer Interaction

Gustavo López, Costa Rica
Sruti Subramanian, Norway

Satellite Events

International Workshop on Energy Aware Systems, Communications and Security

Mauro Migliardi, Italy
Francesco Palmieri, Italy

Program Committee

Adrian Lara	UCR
Adrian Sánchez-Miguel Ortega	Universidad de Castilla-La Mancha
Alberto Morán	UABC

Alejandro Pérez Vereda	Universidad de Castilla-La Mancha
Alessio Merlo	University of Genova
Alireza Souri	Haliç University
Allan Berrocal	Universidad de Costa Rica
Andres Diaz Toro	UNAD
Andrés Oliva	Anáhuac Mayab University
Antonio Robles-Gomez	UNED
Antonio Albín Rodríguez	Universidad de Jaén
Arcangelo Castiglione	University of Salerno
Arfat Ahmad Khan	Department of Computer Science, College of Computing, Khon Kaen University
Beatriz Garcia-Martinez	Universidad de Castilla-La Mancha
Borja Bordel	Universidad Politécnica de Madrid
Bruno Carpentieri	University of Salerno
Carlo Ferrari	University of Padova
Carlos Rovetto	Universidad Tecnológica de Panamá
Carlos Aguilar Avelar	UABC
Carlos E. Galván	Universidad Autónoma de Zacatecas
Carmelo Militello	Italian National Research Council (CNR)
Carmen Martinez Cruz	University of Jaen
Chris Nugent	Ulster University
Colin Shewell	Ulster University
Constantin Cosmin Dobrescu	Universidad de Castilla-La Mancha
Cristiana Pinheiro	University of Minho
Cristina Santos	University of Minho
Cristina Ramirez-Fernandez	TecNM/I.T. de Ensenada
David Carneros-Prado	Universidad de Castilla-La Mancha
David Gil	University of Alicante
Davide Zuccarello	Politecnico di Milano
Dionicio Neira Rodado	Universidad de la Costa
Eduardo Barbará	Anáhuac Mayab University
Elena Navarro	Universidad de Castilla-La Mancha
Ernesto Lozano	CICESE
Ernesto Vera	UABC
Esperanza Johnson	Høgskolen i Innlandet
Fabio Lopes	Universidade Presbiteriana Mackenzie
Fabio Salice	Politecnico di Milano
Federico Cruciani	Ulster University
Federico Botella	UMH
Francesco Palmieri	University of Salerno
Francisco Flórez-Revuelta	University of Alicante
Francisco Javier Cabrerizo	University of Granada

Gabriel Urzaiz	Anahuac University
Gerardo Alonzo	Anahuac Mayab University
Gianni D'Angelo	University of Salerno
Gilberto Borrego Soto	Instituto Tecnológico de Sonora
Gustavo Lopez	University of Costa Rica
Hadi Moradi	School of ECE, UT
Helena Gómez Adorno	IIMAS-UNAM
Higinio Mora	University of Alicante
Ian McChesney	Ulster University, School of Computing
Ian Cleland	Ulster University
Idongesit Ekerete	Ulster University
Ignacio Diaz-Oreiro	Universidad de Costa Rica
Iker Pastor López	University of Deusto
Inmaculada Ayala	Universidad de Málaga
Irvin Hussein Lopez-Nava	Centro de Investigación Científica y de Educación Superior de Ensenada
Iván González Díaz	Castilla-La Mancha University
Jan Havlík	Czech Technical University in Prague
Javier Medina Quero	University of Granada
Javier Ferrandez	University of Alicante
Javier Sanchez-Galan	Universidad Tecnologica de Panama
Jessica Beltrán	Universidad Autónoma de Coahuila
Jesus Fontecha	University of Castilla-La Mancha
Jesus Favela	CICESE
Jesús Peral	University of Alicante
Joana Figueiredo	University of Minho
João Lopes	University of Minho
Joaquin Ballesteros	Universidad de Malaga
Jordan Vincent	Ulster University
Jorge Eduardo Ibarra Esquer	Universidad Autonoma de Baja California
Jorge Rivera	Anáhuac Mayab University
Jose Bravo	Castilla-La Mancha University
José Luis López Ruiz	University of Jaén
Josef Hallberg	Luleå University of Technology
Karan Mitra	Luleå University of Technology
Kåres Synnes	Luleå University of Technology
Kristina Yordanova	University of Rostock
Kryscia Ramírez-Benavides	Universidad de Costa Rica
Laura Villa	University of Castilla-La Mancha
Lilia Muñoz	Universidad Tecnológica de Panama
Long-Hao Yang	Fuzhou University
Luigi Benedicenti	University of New Brunswick

Luis Cabañero	University of Castilla-La Mancha
Luis Quesada	Universidad de Costa Rica
Luis Pellegrin	UABC
Luís Moreira	University of Minho
Luis A. Castro	Instituto Tecnologico de Sonora (ITSON)
Macarena Espinilla	University of Jaén
Manuel Fernández Carmona	Universidad de Málaga
Marcela Rodriguez	UABC
María Martínez Pérez	University of A Coruña
Maria del Pilar Angeles	IIMAS, UNAM
Matias Garcia-Constantino	Ulster University
Matteo Venturelli	Politecnico di Milano
Matthew Burns	Ulster University
Mauro Migliardi	Universita' degli Studi di Padova
Mauro Iacono	Campania University L. Vanvitelli
Mercedes Amor	Universidad de Málaga
Michele Mastroianni	University of Salerno
Muhammad Asif Razzaq	Fatima Jinnah Women University
Nuno Ferrete Ribeiro	University of Minho
Oihane Gómez-Carmona	University of Deusto
Rafael Pastor-Vargas	UNED
Ramón Hervás	University of Castilla-La Mancha
Rene Navarro	University of Sonora
Rosa Arriaga	Georgia Tech
Rubén Domínguez	Anáhuac Mayab University
Saguna Saguna	Luleå University of Technology
Salvatore Source	Università degli Studi di Enna "Kore"
Sandra Nava-Munoz	UASLP
Sara Comai	Politecnico di Milano
Sara Cerqueira	University of Minho
Sergio Ochoa	University of Chile
Solomon Sunday Oyelere	Luleå University of Technology
Sruti Subramanian	Norwegian University of Science and Technology
Tania Mondéjar	University of Castilla-La Mancha
Tatsuo Nakajima	Waseda University
Unai Sainz Lugarezaresti	University of Deusto
Vaidy Sunderam	Emory University
Vincenzo Conti	University of Enna KORE
Vladimir Villarreal	Universidad Tecnológica de Panamá
Wing W. Y. Ng	South China University of Technology
Yarisol Castillo	Universidad Tecnológica de Panamá
Zaheer Khan	University of the West of England, Bristol

Contents

Human-Computer Interaction

About the Editors

Dr. José Bravo is Full Professor in Computer Science in the Department of Technologies and Information Systems at Castilla-La Mancha University, Spain, and Head of the Modelling Ambient Intelligence Research Group (MAmI, mami- lab.eu). He is involved in several research areas such as ubiquitous computing, ambient intelligence, ambient assisted living, context-awareness, Internet of things, mobile computing and m-Health. He is an author of over 37 JCR articles and the main researcher on several projects. H-Index (Scopus). - 22, H-Index (Google Scholar). – 30. Dr. Bravo supervised 7 PhD and over 45 Computer Science undergraduate theses. Since 2003, José Bravo has been the organizer of the International Conference on Ubiquitous Computing & Ambient Intelligence (UCAmI).

Dr. Gabriel Urzáiz received his BS in Computer Engineering at the National Autonomous University of Mexico and his PhD in Advanced Computer Technologies at the Castilla-La Mancha University. His research activity is focused on computer networks, mainly for the integration of heterogeneous networks and their application in ambient intelligence and ubiquitous computing. He is the author of several conference and journal papers, and he has also participated as a reviewer and guest editor. He has been the director of the Computer Science School of the Anahuac Mayab University in Mexico, a research professor and a postgraduate academic coordinator. His current position is as a full-time professor in the Engineering and Exact Sciences Division, primarily focused on teaching and student mentoring.

AmI for Health & (A3L) (Ambient, Active & Assisted Living)

A Computational Model for Agents in a Social Context: An Approach Based on Theory of Mind

Luis Zhinin-Vera⬤, Víctor López-Jaquero⬤, Elena Navarro$^{(\boxtimes)}$⬤, and Pascual González⬤

LoUISE Research Group, University of Castilla-La Mancha, 02071 Albacete, Spain
{Luis.Zhinin, VictorManuel.Lopez, Elena.Navarro, Pascual.Gonzalez}@uclm.es

Abstract. Socialization should be considered in any Ambient Intelligent system (AmI) where several persons interact, and specially in mental health-oriented ones. A crucial aspect of human social interaction and understanding is the Theory of Mind (ToM), which involves the ability to comprehend and predict the mental state of others, being critical for successful social functioning. The hypothesis of this paper is that the use of ToM to develop Multi-Agent Systems (MAS) to support AmI will enable modeling and reasoning about the mind of the people interacting in the system, making the agent that embodies a person more effective, efficient and social-capable than agents lacking such capacity. This paper presents a computational model based on ToM to support the reasoning and decision-making of socio-cognitive agents. The model applies the fundamental principles of ToM and Belief-Desire-Intention (BDI) agent model to develop a multi-agent system adaptable to real-world scenarios. The case study focuses on School Bullying, for which social theories validate the behavior of the proposed socio-cognitive agents. The results provide a suitable framework integrating advanced technologies with the necessary AI capabilities for the development of social-aware Ambient Intelligent systems.

Keywords: social agents · social context · theory of mind · BDI

1 Introduction

Ambient Intelligent systems (AmI) that facilitate human social interaction and understanding have gained significant attention in recent years [8]. These systems aim to create intelligent environments that seamlessly integrate technology with the physical surroundings, enabling enhanced communication, social interaction, and information exchange among individuals. One crucial aspect of successful functioning of social interaction is the ToM, which enables individuals to comprehend and predict the mental states of others [9]. The importance of ToM extends across various domains, including psychology and education. In psychology settings, ToM plays a crucial role in understanding social cognition,

J. Bravo and G. Urzáiz (Eds.): UCAmI 2023, LNNS 835, pp. 3–14, 2023.
https://doi.org/10.1007/978-3-031-48306-6_1

empathy, and ToM development in individuals across different age groups. Moreover, in the educational field, the ability to interpret and anticipate the mental states of students and teachers is essential for fostering effective communication, personalized learning experiences, and collaborative problem-solving.

This paper presents a computational model based on ToM and the BDI agent model, aiming to develop a multi-agent system adaptable to real-world scenarios. The hypothesis driving this research is that incorporating ToM into MAS within a BDI approach can enhance the modeling and reasoning capabilities of socio-cognitive agents, resulting in more effective and socially capable agents. By leveraging the fundamental principles of ToM, which involve understanding others' beliefs, desires, and intentions, the proposed model enables agents to reason and make informed decisions in complex social contexts. The integration of ToM with the BDI approach provides a promising avenue for advancing the capabilities of socio-cognitive agents, fostering more natural and human-like interactions.

The proposed computational model is applied in the context of school bullying. Bullying is a prevalent issue that affects the well-being and social dynamics within educational environments. By applying ToM principles to the development of socio-cognitive agents, this research aims to provide insights into how these agents can detect, prevent, and address instances of bullying. Supporting the proposed model with social theories further enhances its credibility and applicability to real-world scenarios. Moreover, this study sets the stage for future work that will implement the multi-agent system to demonstrate its functionality and effectiveness in combating bullying incidents.

This work is structured as follows. Section 2 provides a background about the main concepts used in this work. Section 3 reviews related works concerning the combination of both approaches. Section 4 presents the proposed computational model that integrates ToM with BDI architecture. Section 5 proposes an application of the model in a case study of a bullying scenario. Finally, Sect. 6 outlines some conclusions and future work.

2   Background

2.1   Theory of Mind (ToM)

The well-known Theory of Mind (ToM) defines a fundamental ability in social interaction that refers to the cognitive ability to attribute mental states, such as *beliefs, desires, intentions,* and emotions, to oneself and others in order to understand and predict behavior [18]. ToM allows individuals to navigate complex social interactions and relationships, and is crucial for human social cognition. This ability begins to develop in the early stages of childhood, being aware that people's knowledge, beliefs, and goals are different from our own. For example, by observing John examining the contents of the refrigerator, we might infer that "John is hungry." Consequently, we might get up and offer him some food. The cognitive ability to empathize with and comprehend the perspectives of others confers a substantial evolutionary advantage to humans. This capacity enhances the adaptive interactions with the environment and fosters more effective cooperation and social cohesion among individuals [9]. This understanding enables individuals to navigate intricate social dynamics, foster mutual comprehension, and enhance collective well-being.

One critical aspect of ToM is the ability to recognize that other people may hold beliefs that differ from our own. This is known as the "false-belief" test, such as the Sally-Anne task, and is commonly used to evaluate ToM in children in different age ranges [2]. In the false-belief test, a child is presented with a story in which a character holds a false belief about a situation, and the child is then asked to predict how the character will behave.

The interplay between ToM and context-awareness is of paramount importance for the strategic modeling of an opponent. The consideration of contextual information becomes crucial in resolving uncertainties about plausible mental states of others. By taking into account the situational factors and the context in which individuals' thoughts, beliefs, and intentions unfold, one can better understand and predict their behavior. This contextual understanding enhances the accuracy of mental state attributions, allowing for more effective interaction and decision-making in social and strategic settings.

Two prominent theories have emerged to attempt to explain ToM: **Theory-Theory** (TT) and **Simulation Theory** (ST) (Sect. 4). These theories aim to elucidate the process by which individuals perceive and interpret the mental states of others. According to TT, individuals construct explicit theories about the minds of others based on observations and inferences [10]. On the other hand, ST posits that individuals simulate the mental states of others by adopting their perspectives and matching them with own mental states [13]. These concepts, by allowing a better understanding of the mental states of others, could prevent complex social situations such as **bullying** (Sect. 5) or facilitate interactions between people with Attention Deficit Hyperactivity Disorder (ADHD) or Autism Spectrum Disorder (ASD).

2.2 BDI Agent Architecture

Belief-Desire-Intention (BDI) models have been widely used for Multi-Agent Systems (MAS) to model the reasoning of agents that interact in an environment with other agents. These models provide a natural way to represent an agent's beliefs, desires, and intentions, which are essential for making decisions and executing actions in a dynamic and uncertain world [19]. In a BDI model, an agent's beliefs represent its knowledge about the world, desires represent its goals or preferences, and intentions represent its plans or strategies for achieving its goals, being all of them used for its reasoning process. A typical BDI agent has a goal base, plan base, plan library, and intentions, and those form the elements of its reasoning [12]. Moreover, the BDI approach allows the integration of new functionalities, such as Artificial Intelligence (AI) techniques, to raise the level of abstraction of agent programming, context sensitivity, flexible decision-making capabilities and a fruitful framework for an adaptation to new technologies [3]. Their popularity is on the rise across multiple fields, including robotics and game theory. In this work, a BDI model is used to represent and reason about mental attitudes mentioned in the previous Section.

3   Related Works

The integration of ToM into computational models has been extensively studied due to its significance for inferring beliefs, predicting behavior, and enhancing reasoning abilities. Some proposals have used different techniques such as Bayesian-ToM in a partially observable Markov decision process (POMDPs), Reinforcement Learning (RL), Inverse Reinforcement Learning (IRL), and MAS [9]. Although these techniques still have limitations, it is notable that they have potential for developing intelligent systems that model complex behaviors of human mind.

One of such ToM integration approaches is that using MAS and BDI model to enhance agents' social behavior capabilities. For instance, Bose et al. [6] have developed a formal model based on BDI concepts to evaluate an agent's reasoning about another agent's reasoning process. This proposal focuses on these concepts at two different levels of architecture. The first one uses BDI concepts within ToM to describe the reasoning of agents, while the second one uses these concepts to describe the metareasoning of one agent about the other agent. It does this using a high-level modeling language called LEADSTO, which describes dynamics in terms of direct temporal dependencies between state properties in successive states. This model has been evaluated in a case study on social manipulation, animal behavior and social anticipation [4,5].

Another proposal discusses the challenges of developing social non-player characters (NPCs) and highlights the importance of a model for character decision making in games [23]. A BDI model is proposed for implementing socially-aware characters based on ToM. Another article describes ToM agents relying on a combination of theory-theory (TT) and simulation theory (ST) [11]. The approach is illustrated as a virtual training system and the goal is to develop agents with a believable social behavior and provide a reasoning about their behavior. A later study compared agents without a ToM, TT-ToM agents with ST-ToM agents [12]. The results revealed that ToM agents outperformed non-ToM agents and that models using ST represent ToM better than TT models.

Recently, a cognitive model has been proposed that enables a robot to regulate its level of collaborative autonomy by integrating Delegation Theory, ToM, and BDI [7]. This decision-making system, implemented using the JaCaMo framework, enables the robot to effectively manage collaborative conflicts and adapt its social autonomy. Another proposal presents a formal semantics for using ToM in agent communication, using the BDI elements (beliefs, desires, intentions, plans) that are applicable with agent-oriented programming languages [16]. Later, this proposal was expanded for providing the agents' with capabilities for uncertainty considerations [22]. They also proposed a high-level approach for modeling deception in MAS using ToM principles, integrating elements from *Interpersonal Deception Theory* and *Information Manipulation Theory 2* to capture the dynamics between agents [21]. The results indicate that unintended deception can arise from certain agent dynamics.

Another proposal presents an algorithm that jointly infers mental states, and action schemas of a BDI agent by maximizing the rationality contained in observations [17]. They evaluated this algorithm on Planning Domain Description Language (PDDL) domains, and constructed a possible probability distribution of plans and observations for actor's model using prior assumptions, beliefs, and observations as evidence. Additionally, a ToM architecture for joint human-robot activities was proposed, which involves mediating joint activities (who will do what) through dialogue to specify commitments between human agent and the robot [24]. The proposed architecture was formalized and illustrated in a scenario where a human and a robot mediate the task of bringing the human medications.

The integration of ToM and BDI models has attracted significant attention in the pursuit of more intelligent and socially adept systems. This integration holds great promise for enhancing system effectiveness and enabling seamless integration with advanced technologies. However, some challenges have been identified, such as the absence of an architecture that correctly represents the ToM, replanning of actions, robustness in reasoning deviations, among others. This paper proposes to tackle the initial challenge by introducing a robust architectural framework that proficiently amalgamates and defines the internal ToM process.

4   Integrating Theory of Mind with the BDI Approach

The integration of ToM with the BDI approach in a computational model offers a promising opportunity to improve the capabilities of agent systems in understanding, simulating, representing and attributing the mental states of others. Within the BDI framework, different elements of ToM can be effectively incorporated, allowing agents to simulate and attribute mental states to other agents. For this reason, in this work an architecture that integrates ToM and BDI is presented to provide agents such facilities. Figure 1 depicts how this architecture adapts the BDI-approach including a *Theory of Mind Processing* component. As this Figure illustrates, the architecture enables the interplay between the key components of BDI and the cognitive processes associated with ToM, highlighting how these elements contribute to agents' enhanced ability to simulate, infer, and reason about the beliefs, desires, and intentions of their counterparts. Following, the elements that make up the architecture are presented:

- **Mental State:** This definition plays a fundamental role within the framework of ToM. It refers to internal representations of cognitive and emotional processes occurring in an individual's mind. Mental states attributed to other agents are conceived as unobservable, theoretical assumptions invoked to explain and predict behavior. In order to represent such Mental State, the BDI approach has been used as follows:
    - **Beliefs:** In ToM, beliefs are fundamental for comprehending the perspectives and expectations of others. To incorporate this concept into the BDI approach, agents maintain a belief base that includes not only their own beliefs but also beliefs attributed to other agents. This enables agents to reason about the mental states of others and anticipate their behaviors based on their beliefs.

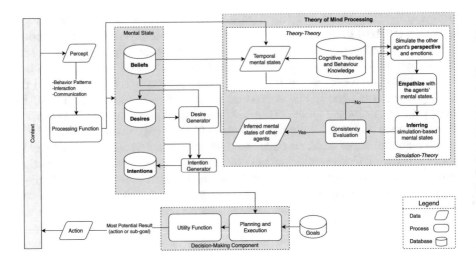

Fig. 1. The general BDI-ToM architecture

- **Desires and Intentions:** ToM recognizes desires and intentions as essential components for understanding the actions of others. BDI agents can address such understanding by having their own desires and intentions, as well as attributing mental states to other agents. This integration allows agents to better grasp the motivations and goals driving the behaviors of other agents, leading to more accurate modeling and decision-making.

- **Theory of Mind Processing Component:** The process of ToM is dynamic and involves revision as new information is acquired. This reflects a learning process that occurs through interactions between the mind and the environment. Theory-Theory (TT) and Simulation-Theory (ST) have been proposed for explaining ToM [1]. The integration of ToM and BDI paves the way for a more comprehensive understanding of how agents can model and interact with the mental states of others. For this reason, it is important to detail how the *Theory of Mind Processing* component can attribute mental states to support ToM. This component combines ST and TT in order to create a *hybrid model* in the processing component, aimed at obtaining a more comprehensive representation of the mental states of other agents. The general process is outlined in Algorithm 1 and is described as follows:

  - Theory-Theory. This explains that understanding other minds involves constructing an explicit symbolic theory with axioms and inference rules [10] where mental states are represented as inferred constructs derived from a naive theory. Through observation, and focusing on the behaviors of others, individuals are able to infer their possible mental states. Therefore, as Algorithm 1 states, TT uses cognitive theories and social behavioral knowledge, also incorporating the perceptions of the context and the initial beliefs of the agent, to generate possible mental states attributed to the other agents. As perceiving the context is an important part of the model, it is proposed to use a *Processing Function* as a context-aware system under the characteristics of an AmI system to obtain this context information [20].

- Simulation-Theory. It proposes that individuals simulate the mental states of others by using their own thoughts and feelings as models. They strive to *infer*, within their mental state, and *perspectives* of others, enabling them to better understand those mental states and consider them when making decisions. ST suggests that this is possible by adopting their perspective, tracking or matching their states with *resonant states* of one's own, being the later simulated mental states experienced by an observer while attempting to understand the mental states of another individual. This simulation process facilitates *empathy* and understanding of others' behaviors in terms of their own mental states and enhances the agents' ability to infer the intentions behind others' actions and react appropriately. Subsequently, as Algorithm 1 depicts, the output of TT acts as input for ST, which simulates, empathizes, and infers the definitive mental states of the agents. These inferred mental states are evaluated and then integrated into the agent's beliefs, ultimately updating their mental state.

- Consistency Verification. It validates the coherence of the generated mental states with observed sensory input and contextual cues, assessing their alignment with the goals of the agent using predefined parameters. It iterates to refine simulations and adjusts mental states if inconsistencies are detected. Otherwise, the generated mental states become definitive and are incorporated into agent's beliefs.

- **Decision-Making Component:** The agent's mental state, updated through inference of other agents' mental states carried out by the *Theory of Mind Processing* component, is integrated into a planning and execution mechanism. This mechanism employs system goals and a utility function to determine the actions or sub-goals necessary for goal satisfaction.

5   Case Study: Bullying Scenario

The following case study showcases the practical application of the presented architecture in the context of bullying, emphasizing its relationship with ToM. It has been detected that "pure" bullies exhibit advanced ToM skills, allowing them to effectively plan and execute their actions, or conversely, victims of bullying may exhibit deficits in these skills which perpetuates the cycle of victimization [14]. Therefore, applying ToM as an approach is viable to understand the bullying process and gain insights into how children perceive the behavior of individuals involved in the case study.

For developing this case study, the **Prometheus Methodology** [15], was used for designing a MAS for bulling intervention, as this methodology enables the development of socio-cognitive agents. It has been adapted for providing such

---

Algorithm 1: Theory of Mind Processing component

Input   : Agent's own Beliefs and Percepts
Output: Beliefs, desires, and intentions to the other agents based on the
           inferred mental states.
begin
   TT Subprocess:
      Inputs: context percepts, initial beliefs of the agent, cognitive
      theories, and knowledge about social behaviors.
      Output: temporary mental states of other agents.
   ST Subprocess:
   while *mental states are not consistent* **do**
           - Internal Simulation Section:;
              Inputs: temporary mental states from TT subprocess;
              Output: perspectives and emotions simulated.
           - Empathy Section:
              Inputs: internal information from other agents.
              Output: empathetic understanding of the mental states.
           - Inference Section:
              Inputs: understanding of TT's mental states.
              Output: refined possible mental states of other agents.
           - Consistency Evaluation:
              Inputs: mental states of other agents.
              Output: evaluation of consistency.
  return Inferred Mental States

---

agents with ToM capabilities, enhancing their ability to address bullying incidents effectively. The case study revolves around a bullying incident involving the teacher and student agents. It highlights the importance of timely intervention and a safe environment. It is also described how, by integrating the BDI and ToM framework, the student agent understands their own mental states, and empathizes with the victim. This allows for assessment, intent to help, and decision making that leads to teacher intervention and victim support. Due to the limited space constraints, some design details have been omitted in this paper.

5.1   System Specification

According to Prometheus, the first step is to identify the main goals of the MAS to capture the interactions and behaviors of its different agents. For this case study, the primary goals are to detect bullying and implementing interventions to mitigate their occurrence (Fig. 2). This ultimately aims to improve the understanding of the underlying mechanisms of bullying and contribute to the development of effective prevention and intervention strategies. Then, once the goals were identified, the *Context* of the MAS was defined as follows:

– A school setting with multiple classrooms, a playground, and common areas.
– Students with different backgrounds interacting in social groups.

5.2 Architectural Design

The *Agents* that make up the MAS were identified during this phase. One of the agent types is *Student* that may play 'Victim', 'Bully', and 'Observer' roles. The other agent type is *Teacher* who plays 'Intervener' role (Fig. 3). An explanation of the roles of the agents and their actions is detailed in Table 1. In the following, the computational model for the student agent with the role of Observer (Fig. 4) is designed. This agent perceives the behavior of other agents and, once understands their motivations, chooses its future actions based on the defined goals.

Table 1. Roles and actions

Role	Description	Actions
Victim(V)	A student who may become a target of bullying	Participates in school activities, interacts with peers, and responds to the behaviors of others
Bully(B)	A student engaging in bullying behaviors	Displays aggressive behaviors such as teasing, physical intimidation, exclusion, or spreading rumors
Observer(O)	An agent with a role in perceiving bullying incidents	Observes social interactions, monitors V and B behaviors, and notifies when bullying is perceived to be occurring
Intervener(I)	An agent responsible for addressing bullying incidents	Receives notification from the O and intervenes when necessary to address bullying and provide support to the V

5.3 Detailed Design

In this phase it is important to define the capabilities of the agents and the actions for their assigned roles. For this reason, the *Capability Overview* and the *ToM Processing Component* were defined as detailed next.

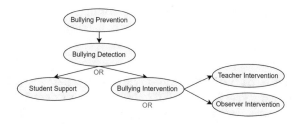

Fig. 2. Main Goals of the system

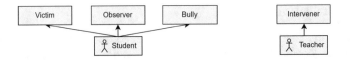

Fig. 3. Assignment of roles to agents of the MAS

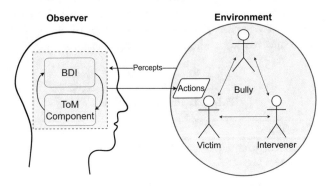

Fig. 4. Overview of the Agent with Observer Role

Capability Overview. It defines the plans and actions of the agents using the BDI architecture. For the bullying case study, it was defined as follows:

- Student Bully approaches Victim student, and starts verbally harassing or physically intimidating them.
- Victim student, feeling targeted and threatened, responds with signs of distress or attempts to defend themselves.
- Observer student, with *ToM Processing Component* capabilities, notices the interaction between Victim and Bully students.
- Applying ToM, Observer student understands that Victim student is experiencing distress due to the bullying behavior of Bully student.
- Observer student empathizes with Victim student, recognizing their perspectives and updates their own beliefs and desires about the situation.
- Based on the inferred mental states, Observer decides the appropriate interventions to address the bullying through a decision-making component to fulfill the simulation goals.
- Observer interprets and empathizes with Bully's behavior, addressing the issue while supporting Victim and reporting incidents to teacher agent when needed.

ToM Processing Component. This component is part of the capabilities of Observer, which runs the TT and ST subprocesses to detect bullying. In the following the subprocesses that made it up, as well their inputs and output, are detailed for the bullying case study:

1. *TT_subprocess(Percepts, Beliefs, Cognitive Theories) → Mental_states_temp*

- **Input:** *Percepts:* Sensory information, including observable behaviors of Victim and Bully. *Beliefs:* Initial beliefs about Victim and Bully's backgrounds and attitudes. *Cognitive Theories:* Behavior patterns based on cognitive theories.
- **Output:** Generate temporal Mental States for Victim and Bully, such as beliefs, intentions, and emotions.

2. *ST_subprocess(Mental_states_temp) → Inferred_Mental_States*

- **Input:** Receive the output of the TT subprocess including temporary mental states for Victim and Bully.
- **Output:** Infer Mental States by:
  - Simulating the perspectives of Victim and Bully within Observer's mental state.
  - Empathizing with Victim and Bully by mentally adopting their perspectives and experiencing resonant states.
  - Understanding the reasons why Agent Bully is bullying and why Agent Victim is susceptible to being bullied.

6   Conclusions

In conclusion, this work has demonstrated the potential of incorporating ToM into MAS for supporting social-aware AmI environments. The computational model presented in this study successfully applies the principles of ToM and the BDI agent model to develop socio-cognitive agents capable of reasoning and decision-making in real-world scenarios. The case study focused on bullying has shown the possibility of incorporating this computational model to promote social behaviors with the aim of reducing the presence of this problem in schools.

Moreover, as part of our future work, the integration of ToM and the BDI approach presented in this study will be implemented in a modeling and simulation framework for creating explicit agent-based simulations to demonstrate its functionality and effectiveness. This is important for researching and comparing the behavior of agents who do or do not have a ToM. These areas present exciting opportunities for advancing the capabilities of socio-cognitive agents and enhancing their social interaction and understanding, which would contribute to the development of more sophisticated and social-aware Ambient Intelligent systems, enabling them to effectively navigate and participate in various social contexts with human-like reasoning and decision-making abilities.

Acknowledgements. This paper is part of the R+D+i projects PID2019-108915RB-I00 and PID2022-140907OB-I00 and the grant PRE2020-094056 funded by MCIN/AEI/10.13039/501100011033. Moreover, it has been funded by the University of Castilla-La Mancha as part of the project 2022-GRIN-34436 and by 'ERDF A way to make Europe'.

References

1. Bach, T.: Structure-mapping: directions from simulation to theory. Philos. Psychol. **24**(1), 23–51 (2011)
2. Baron-Cohen, S., Leslie, A.M., Frith, U.: Does the autistic child have a "theory of mind"? Cognition **21**(1), 37–46 (1985)
3. Bordini, R.H., et al.: Agent programming in the cognitive era. In: AAMAS 2021, pp. 1718–1720 (2021)
4. Bosse, T., et al.: Modelling animal behaviour based on interpretation of another animal's behaviour. In: ICCM 2007, pp. 193–198 (2007)
5. Bosse, T., et al.: A recursive BDI-agent model for theory of mind and its applications. Appl. Artif. Intell. **25**, 1–44 (2011)
6. Bosse, T., et al.: A two-level BDI-agent model for theory of mind and its use in social manipulation (2007)
7. Cantucci, F., Falcone, R.: A computational model for cognitive human-robot interaction: an approach based on theory of delegation. In: WOA (2019)
8. Costa, A., et al.: Emotions detection on an ambient intelligent system using wearable devices. Future Gener. Comput. Syst. **92** (2018)
9. Cuzzolin, F., et al.: Knowing me, knowing you: theory of mind in AI. Psychol. Med. **50**(7), 1057–1061 (2020)
10. Gopnik, A., Wellman, H.: Why the child's theory of mind really is a theory. Mind Lang. **7**(1–2), 145–171 (1992)
11. Harbers, M., et al.: Agents with a theory of mind in virtual training. In: Multi-Agent Systems for Education and Interactive Entertainment, p. 172 (2010)
12. Harbers, M., et al.: Modeling agents with a theory of mind: theory–theory versus simulation theory. WIAS **10**, 331–343 (2012)
13. Harris, P.L.: From simulation to folk psychology: the case for development. Mind Lang. **7**(1–2), 120–144 (1992)
14. Kellij, S., et al.: The social cognitions of victims of bullying: a systematic review. Adolesc. Res. Rev. **7** (2022)
15. Padgham, L., Winikoff, M.: Prometheus: a methodology for developing intelligent agents. In: Giunchiglia, F., Odell, J., Weiß, G. (eds.) AOSE 2002. LNCS, vol. 2585, pp. 174–185. Springer, Heidelberg (2003). https://doi.org/10.1007/3-540-36540-0_14
16. Panisson, A.R., et al.: On the formal semantics of theory of mind in agent communication. In: International Conference on AT (2018)
17. Persiani, M., Hellström, T.: Inference of the intentions of unknown agents in a theory of mind setting. In: Advances in PAAMS, pp. 188–200 (2021)
18. Premack, D., Woodruff, G.: Does the chimpanzee have a theory of mind? Behav. Brain Sci. **1**(4), 515–526 (1978)
19. Rao, A.S., Georgeff, M.P.: BDI agents: from theory to practice. In: International Conference on Multiagent Systems (1995)
20. Roda, C., et al.: A multi-agent system for acquired brain injury rehabilitation in ambient intelligence environments. Neurocomputing **231** (2017)
21. Sarkadi, C., et al.: Modelling deception using theory of mind in multi-agent systems. AI Commun. **32**(4), 287–302 (2019)
22. Sarkadi, Ş., et al.: Towards an approach for modelling uncertain theory of mind in multi-agent systems. In: Agreement Technologies, pp. 3–17 (2019)
23. Sindlar, M., et al.: BDI-based development of virtual characters with a theory of mind, pp. 34–41 (09 2009)
24. Tewari, M., Persiani, M.: Towards we-intentional human-robot interaction using theory of mind. In: 5th Conference on CCHIRA. Sitepress (2021)

A Reinforcement Learning Algorithm for Improving the Generation of Telerehabilitation Activities of ABI Patients

Luis Zhinin-Vera[1]($\boxtimes$) (ID), Alejandro Moya[1] (ID), Elena Navarro[1]($\boxtimes$) (ID), Javier Jaen[2] (ID), and José Machado[3] (ID)

[1] LoUISE Research Group, University of Castilla-La Mancha, 02071 Albacete, Spain
{Luis.Zhinin,Alejandro.Moya,Elena.Navarro}@uclm.es
[2] Instituto Universitario Mixto de Tecnología de Informática, Universitat Politècnica de València, València, Spain
fjaen@upv.es
[3] Centro Algoritmi/LASI, University of Minho, Braga, Portugal
jmac@di.uminho.pt

Abstract. Acquired Brain Injury (ABI) is a condition caused by an injury or disease that disrupts the normal functioning of the brain. In recent years, there has been a significant increase in the incidence of ABI, highlighting the need for a comprehensive approach that improves the rehabilitation process and, thus, provides people with ABI with a better quality of life. Developing appropriate rehabilitation activities for these patients is a major challenge for experts in the field, as their poor design can hinder the recovery process. One way to address this problem is through the use of smart systems that generate such rehabilitation activities in an automatic way that can then be modified by therapists as they deem appropriate. This automatic generation of rehabilitation activities uses experts' knowledge to determine their suitability according to the patient's needs. The problem is that this knowledge may be ill-defined, hampering the rehabilitation process. This paper investigates the possibility of applying Deep Q-Networks, a Reinforcement Learning (RL) algorithm, to evolve and adapt that information according to the outcomes of the rehabilitation process of groups of patients. This will help minimize possible errors made by experts and improve the rehabilitation process.

Keywords: reinforcement learning · deep q-networks · acquired brain injury · multimodal gesture interaction

1 Introduction

An increase in the number of people with disabilities has occurred during the last decade, prompting agencies and governments around the world to pay more attention to their care. This fact was noted in the 2022 Convention on the Rights of Persons with Disabilities (CRPD) [15], that emphasizes the need for protection

J. Bravo and G. Urzáiz (Eds.): UCAmI 2023, LNNS 835, pp. 15–26, 2023.
https://doi.org/10.1007/978-3-031-48306-6_2

and care for persons with disabilities. Among this group, people with acquired brain injury (ABI) are of particular interest. ABI is caused by a bump, blow, or jolt to the head or a penetrating head injury, or including brain tumors and vascular degeneration [12]. In particular, the World Health Organization (WHO) expects ABI to become a trend, with some 10 million cases annually [7]. According to the Castilla-La Mancha Acquired Brain Injury Association (ADACE) [1], one of the biggest challenges in the rehabilitation of ABI patients is the wide range of disabilities they may suffer, People with ABI may suffer up to 10 different physical or motor deficits, 36 different cognitive or intellectual deficits, and 5 different behavioural or emotional deficits with different acuteness depending on the area of and severity of the damage (see [10] for a detailed description). There is a wide diversity of types of rehabilitation activities for the recovery of people with ABI, that are usually classified according to the main area they rehabilitate. In a previous work [10], it was presented that activities for rehabilitating cognitive functions can be grouped into Front executive, Attentional, Mnesic, Constructive and Linguistic [10]. In this work, it was also detailed which features are necessary for the design of the rehabilitation activities.

In order to ensure a good design of rehabilitation activities for ABI patients, the use of tools that automate the process is a must. In a previous study, Moya et al. [11], based on the ideas presented in [10], presented a proposal for automatically generating *association activities* which are used, among others, to rehabilitate front executive functions. An association activity consists of asking patients to associate elements or ideas that have an implicit relationship among them. The association activities generated can be run using the MRehabilitation tool [13]. This includes the typical features of any association activity, such as which elements to use (textual, graphical, etc.), the size, the rotation, etc. However, MRehabilitation was developed to offer a Kinect-based multimodal gesture interaction in order to simultaneously rehabilitate physical and cognitive deficits. For this reason, when creating an association activity with MRehabilitation other features are also required, such as the location in the space or which body parts will be used for the interaction. This means that all these features must be considered while designing a rehabilitation activity in order to define its most suitable configuration for the rehabilitation of a specific patient, depending on his or her deficits, needs, etc.

Moya et al. [11] developed a genetic algorithm that generates association rehabilitation activities for patients with ABI, taking into account their physical/cognitive characteristics. In order to be able to assess the suitability of each activity generated by the algorithm, expert knowledge was used and defined in what is called a *contribution map*. This determines how each feature that must be set while creating an association activity can contribute either positively or negatively to rehabilitating each deficit that people with ABI may suffer. However, this expert's knowledge may be subject to error, as it may indicate that a certain feature is good for a specific deficit when this is not correct. This led us to define the following research question: **RQ**: *is it possible to develop a Reinforcement Learning algorithm able to learn and evolve such knowledge according* *to the outcomes of the rehabilitation process?* This would facilitate that, as the rehabilitation process of the patients progresses, such knowledge would change to adapt accordingly to the generation of activities. As far as we know, this is the first proposal that is both able to generate automatically rehabilitation activities according to the needs of people with ABI, and evolve this generation process.

The paper has been structured as follows: Sect. 2 describes the main concepts used in this work; Sect. 3 presents the proposed algorithm to generate the telerehabilitation activities; Sect. 4 details the experiments performed and results obtained; and finally, Sect. 5 outlines some conclusions and future work.

2   Background: Reinforcement Learning

Reinforcement Learning (RL) is a type of *Machine Learning* method that guides an agent in learning how to behave in an environment to maximize its performance. Unlike supervised and unsupervised learning, which rely on specific datasets, RL operates on the basis of rewards and penalties received by the agent for its actions. The agent's goal is to learn the optimal policy, which determines the best action to take in each state to obtain the maximum reward [14]. RL uses two key concepts: exploration and exploitation [8]. Exploration allows the agent to discover new information from the environment, while exploitation uses this information to find the optimal policy, refining its decision-making process. Through iterative interactions and learning from rewards and penalties, RL enables agents to continuously improve their performance on a given task.

The underlying mathematical framework in RL is the **Markov Decision Process** (MDP). This is used for modeling agents' decision-making when the environment is fully observable. The MDP [5] is defined as a tuple $(\mathcal{S}, \mathcal{A}, \mathcal{P}, \mathcal{R})$ where:

- $\mathcal{S}$ is a finite set of states $\mathcal{S} = \{s_1, s_2, ..., s_n\}$, $s_t \in \mathcal{S}$ the state at time t.
- $\mathcal{A}$ is a finite set of actions $\mathcal{A} = \{a_1, a_2, ..., a_n\}$, being $a_t \in \mathcal{A}$ the action executed at time t.
- $\mathcal{P}$ is a transition function $P(s, a, s')$ that determines the probability of transition from state s to state s' after performing action a, being $s, s' \in \mathcal{S}$.
- R is a reward function $R(s, a)$ that determines the reward of executing action a while in state s, being r_t the result of the reward function obtained at time t.

In the domain of RL, an *optimal* policy is defined as the policy that maximizes (or minimizes) a pre-defined objective function. Thus, RL algorithms aim to discover the optimal policy, $\pi^*(s)$, that maps states to actions, enabling agents to select the actions that maximize the overall expected reward.

One of the most widely-used RL algorithms is **Q-learning**. It is a model-free approach that allows an agent to learn an optimal policy through interaction with an unknown environment [14]. This algorithm, proposed in 1989 by Watkins [16], is defined as follows:

$$Q_{t+1}(s_t, a_t) \leftarrow Q_t(s_t, a_t) + \alpha \left[r_{t+1} + \gamma \max_a Q_t(s_{t+1}, a) - Q_t(s_t, a_t) \right] \quad (1)$$

At its core, Q-learning estimates the value of taking a specific action in a particular state, known as the Q-value. The Q-value represents the expected cumulative discounted reward that an agent can obtain by following a particular action policy. The Q-learning algorithm iteratively updates estimated Q-values based on observed rewards and expected future rewards to discover the optimal policy [2]. In each iteration, γ is used as a discount factor that determines the importance of future rewards compared to immediate rewards in the agent's decision-making process. This iterative process, known as Q-value iteration, refines the Q-values towards their true values. By selecting actions with the highest Q-values, Q-learning enables the agent to discover optimal policies.

While Q-learning is powerful in solving RL problems, challenges arise in complex applications, such as memory limitations, high-dimensional inputs, and convergence issues. Deep Q-Networks (DQNs) have emerged as a promising solution to address these challenges [9]. Neural networks and deep learning approaches possess powerful universal approximation capabilities, allowing them to approximate complex functions and capture intricate patterns in data [4]. For high-dimensional state-action problems in RL, **DQNs** arise from combining the power of Deep Neural Networks with the Q-learning algorithm.

In the DQN approach, a deep neural network replaces the traditional method of generating Q-value pairs. The DQN model, represented by a parameterized function $Q(s, a; \theta)$, maps states and actions to Q-values in the Q-learning algorithm [3]. By optimizing the model parameters θ through backpropagation, the DQN learns to approximate the optimal Q-values. This iterative process minimizes the discrepancy between predicted and actual Q-values, enabling the discovery of the optimal policy for the RL problem.

DQNs have succeeded in solving a wide variety of challenging problems, from playing an Atari game to controlling robotic systems [2,6]. By exploiting deep neural networks representational power, DQNs can effectively approximate the optimal action value function and enable agents to make intelligent decisions in complex, dynamic environments.

3   A Reinforcement Learning Algorithm for Generation of Telerehabilitation Activities

For developing a comprehensive understanding of the RL algorithm, the study introduces fundamental mathematical formulations to define the crucial components of the DQN. These mathematical expressions serve as the fundamental backbone of the model, allowing for precise definitions and the establishment of a robust theoretical framework. The following notations are defined:

F: set of features $F = \{F_1, F_2 \ldots, F_p\}$, being p the number of features that the association telerehabilitation activities may have.

F$_i$: set of feature values $F_i = \{f_{i1}, f_{i2}, \ldots, f_{iq_i}\}, being q_i$ is the number of values of feature F_i.

Ω: set of weights $\Omega = \{\omega_{F_1}, \omega_{F_2}, \ldots, \omega_{F_p}\}$ assigned by the therapist to each feature F_i.

AC: an association telerehabilitation activity configuration $AC = \{f_{1j_1}, f_{2j_2}, \ldots, f_{pj_p}\}$, $1 \leq j_i \leq q_i; 1 \leq i \leq p$

Cont$_{ij}$ contribution of feature value f_{ij}:

    1 strongly positive contribution

    0.5 positive contribution

    0 neutral contribution

    −0.5 negative contribution

    −1 strongly negative contribution

The overall benefit of an association telerehabilitation activity configuration **AC** is computed as:

$$\sum_{i=1}^{p}\sum_{j=1}^{q_i}\omega_{F_i} \cdot \mathbf{Cont_{ij}} \tag{2}$$

In order to effectively develop the DQN algorithm, it is crucial to define and understand the key components that are the foundation of the algorithm. These components lay the groundwork for the algorithm's execution and play a vital role in achieving optimal performance. The key components are outlined below:

– **States:** Set of all possible association telerehabilitation activity configuration $S = \{AC_1, AC_2, ..., AC_k\}$, being k the number of combinations. For this work, AC_k are generated using the features, and their corresponding values, that are shown in Table 1, as well as the contributions defined above.

– **Actions:** Set of possible actions $A = \{(i, v)|i, v \in Cont_{ij}]\}$. Each action (i, v) involves exchanging a certain contribution value i for a new value v, resulting in a transition to a new state.

– **Rewards:** The reward structure is designed to encourage the agent to move towards the optimal state. In our case, the reward has been computed as the Euclidean distance between the current state s_t and the optimal state s_o. If the states are identical, the agent receives the maximum reward.

$$r_t = 100 - (distance(s_t, s_o) * 10)$$

– **Neural Network Architecture:** The Q-network takes the state s_t as input and outputs the Q-values for all possible actions. The architecture includes the input layer, the hidden layers with a 192-dimensional fully connected layer, and the output layer (Fig. 1). All hidden layers are activated by rectified linear units (ReLUs). The parameters are updated by gradient descent.

– **Experience Replay:** Experience replay is used to enhance the learning process. It involves storing the agent's experiences, including transitions of states, actions, rewards, and next states, in a memory buffer.

– **Exploration-Exploitation Strategy:** An ε-greedy exploration strategy is employed, where the agent chooses a random action with a certain probability ϵ to explore, otherwise, the agents exploits the action with the highest Q-value.

Table 1. Set of features and values

Name | Value	0	1	2	3
Size	Little	Medium	Big	
Yaw	0°	90°	180°	270°
Distance	Alongside	Close	Far	
Transparency	25%	50%	75%	100%
Type of sign	Image	Textual	Iconic	
Highlight	False	True		
Draggable	False	True		
Hide	False	True		
Time	Short	Normal	Long	
Errors	Few	Normal	Many	
NumberPairs	Few	Normal	Many	

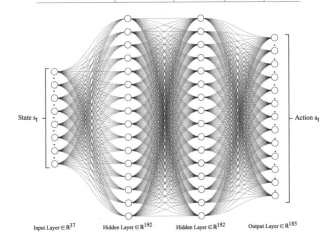

State s_t — Action a_t

Input Layer $\in \mathbb{R}^{37}$ Hidden Layer $\in \mathbb{R}^{192}$ Hidden Layer $\in \mathbb{R}^{192}$ Output Layer $\in \mathbb{R}^{185}$

Fig. 1. Architecture of the proposed Deep Q-Network

By defining and configuring these components according to our problem domain, the DQN algorithm can learn to optimize the state array configuration and achieve the maximum reward. These components collectively shape the learning process, enabling the agent to make informed decisions and converge towards the desired optimal state. The overview of the implementation of the DQN algorithm is presented in Algorithm 1. It uses a neural network as its action-value function approximator. The network was trained using a combination of a loss function and the stochastic *Gradient Descent* optimization algorithm. The loss function employed in our approach quantifies the disparity between the predicted Q-values and the target Q-values. Specifically, we utilize the Mean Squared Error (MSE) loss as our chosen loss function. Through gradient descent, the network's weights are iteratively adjusted to minimize this discrepancy, aligning the predicted Q-values more closely with the target values.

In order to strike a balance between exploration and exploitation, we implemented an epsilon-greedy strategy during the training process. Under this strategy, at each step, the agent has two possible courses of action. It may either select a random action with a probability represented by ϵ, or opt for the action with the highest Q-value, based on the current policy.

The crucial factor that determines the agent's decision-making process lies in the value of ϵ. When ϵ is set to a higher value, the agent is more inclined to explore and take random actions, thereby fostering a broader search of the action space. Conversely, when ϵ is lowered, the agent prioritizes exploitation by selecting actions with the highest Q-values, which are considered more likely to yield immediate rewards. Therefore, the parameter ϵ acts as a tuning knob, allowing us to control the degree of exploration versus exploitation within the learning process.

Algorithm 1: DQN for Generation of Telerehabilitation Activities

---

Initialize replay memory RM
Initialize action-value function Q with random weights
Initialize target action-value function Q_target with the same weights as Q
begin
  for *episode = 1 to M* **do**
    Initialize state $s_1 \in$ S
    for *t = 1 to T* **do**
      Select an action a_t based on the ε-greedy policy derived from Q
      Execute action $a_t \in$ A
      Observe the new state s' and reward r_t
      Observe whether the episode has terminated
      Store the transition (s_t, a_t, r_t, s') in replay memory RM
      Sample a random mini-batch of transitions (s_j, a_j, r_j, s'_j) from RM
      if *s' is a optimal state (s_o)* **then**
        $Q\_target(s_t, a_t) = r_j$
      else
        $Q\_target(s_t, a_t) = r_j + \gamma max_{a'} Q(s', a')$
      Train the Q network by minimizing the loss between Q and Q_target using Gradient Descent
      Every C steps, update Q_target with the weights from Q
      Set current state $s_{t+1} = s'$
    Decay ε according to the chosen schedule
  Save the current Q network weights
  return Trained Q network

As it can be observed in Algorithm 1, during training, the agent observed the current state s_t, selected an action a_t based on its policy, and received a reward r_t and the next state s'. These experiences were stored in a replay memory (RM), which allowed the agent to learn from past interactions, speeding up the process. The training process aimed to optimize the Q-values so that the agent could make informed decisions that maximized long-term rewards. As the algorithm progressed through episodes, it gradually learned to balance exploration and exploitation, resulting in improved performance over time.

4   Experiments and Results

The DQN algorithm proposed in this work (see Algorithm 1) is a non-interactive approach, that is, it does not generate contribution maps until a sufficient number of results have been accumulated to ensure the generation of reliable outcomes. To evaluate the algorithm's performance and refine its decision-making abilities, a significant amount of feedback is necessary. As the research question driving this proposal is to determine whether a RL algorithm may help evolve the contribution map, it was decided to evaluate the algorithm under a stressful condition. Consequently, the following null hypothesis was formulated to assess the algorithm's performance and validate its effectiveness:

H_0: *Is a reinforcement learning algorithm able to evolve a contribution map defined by experts with a low rate of errors (10%) but having a high impact on the generated rehabilitation activities?*

As it can be observed, only a 10% of errors were defined in the hypothesis, as experts are expected to have a proper background in their area. However, it was also defined in the hypothesis that the impact on the outcome was high in order to evaluate the proposed algorithm under unfavorable conditions where the rehabilitation activities proposed initially with the information given by experts are far from optimal for the ABI patients under consideration.

First, the proposed algorithm (see Algorithm 1) was trained by iteratively updating the Q-values of the agent's action-value function. This method was trained for a total of 500 episodes, with each episode consisting of a series of iterations until convergence was reached. Notably, the number of iterations gradually decreased over the course of training, indicating an improvement in the algorithm's performance as it learned to make more efficient decisions (Fig. 2). The hyperparameters used in the training process are summarized in Table 2.

Hence, the trained algorithm is executed 50 times using diverse initial states, in accordance with the formulated hypothesis. The aim of these iterations is to ascertain whether the algorithm can converge towards the optimal contribution maps within the specified maximum number of iterations. Additionally, the algorithm's computational complexity results in an average runtime of 120 min to converge to the optimal solution.

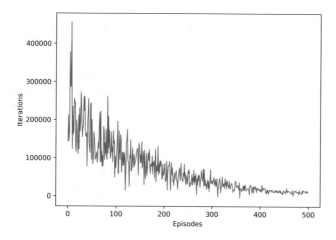

Fig. 2. Number of iterations in each episode to reach the maximum reward.

Table 2. Hyper-parameters used in the Deep Q-Learning Algorithm

Hyper-Parameter	Value
Episodes (M)	500
Replay Memory (RM) size	300
Minibatch size	64
Discount factor	0.1
Learning rate	0.01
Initial exploration rate	1.0
Final exploration rate	0.01
Number of hidden layers	2
Number of neurons in each layer	192
Max. Iteration	500000

After performing multiple runs, in order to evaluate the performance of the algorithm, we adopted the Euclidean distance as the evaluation metric. This measures the difference between the optimal state (the state that would best fit the condition of the ABI patients under consideration) and the resulting state after each of the 50 runs of the algorithm. Then, $D = \{d_1, d_2, ..., d_{50}\}$ is the set of the 50 differences between the resulting states and the optimal states. If $d_i = 0$, it means that the algorithm was able to offer an optimal solution. A descriptive analysis of the results is shown in Table 3, while Fig. 3 shows the frequency of the results.

In this work, the use of a statistical test is crucial to evaluating the effectiveness and significance of the results obtained from the application of the Deep Q-Learning algorithm in the rehabilitation process. By performing a statistical test, it is possible to determine whether the differences observed between the results of the algorithm and the expected results are statistically significant or whether these differences could be attributed to random chance.

The non-parametric Wilcoxon test is chosen because it does not assume a specific distribution of the data, making it suitable for analyzing non-normally distributed data. Given the complex nature of our rehabilitation outcomes and the potential for non-normal distributions, the Wilcoxon test provides a robust statistical approach to assessing the significance between groups. The test was performed using SPSS (Statistical Package for the Social Sciences), a set of data processing tools for statistical analysis. This test measures the magnitude of the differences and their directionality. The null hypothesis (H_0) assumes that the median of the differences is zero, implying no significant difference between the resulting states and the optimal states.

Table 3. Descriptive Statistics of set D

	N	Range	Minimum	Maximum	Mean	Mean	Std. Deviation	Variance
	Stat.	Stat.	Stat.	Stat.	Stat.	Std. Error	Stat.	Stat.
Distance	50	.5	.0	.5	.030	.0170	.1199	.014
Valid N	50							

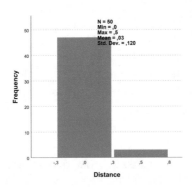

Fig. 3. Frequency of results from the set D

The results of our statistical analysis using the Wilcoxon test on the distances between states yielded a p-value of 0.083, indicating that we retain the null hypothesis. This suggests that there is no significant difference between the mean distance and zero. While the p-value is slightly above the conventional significance level ($\alpha = 0.05$), it is important to note that our findings provide suggestive evidence that the mean distances are close to zero. These results indicate a trend towards achieving optimal states in our algorithm. A summary of the Hypothesis test is shown in Table 4.

Table 4. Hypothesis Test Summary

Null Hypothesis	Test	Sig.[a,b]	Decision
The median of Distance equals .0	One-sample Wilcoxon Signed Rank Test	.083	Retain the null hypothesis

a. The significance level is .050

b. Asymptotic significance is displayed.

Finally, the analysis of the results and the application of the Wilcoxon test offer a robust and scientific approach to assessing the algorithm's performance. This statistical analysis enhances our understanding of the DQN algorithm's convergence capability and its potential to achieve an optimal state within the specified number of iterations.

5   Conclusions

This study's main goal was to look into how Deep Q-Learning, an RL algorithm, could be used to deal with the problems that come with making rehabilitation activities for people with ABI. The results obtained were highly promising and demonstrated the potential of using RL to evolve and adapt expert knowledge based on the outcomes of the rehabilitation process.

The purpose of the study was to address the issue of poorly defined knowledge in rehabilitation activities, which can hinder ABI patients' recovery. By utilizing Deep Q-Learning, the algorithm was able to learn and evolve the knowledge underlying the generation of activities, thereby reducing the number of mistakes made by experts and improving the rehabilitation process.

By evaluating the algorithm under extreme conditions, it was hypothesized that the reinforcement learning algorithm would provide contribution maps with mean scores close to the optimal when experts make few errors that have a significant impact on the outcome. This hypothesis seeks to ensure that the algorithm can effectively adapt the generation of activities as the rehabilitation process progresses, so that they remain aligned with the patients' evolving needs.

Deep Q-Learning's successful implementation in this context holds great promise for the field of ABI rehabilitation. The ability of the algorithm to learn from experience and continuously adapt the generation of activities paves the way for new personalized and effective rehabilitation interventions. Future research may explore and refine the use of reinforcement learning algorithms to optimize the rehabilitation process and enhance the quality of life for patients with gathered brain injury.

Acknowledgements. This paper is part of the R+D+i projects PID2019-108915RB-I00 and PID2022-140907OB-I00, and the grant PRE2020-094056 funded by MCIN/AEI/ 10.13039/501100011033. It has also been funded by the University of Castilla-La Mancha as part of the project 2022-GRIN-34436, by 'ERDF A way to make Europe' and thanks to the PhD scholarship 2019-PREDUCLM-10772 (co-financed by the FSE Operational Programme 2014–2020 of Castilla-La Mancha through Axis 3).

References

1. ADACE CLM: ADACE - Association of ABI of Castilla - La Mancha. https://www.adaceclm.org/
2. Agostinelli, F., Hocquet, G., Singh, S., Baldi, P.: From reinforcement learning to deep reinforcement learning: an overview. In: Rozonoer, L., Mirkin, B., Muchnik, I. (eds.) Braverman Readings in Machine Learning. Key Ideas from Inception to Current State. LNCS (LNAI), vol. 11100, pp. 298–328. Springer, Cham (2018). https://doi.org/10.1007/978-3-319-99492-5_13
3. Bidgoly, A.J., Arabi, F.: Robustness evaluation of trust and reputation systems using a deep reinforcement learning approach. Comput. Oper. Res. **156**, 106250 (2023)
4. Hornik, K., Stinchcombe, M., White, H.: Multilayer feedforward networks are universal approximators. Neural Netw. **2**(5), 359–366 (1989)
5. Howard, R.A.: Dynamic Programming and Markov Processes. John Wiley (1960)
6. Huang, Y.: Deep Q-networks. In: Dong, H., Ding, Z., Zhang, S. (eds.) Deep Reinforcement Learning, pp. 135–160. Springer, Singapore (2020). https://doi.org/10.1007/978-981-15-4095-0_4
7. Hyder, A.A., Wunderlich, C.A., Puvanachandra, P., Gururaj, G., Kobusingye, O.C.: The impact of traumatic brain injuries: a global perspective. NeuroRehabilitation **22**(5), 341–53 (2007)
8. Kaelbling, L.P., Littman, M.L., Moore, A.W.: Reinforcement learning: a survey. J. Artif. Intell. Res. **4**, 237–285 (1996)
9. Mnih, V., et al.: Human-level control through deep reinforcement learning. Nature **518**(7540), 529–533 (2015)
10. Montero, F., López-Jaquero, V., Navarro, E., Sánchez, E.: Computer-aided relearning activity patterns for people with acquired brain injury. Comput. Educ. **57**(1), 1149–1159 (2011)
11. Moya, A., Navarro, E., Jaén, J., López-Jaquero, V., Capilla, R.: Exploiting variability in the design of genetic algorithms to generate telerehabilitation activities. Appl. Soft Comput. **117**, 108441 (2022)
12. Network, T.A.: Definition of ABI (2019). http://www.abinetwork.ca/definition
13. Oliver, M., Teruel, M., Molina, J., Romero-Ayuso, D., González, P.: Ambient intelligence environment for home cognitive telerehabilitation. Sensors **18**(11), 3671 (2018)
14. Sutton, R.S., Barto, A.G.: Reinforcement Learning: An Introduction. MIT Press (2018)
15. UN: Convention on the Rights of Persons with Disabilities (2022). https://www.un.org/development/desa/disabilities/convention-on-the-rights-of-persons-with-disabilities.html
16. Watkins, C.J.C.H.: Learning from delayed rewards. King's College, Cambridge United Kingdom (1989)

Monitoring Environments with New Generation Devices

Alicia Montoro Lendínez[1]([✉])⬤, José Luis López Ruiz[1]⬤,
María Paz Barbero Rodríguez[1], David Díaz Jiménez[1]⬤, Chris Nugent[2]⬤,
and Macarena Espinilla Estévez[1]⬤

[1] Department of Computer Science, University of Jaén, Jaén, Spain
amlendin@ujaen.es
[2] School of Computing, Ulster University, Belfast, UK

Abstract. Currently, a number of studies have found that high levels of carbon dioxide in an enclosed space increase the probability of COVID-19 infection and can have adverse health effects. The control of the levels of certain gases, such as carbon dioxide, has become significantly relevant during the period of the COVID-19 pandemic. Therefore, this study presents a system for indoor air quality monitoring and counting in an enclosed space. Furthermore, the system has been evaluated in a case study with different scenarios and it is analysed how the ventilation affects the air quality levels in the enclosed space. For example, the raw carbon dioxide value obtained from the MQ135 sensor is 63 without any person in the room compared to the raw value of 148 when there are 4 people. All this is done using the Internet of Things paradigm and the implementation of intelligent ambients.

Keywords: Intelligent ambients · Internet of Things · Monitoring system · Low-cost development board

1 Introduction

COVID-19 has been a major focus of global attention since its emergence. From its early stages, research focused on how the virus was transmitted in order to prevent its massive spread. These studies have concluded that a healthy person can contract the disease through the inhalation of aerosols containing the virus, as they remain suspended in the air for extended periods of time [20].

The higher the concentration of aerosols in an enclosed space, the higher the viral load and therefore the greater the possibility of infection. Therefore, good ventilation, together with maintaining an adequate safety distance in enclosed spaces, has become a key measure to control the percentage of aerosols suspended in the air and thus reduce the spread of the virus [2].

From another perspective, several studies have shown that high levels of carbon dioxide (CO_2) in the environment can have negative effects on health,

Grant PID2021-127275OB-I00 funded by MCIN/AEI/10.13039/501100011033 and by "ERDF A way of making Europe".

affecting both personal performance and physical fitness. In particular, the study by Satish et al. [18] looked at the impact on decision-making and physical health (such as headaches and dizziness) or physiological health (such as increased blood pressure). Therefore, it is essential to maintain adequate ventilation in enclosed spaces to prevent the accumulation of CO_2. However, in cases such as offices or classrooms with a high concentration of people, keeping windows permanently open may not be feasible due to energy savings in heating or air conditioning [6].

Consequently, indoor air monitoring has a dual objective: to control the quality through CO_2 saturation and viral load. For CO_2 monitoring, the use of new technologies to obtain, analyse and visualise the data is important. Accordingly, the aim of this work is to address the current problem in relation to air quality and propose a potential solution through the monitoring of intelligent ambients by leveraging the IoT paradigm.

The paper is structured as follows. Section 2 presents a review of existing works on current technologies related to sensorisation in gases and intelligent ambients. Section 3 presents the system architecture and data collection. Section 4 shows the case studies carried out and their corresponding results. Finally, Sect. 5 presents the conclusions and the contribution provided.

2   Related Works

This section reviews intelligent ambients and gas monitoring, both of which are considered fundamental applications of the IoT and are the foundation of this paper.

It should be noted that the concept of IoT emerged in 1999, coined by computer engineer Kevin Ashton, who at that time was working in the field of radio frequency identification (RFID) [16]. According to Gartner, IoT is defined as a network of physical objects that incorporate technology to communicate, sense or interact with their internal states or the external environment [1]. On the other hand companies such as Cisco point out that IoT refers to the time when there were more objects connected to the internet than people between 2008 and 2009 [7]. However, both definitions highlight the idea of things connected to the network with embedded technology for connectivity and data transmission.

The future of IoT is being realised in the concept of the Internet of Everything (IoE), which is defined as the networked interconnection of people, processes, data and objects, which benefits from the value generated by this extensive connection as everything is integrated online. This definition goes one step further, because while IoT is considered one technology transition, IoE encompasses multiple technology transitions, including IoT, and is emerging as one of the most important trends in technology today [8, 15]

Both IoT and IoE have applicability in various fields, such as domestic, business, industrial, health, education, among others [17]. In the context of this paper, the topic of intelligent ambients is addressed. Intelligent ambients are electronic environments that are sensorised and responsive to the presence of

a user, who remains unnoticed in the environment, with only the user interface being perceptible. This paradigm, known as Ambient Intelligence (AmI), is
based on persuasive or ubiquitous computing, profiling, context awareness and
human-centred interaction design [14]. An example of intelligent ambients is the
Smart Lab of the University of Jaén. This laboratory is characterised by the
integration of a wide range of sensors at various points, which allows its use in
multiple projects and applications [10].

An important component of intelligent ambients is gas monitoring, specifically in this work on CO_2. Gas monitoring involves the use of sensors for gas
detection. Each sensor has its own characteristics, such as sensitivity, accuracy,
ability to select between various gases, detection ranges, response time and recovery time, which determine the performance of the sensor in question.

In a literature review, it is possible to find several types of gas sensors, among
which, according to Z. Yunusa et al. [4]:

- Catalytic sensors. Detect toxic, flammable or combustible gases. They operate
  by a chemical reaction in which gases interact with a catalyst, generating an
  electrical current which is measured to determine the gas concentration [19].
- Thermal conductivity sensors. Detect the presence of gas by measuring the
  amount of heat that the gas transfers to the sensor. This device is composed
  of two elements: one is kept at a constant temperature and the other cools in
  response to the gas flowing over it. The temperature difference between the
  two elements is proportional to the thermal conductivity of the gas, which
  allows the concentration of the gas to be determined. They are simple but
  robust sensors, however, they require a heating wire [5].
- Electrochemical sensors. Work by generating an electrical current from a
  chemical reaction between the gas to be detected and an electrode. The electrode is coated with a gas-sensitive material that acts as a catalyst for the
  chemical reaction. As the gas comes into contact with the sensitive material,
  an electrochemical reaction takes place which produces an electrical current
  proportional to the gas concentration [12].
- Optical sensors. Measure the absorption of light at a specific wavelength by
  the gas to be detected. The sensor contains a light source and a light detector, and the gas is placed between the two. When light is emitted from the
  source, it passes through the gas and reaches the detector. The amount of
  light reaching the detector is reduced by the amount of gas present, as the
  gas absorbs some of the light. The decrease in the amount of light detected is
  converted into an electrical signal that can be measured to determine the gas
  concentration. This type of sensor is commonly used in gas detection applications, such as measuring carbon monoxide in air. Its main disadvantage
  is the influence that ambient light can have, however, it is not sensitive to
  electromagnetic interference and the detection area is very large [21].
- Infrared sensors. These are a sub-type of the optical sensors mentioned in the
  previous point. In this case, they are based on the emission of an infrared
  light source. These sensors are highly sensitive and can detect gases even at
  low concentrations [11]. And there is variety on the market [3].

– Semiconductor sensors. Based on a chemical reaction between the gases and the semiconductor material, which is usually a tin oxide layer, producing a change in the electrical resistance of the sensor. This change is measured and used to determine the gas concentration. Advantages include speed of response, low cost, high sensitivity and ease of use [9].
– Acoustic wave sensors. Work by measuring the resonant frequency of an acoustic wave propagating along a piezoelectric crystal. When gases bind to the crystal surface, they change their resonant frequency and this measurement determines the gas concentration. In addition, they have high sensitivity, selectivity and stability, as well as fast response and low power consumption. They are widely used in gas monitoring and environmental analysis applications [13].

In this work, a semiconductor sensor of the MQ series has been used, due to the great advantages of low power consumption, speed of response and ease of use And above all because we are looking to design a low-cost device and this sensor has a very low price. However, a disadvantage is the cross-sensitivity which in this case in ideal conditions (closed room) would not be a problem.

3   Materials and Methods

This section presents the architecture of the system and the procedure carried out for data collection.

3.1   System Architecture

The architecture of the prototype system is presented in Fig. 1. The hardware part is composed of:

– NodeMcu V3 (ESP8266) development board. It has a 5V pin that allows powering the gas sensor and accelerometers. Through its analogue-to-digital converter (ADC) pin, the raw values of the gas sensor can be obtained in a range from 0 to 1023 unsigned integers (10bits). In addition, this small-sized board incorporates Wi-Fi connectivity, which will be used to transmit the data obtained during the measurement.
– Raspberry Pi 3 Model B+ development board. It is compact in size but has considerable computing power. Thanks to its ability to manage the flow of information between the ESP8266 boards, the database and the IoT platform, it becomes a key part of the correct working of the system.
– MQ135 sensor. It is highly sensitive and accurate, with fast response and low power consumption and is compatible with different development platforms, making it ideal for use in IoT projects and ambient monitoring. It detects the presence of various toxic gases, such as ammonia, benzene, carbon dioxide, carbon monoxide and sulphur dioxide.

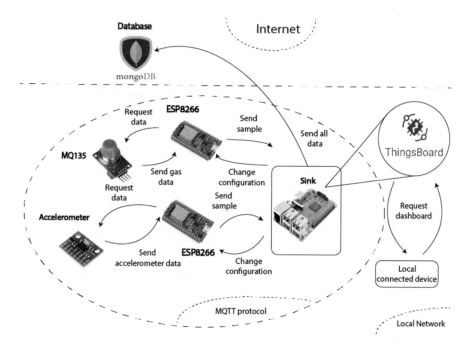

Fig. 1. Initial system prototype diagram.

- Voltage divider. It is an intermediate stage between the MQ135 sensor and the ADC pin of the ESP8266. Its function is to adapt the electrical signal generated by the gas sensor, which has a range of 0–5 V, to the reading range of the ESP8266 ADC pin, which is 0–3.3 V. In this way, the voltage divider allows a suitable connection between the gas sensor and the development board, guaranteeing the correct capture and analysis of the data obtained.
- MPU6050 sensor. It is an inertial measurement device that integrates a three-axis gyroscope and a three-axis accelerometer on a single chip. It is used to determine the open/closed status of door and window.
- Smartphone. It is used to count the presence of nearby people through Wi-Fi technology.

On the other hand, the software part is composed of:

- Visual Studio Code. Development tool used to program ESP8266 development boards.
- Message Queuing Telemetry Transport (MQTT). It is the communication protocol based on subscription and publication that has been used to communicate and send data between the different devices in the system.
- MongoDB. It is the database used to persistently store all the data coming from the MQ135 and MPU6050 sensor.
- ThingsBoard. The open source IoT platform that enables device management and data collection, processing and visualization. It offers paid and Community Edition versions, which will be used in this project and is compatible with

various architectures. ThingsBoard offers features such as horizontal scalability, customization and communication via MQTT, HTTP and LrM2M.
- ThingsBoard Gateway. It is used for integrating devices connected to legacy and third-party systems with ThingsBoard IoT. This allows the extraction of data from devices connected to external MQTT brokers, OPC-UA servers, Sigfox Backend, Modbus slaves or CAN nodes.

3.2 Data Collection

In this section, the data collection of the system is discussed. For this, an ESP8266 board connected to an MQ135 sensor is placed near the main door of the room. In the same way, another board with the same sensor is deployed near the window to measure possible differences. The connection between the board and the sensor and the applied voltage divider is shown in Fig. 2.

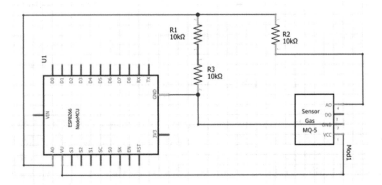

Fig. 2. Schematic of MQ135 and ESP8266 sensor assembly near window and door.

On the other hand, another ESP8266 board connected with a MPU6050 sensor is deployed and attached to the window. Similarly, another board and sensor is mounted on the main door to check the open/closed status of the window and the door, as shown in Fig. 3.

Finally, a local Wi-Fi network created by one of the ESP8266 boards is established. This device is based on the principle of communication between networks and nearby mobile devices to obtain basic data such as the network identifier and the Received Signal Strength Indicator (RSSI). The RSSI is used to establish the inflow based on a threshold value defined by experts.

The data collected by the 4 ESP8266 boards plus the Wi-Fi based flow information from the mobile devices is collected and sent to a remote database for

Fig. 3. Device installed on the door and window.

persistence by a central element. In addition, an own dashboard has been created on the ThingerBoard IoT platform for real-time visualisation on a designed dashboard shown in Fig. 4.

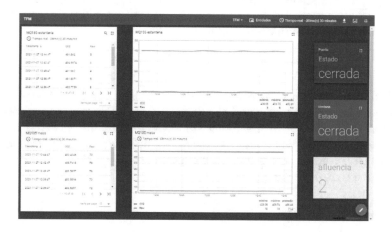

Fig. 4. Our complete dashboard.

4   Case Study

In this section, the system is evaluated through a case study with three different scenarios that have been carried out to observe the progress of the CO_2 gas variable. The closed space in which the three case studies are presented is given by Fig. 5. Its total area is estimated to be $12m^2$.

The variables to be studied are the following: the number of people in the room at each instant, the opening of both the window and the door at each

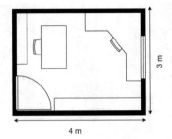

Fig. 5. Map of the room where the case studies are implemented.

instant, as well as the differences between the values recorded by the CO_2 meter located near the window and the one located near the door. The way to collect data is as follows: close both the door and the window and keep as many people as necessary inside, depending on the scenario you want to experiment with. Then, you start collecting data which is visualised from the graphs on the dashboard that has been generated. After an extended period of time, which can vary depending on the scenario between one hour and two hours, both the window and the door are opened and the data collection continues, this time, ventilating to observe the changes that may occur.

4.1   First Scenario

In this first scenario, a session was conducted without the participation of any person and both the window and the door were closed, in order to observe the minimum values of CO_2 in the absence of any external influence. Over a period of one hour, stable values around 63 were recorded. These values indicate low and constant CO_2 levels in the passenger compartment, suggesting good air quality under these controlled conditions.

4.2   Second Scenario

In this second experiment, one session was carried out with the participation of two people, with an approximate duration of two hours. The gas recording started at 11:53 h and ended at 14:10 h.

Figure 6a shows the raw values obtained from the MQ135 gas sensor for the measurement of CO_2. These values are in the range of 0 to 1023 unsigned integers (10 bits). During the measurement, a first peak of high values is observed, which corresponds to the calibration process of the sensor. This process starts with a stabilisation around the value of 123, until it reaches a new maximum close to 138 at 13:50 h, at which time the device had been measuring for two hours without ventilation. Subsequently, a sharp decrease in the values recorded can be seen, just after the opening of the door and the window at 14:00 h. This decrease is kept until the last recorded value at 14:10 h, where a minimum of 116 is reached after a short period of ventilation.

Figure 6b shows the results obtained from the MQ135 gas sensor located near the window. In this case, it can be seen that the maximum peak of 89 obtained occurred at 13:44 h, which coincides with the maximum recorded in the gas sensor mentioned above. After the ventilation period, a minimum peak of 66 is obtained at 14:04 h. It should be noted that this minimum value corresponds to a much lower CO_2 level than the previous maximum value, indicating that the ventilation has been effective in reducing the CO_2 levels in the passenger compartment.

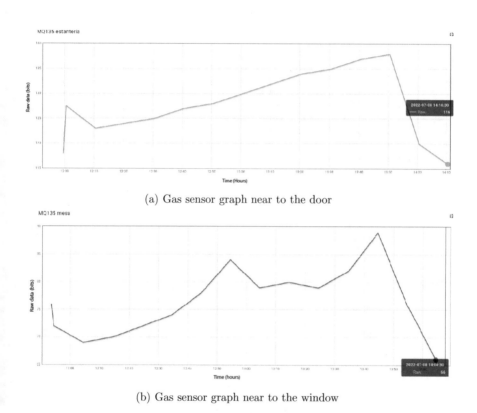

(a) Gas sensor graph near to the door

(b) Gas sensor graph near to the window

Fig. 6. Second scenario

4.3   Third Scenario

In the third scenario, a session was held with the participation of four people, lasting approximately two hours.

Figure 7a shows the raw values obtained from the MQ135 gas sensor located near the door. During the measurement, a first global minimum of 125 is observed. After a period of 110 min without ventilation, a maximum peak of

168 is observed. Subsequently, when ventilation is started, the values decrease quickly.

In Fig. 7b the results obtained from the MQ135 gas sensor located near the window are presented. In this case, it is observed that the maximum peak of 148 also occurs after 110 min without ventilation, as in the previous case. Finally, at the beginning of the ventilation period, the values decrease sharply in only 10 min, reaching 89.

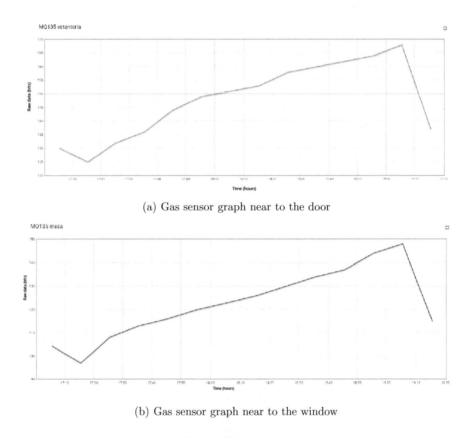

(a) Gas sensor graph near to the door

(b) Gas sensor graph near to the window

Fig. 7. Third scenario

Thus, it can be observed that as the number of people in the room increases, in this case 4 people, there are changes in the values recorded by the MQ135 sensor near both the window and the door. In the case of the sensor near the window, the maximum peak increases from 138 bits from the second scenario to 168 bits. For the sensor near the door, the maximum peak increases from 89 bits to 148 bits.

5   Conclusions

This work has focused on the design, implementation and experimentation of a low-cost IoT-based system to observe and record CO_2 measurements in an enclosed space, using two sensors, as well as collecting additional information from the ambient environment, such as door and window status and affluence.

The results obtained from the case studies have shown favourable conclusions (refer to Table 1). Over the course of the time without ventilation, a gradual increase in gas concentrations has been observed, especially in situations of increased human traffic in the studied ambient. However, a significant decrease in gas concentrations was observed when both the door and the window were opened to allow suitable ventilation.

Table 1. Summary of the results of the different scenarios.

$N^{\underline{o}}$ Scenario	$N^{\underline{o}}$ Scenario	Max raw data of carbon dioxide
1	0	63
2	2	138
3	4	168

These results conclusively demonstrate the efficacy of ventilation as an effective measure to reduce gas concentrations in the analysed ambient.

5.1   Limitations and Future Works

This work aimed to raise knowledge of the benefits of using smart gas devices to monitor environments. However, some improvements that could be made in the future would be to calibrate the MQ135 sensor to provide ppm values and not just raw values, to use contact sensors for opening and closing doors and windows and reduce the computational load on the development board, generate a statistical analysis including significance tests on the differences that are observed and finally, to extend the study to new environments and gases.

References

1. Definition of Internet Of Things (IoT) - IT Glossary — Gartner. https://www.gartner.com/en/information-technology/glossary/internet-of-things#: :text=The%20Internet%20of%20Things%20(IoT,states%20or%20the%20external%20environment., (Accessed on 05/11/2023)
2. El gobierno publica recomendaciones sobre el uso de sistemas de climatización y ventilación para prevenir la expansión del covid-19. https://www.miteco.gob.es/es/ministerio/medidas-covid19/sistemas-climatizacion-ventilacion/default.aspx#: :text=se%20recomienda%20desconectarlos%2C%20dando%20prioridad,dos%20horas%20despu%C3%A9s%20del%20cierre. Accessed 05 Nov 2023

3. Rs pro through beam emitter photoelectric sensor, barrel sensor, 20 m detection range — rs. https://uk.rs-online.com/web/p/photoelectric-sensors/2043997?cm_mmc=UK-PLA-DS3A-_-google-_-CSS_UK_EN_Automation_%26_Control_Gear_W hoop-_-Photoelectric+Sensors_Whoop+(2)-_-2043997&matchtype=&pla-33404615 5129&gclid=CjwKCAjwjaWoBhAmEiwAXz8DBaTFtNkc8eKG0OAaT_UMGZ6t oicwNTiwc6gpx6Lcw1nSoJNecZd8vhoCfb4QAvD_BwE&gclsrc=aw.ds (Accessed on 09/19/2023)
4. Awang, Z.: Gas sensors: a review. Sens. Transducers **168**(4), 61–75 (2014)
5. Berndt, D., Muggli, J., Wittwer, F., Langer, C., Heinrich, S., Knittel, T., Schreiner, R.: Mems-based thermal conductivity sensor for hydrogen gas detection in automotive applications. Sens. Actuators, A **305**, 111670 (2020)
6. Cuerdo-Vilches, T.: El aire acondicionado y la covid-19 en espacios interiores: recomendaciones para generar entornos más seguros (2020)
7. Dave, E., et al.: How the next evolution of the internet is changing everything. The Internet of Things, p. 2011 (2011)
8. DeNardis, L.: The Internet in everything. Yale University Press (2020)
9. Dey, A.: Semiconductor metal oxide gas sensors: a review. Mater. Sci. Eng. B **229**, 206–217 (2018)
10. Espinilla, M., Martínez, L., Medina, J., Nugent, C.: The experience of developing the ujami smart lab. Ieee Access **6**, 34631–34642 (2018)
11. Goldenstein, C.S., Spearrin, R.M., Jeffries, J.B., Hanson, R.K.: Infrared laser-absorption sensing for combustion gases. Prog. Energy Combust. Sci. **60**, 132–176 (2017)
12. Jadon, N., Jain, R., Sharma, S., Singh, K.: Recent trends in electrochemical sensors for multianalyte detection-a review. Talanta **161**, 894–916 (2016)
13. Jakubik, W.P.: Surface acoustic wave-based gas sensors. Thin Solid Films **520**(3), 986–993 (2011)
14. Korzun, D., Balandina, E., Kashevnik, A., Balandin, S., Viola, F.: Ambient intelligence services in iot environments: Emerging research and opportunities: Emerging research and opportunities (2019)
15. Miraz, M.H., Ali, M., Excell, P.S., Picking, R.: A review on internet of things (iot), internet of everything (ioe) and internet of nano things (iont). In: 2015 Internet Technologies and Applications (ITA), pp. 219–224 (2015)
16. Rose, K., Eldridge, S., Chapin, L.: The internet of things: an overview. Internet Soc. (ISOC) **80**, 1–50 (2015)
17. Salazar Soler, J., Silvestre Bergés, S.: El mundo internet of things (iot) (2019)
18. Satish, U., Mendell, M.J., Shekhar, K., Hotchi, T., Sullivan, D., Streufert, S., Fisk, W.J.: Is co2 an indoor pollutant? direct effects of low-to-moderate co2 concentrations on human decision-making performance. Environ. Health Perspect. **120**(12), 1671–1677 (2012)
19. Sturm, H., Brauns, E., Seemann, T., Zoellmer, V., Lang, W.: A highly sensitive catalytic gas sensor for hydrogen detection based on sputtered nanoporous platinum. Procedia Eng. **5**, 123–126 (2010). https://doi.org/10.1016/j.proeng.2010.09.063, eurosensor XXIV Conference
20. Wang, C.C., Prather, K.A., Sznitman, J., Jimenez, J.L., Lakdawala, S.S., Tufekci, Z., Marr, L.C.: Airborne transmission of respiratory viruses. Science **373**(6558), eabd9149 (2021). https://doi.org/10.1126/science.abd9149
21. Zhou, H.y., Ma, G.m., Wang, Y., Qin, W.q., Jiang, J., Yan, C., Li, C.r.: Optical sensing in condition monitoring of gas insulated apparatus: a review. High Voltage **4**(4), 259–270 (2019)

Using Reference Points for Detection of Calcifications in Mammograms for Medical Active Systems

Francisco E. Martínez-Perez$^{(\boxtimes)}$ ⓘ, César A. Ramírez-Gámez ⓘ,
Alberto Núñez-Varela ⓘ, Sandra Nava-Muñoz ⓘ, José Ignacio Núñez-Varela ⓘ,
Héctor G. Pérez González ⓘ, Pedro David Arjona-Villicaña ⓘ,
and Francisco Javier Ramírez-Aguilera

Universidad Autónoma de San Luis Potosí, Av. Dr. Manuel Nava No. 8 Zona Universitaria,
078290 San Luis Potosí, Mexico
{eduardo.perez,crgamez,alberto.nunez,senavam,jose.nunez,
hectorgerardo,david.arjona,javier}@uaslp.mx

Abstract. Context-awareness is one of the main features of any AmI system, particularly related to healthcare systems. To develop this kind of active systems it is necessary to create methodologies where it can be observed how the experts solve specific problems or situations. This paper focuses on the problem of breast cancer detection and classification, and presents a methodology based on a case study where radiologists and medical doctors were involved and their knowledge was extracted in order to define a model that could be used to develop a breast cancer automatic interpretation system, with the goal of helping the medical personnel in their decision-making process. As a first step towards the development of such active system, this paper presents the detection of calcifications, a type of finding in the breasts that could be benign or malign, depending on certain features. The detection is made by applying composite correlation filters, and even though the calcifications used for testing cover all the categories defined by a medical taxonomy (BI-RADS system), the accuracy of such detection is promising, where most categories have a detection accuracy of 80% or above.

Keywords: Breast Cancer · Calcifications Detection · Composite Correlation Filters

1 Introduction

The paradigm of Ambient Intelligence (AmI) involves the integration of multiple sensors in a distributed interface in order to observe and interpret the activities and intentions of the user, with the purpose of allowing the user to interact with the environment in a passive and transparent way [1]. To achieve this, it is necessary to represent the information and knowledge that is provided by the environmental context of specific situations [2].

J. Bravo and G. Urzáiz (Eds.): UCAmI 2023, LNNS 835, pp. 39–50, 2023.
https://doi.org/10.1007/978-3-031-48306-6_4

Current systems should be aware of the context related to the problem they try to solve, where collaborative processes could be involved for the interpretation of events in terms of some patterns. Therefore, to get context awareness it is necessary to create algorithms that consider taxonomies from several fields of study; such is the case of breast cancer detection, where time becomes a key factor to act and save lives. Therefore, to create healthcare active systems it is important to extract knowledge from medical taxonomies such as Breast Imaging Reporting and Data System (BI-RADS) [3], to allow these systems to help in the decision-making process of healthcare personnel. The creation of active systems also requires the development of algorithms for data gathering, collaborative processes, data fusion, interpretation of events, context extraction, behavior modeling, among others [4]. Nowadays, technology has increased in power, for instance, the equipment to obtain a mammogram has improved its image acquisition capabilities. They have gone from 12 bits to 14 bits in depth, allowing a higher resolution in possible findings (such as calcification, masses) within the image. Current mammographers are considered passive systems in the sense that they only highlight some findings with no notification of possible anomalies.

The purpose of this work is to present a case study of the understanding of a medical taxonomy and its adaptation to a computational language where it would be possible to detect masses, calcifications, and other findings using our PREVEMM (Pattern recognition to Evaluate MaMmograms) conceptual model. As an example of this understanding, results of the detection of calcifications are presented, considering the real size of the images of mammogram studies ranging from 2736x3580, 3328x4084 and 2560x3328 pixels. Three public datasets were evaluated that were manually tagged for a total of 2,015 processed images. The detection of these findings was made through a composite of correlation filters using 13 reference patterns. Those processed images used a resolution of 7200x7200 and 4100x4100 pixels, with a depth of 8, 14 and 16 bits, depending on the dataset. These results may allow computers to automatically classify findings using the BI-RADS [3], which could help to speed up the diagnostic process and to create active systems capable of transmitting contextual information based on the BI-RADS medical taxonomy for early awareness. With this understanding, it will be possible to provide better results since they closely follow the medical taxonomy, unlike most of the current systems where only binary results are provided (cancer/not cancer) [5].

The rest of this paper is organized as follows. Section 2 introduces the methodology and results of the case study conducted in this research. Section 3 explains the results using the three datasets. Section 4 presents the conclusions and directions for future work.

1.1 Related Work

Breast cancer is a disease which needs to be detected as early as possible [6]. Mordang et al. show that less than of 1% of a total of 63,895 cases present possible malign findings (only 413 cases). Currently, most researchers have proposed the use of AI techniques for cancer detection [7–10], particularly Convolutional Neural Networks (CNNs) for the recognition of calcifications. One main drawback is that a large number of images for each category that has to be detected, is needed for training. Another important thing to mention is that these datasets and research works do not consider the knowledge from radiologists and medical personnel that is used for the interpretation of mammogram studies.

One of the main goals of our research is to propose the use of this knowledge for the creation of a methodology to analyze and apply in all the problems related to breast cancer detection. In this sense, it is also important to apply this knowledge for the creation of public datasets that could consider details of findings (calcifications, masses, ganglions, among others), such as, asymmetries, density, measurements, relationships among each finding, etc.

Also, previous works based on image processing techniques make a reduction of the image size of the four projections generated by the mammogram study. These reductions are of 640x640 [7], 1024x 1024 [8], or 800x800 [11]. Additionally, these works use an image depth of 8 bits. In our paper, we present an example of calcifications detection considering the real size of the mammogram studies projections with depths of 8, 12 and 14 bits. By using the real size and depth, it is expected to obtain a more accurate classification of findings in terms of the BI-RADS classification systems.

2 Materials and Methods

The main goal of this research is to employ the experience of health professionals to create a methodology with the purpose of correctly identify calcifications, masses, and other breast cancer findings, through pattern recognition techniques. This case study was conducted in a public hospital in the city of San Luis Potosí, Mexico, where more than 500 h were dedicated in understanding the needs of the professionals in charge of analyzing mammograms.

2.1 Subjects

A collaborative work group was established with four medical doctors from a public healthcare hospital: A surgical oncologist with ten years of experience, and three radiologists with twelve, fifteen and twenty years of experience.

Fig. 1. Think-aloud technique performed by radiologist on visualization lab.

2.2 Methodology

Work meetings were organized with the medical doctors, where they described how to identify calcifications, masses, and other formations in mammograms, and to understand the way in which they perform such interpretations (as shown in Fig. 1). At the same time, the team documented themselves on other research that have tried to solve similar problems to the ones presented here. As the meetings progressed, the medical doctors explained how breast formations are classified according to the ACR BI-RADS Atlas [3]. Additionally, an intern radiologist joined the team and helped to label the calcifications in some mammogram studies. In this paper, only the interpretation for calcifications is described.

2.3 Results of the Analysis

Understanding the characteristics and labels that radiologists employ to classify and diagnose calcifications in a mammogram is a laborious task because it requires to understand how physicians classify images and select the ones that better describe the formations that the algorithm would need to search.

As shown in Fig. 1, the radiologists were asked to perform a think-aloud technique to understand how the interpretation is performed on the whole mammogram, and the terminology used for classification. They start their analysis by observing the mammogram images and they conclude with the creation of the interpretation.

Figure 2 illustrates the conceptual model describing the complete process of how a mammogram study is analyzed. This model has three main features that are described next:

a) Conditions: It is necessary to have a complete mammogram study, composed of four projections corresponding to: two mediolateral oblique projections (MLO left L and right R), and two craniocaudal projections (CC left L and right R). Each projection could contain none, one or more findings.

b) Action and Interaction Strategies: The radiologist performs a zig-zag visual review for each projection, where it is important to identify information related to i) density and type findings, ii) calcifications, or iii) masses. There are two characteristics related to asymmetries and distortions that were not considered in this research. Density is the only characteristic that all projections contain. Findings cannot exist or appear in an isolate or jointly way. Findings are described by laterality (right and left breast). The action strategy is to store the information of each projection based on detected features as shown in Fig. 2. A calcification could have a shape as those indicated in Table 1. A mass could have an oval, round, or irregular shape, as well as other specific features. There could be only calcifications, masses, or both at the same time. All this information should be fused together to create the necessary context established by the BI-RADS classification system. Thus, the information obtained for the findings of each projection should be fused together, compared, and analyzed.

c) Consequences: Based on the previous analysis and information fusion of the four projections, the BI-RADS classification is obtained as shown in Fig. 2 to obtain a possible interpretation.

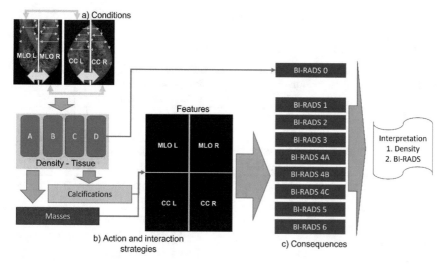

Fig. 2. PREVEMM conceptual model to create an interpretation from a mammogram study.

This proposed conceptual model, called PREVEMM (Pattern recognition to Evaluate MaMmograms), presents some challenges that need to be solved to create healthcare active systems that are able to analyze, gather and fuse information, and send notifications related to possible found findings to help medical doctors and radiologists in their decision-making process. Some challenges are: i) calcifications detection, ii) masses detection, iii) tissue density definition, and iv) the definition of rules and knowledge base of the relationship between features of challenges i) and ii) with the goal of obtaining an accurate BI-RADS classification. This also considers data fusion techniques for the creation of context considering the features present in calcifications and masses. Our research team has started tackling these challenges at different levels. In this paper, the detection of calcifications is described, as a first step for the creation of an automatic breast cancer classification system.

Calcifications Detection

To create an automatic breast cancer classification system to help in the decision-making process of radiologist and other medical personnel, a first step is the detection of findings in general, and calcifications (particularly for this paper). According to the radiologists' experience there is a strong relationship between the calcification's morphology and their distribution in an image. This relationship is not documented in the ACR BI-RADS Atlas [3], however, it needs to be considered for the final decision of a diagnostic. Moreover, this relationship contributes to the set of rules that need to be considered when implementing a computational algorithm.

Table 1 shows the final classification provided by the radiologists. Rows or cells with asterisk(*) describe diagnostics defined from the radiologists' experience. Once the data in this Table and the medical terminology were validated, the preliminary analysis of images started. Two applications were developed to identify and organize the preliminary image analysis, but these are not included in this paper.

Table 1. Classification of calcifications.

Distribution in the image	Morphology	Shape	BI-RADS classification
Ductal*	Skin	Polygonal, round, with radiolucent center	2
Grouped/Difusse	Vascular	Parallel lines in a serpentine path like curves	2
NA	Coarse or "popcorn-like"	Several circular and curves joined	2
Regional/Segmental	Large rod-like	Linear, smooth calcifications	2
Segmental	Round	Round, several sizes, greater than 0.5 mm	2
NA	Rim	Round with radiolucent centered	2
NA	Dystrophic	Thick calcifications, with radiolucent areas, irregular polygonal	2
NA	Milk of calcium	Crescent upper concavity shaped or linear	2
NA	Suture	Very thick and bright. Linear and curves	2
Grouped*	Amorphous* (Big)		2*
NA	Coarse heterogeneous		3
Linear*	Fine-linear*		3*
Grouped/Diffuse/Linear	Fine pleomorphic		4A/4C
Grouped/Diffuse	Amorphous (small)		4B
Grouped	Branching		4C
Grouped*	Fine pleomorphic + mass*		5*
Grouped*	Branching + mass*		5*

Computational Understanding

To create the necessary context for calcification detection and recognition it is important to describe some particular features that could be found in calcifications [12]:

- These are small deposits of calcium, appearing as bright white specks or dots on the soft tissue background of the breasts.

- When these are more typically benign: a) They are larger than 0.5 mm (also known as macrocalcifications); b) They have well defined edges and standard shapes; c) They are not clustered in one area.
- When these have suspicious morphology: a) They are smaller than 0.5 mm each (also known as microcalcificactions); b) They have variable size and shape; c) They are clustered in a specific area.

Among these features, the calcifications' morphology and distribution could provide further information. Table 1 presents four columns that are related to those features and are described as:

1. Morphology: The concepts showed in this column are medical terms. For practical purposes, these terms are similar to some geometric forms. For example, a skin calcification, means that it is found more frequently in the inframammary fold, parasternal region, armpit or areola as shown in [13]; this kind of calcification can be found as a polygonal form, some round, with a radiolucent center in the skin.
2. Shape: Calcifications can be shaped as round, linear, polygonal, and curved. Nevertheless, the label to be assigned to a calcification does not depend on its shape only, but also on the location within the breast. For instance, the skin calcification, which is found in a superficial location in the skin, has a round form.
3. Distribution in the image: This is an important descriptor, which refers to the arrangement of the calcifications. The distribution can be grouped, regional, diffuse, segmental and linear. To identify this distribution, it is possible to apply the Euclidean distance or the Mahalanobis distance between different calcifications. Calcifications can be found in isolation or with some specific structure [13].
4. BI-RADS classification: The American College of Radiology (ACR) has established a classification system known as BI-RADS [3]. It contains nine categories, from benign findings (B1, B2 and B3) to malign findings (B4A, B4B, B4C, B5 and B6), and even a need for additional examinations (B0). This classification is more specific than just the commonly used binary classification of cancer and not cancer.

Reference Points for the Detection of Calcifications

As mentioned before, a first step is the detection of findings so that they can be further analyzed for their proper classification. For this research, correlations are employed in order to detect calcifications. The role of correlations is to measure how similar or different a test object is from training objects. Correlation works best when the test object matches well with the training set. Typically, a reference image is correlated with a test image (also called a scene) to detect and locate the reference image in the scene. Thus, a correlation operation may be considered as a system with an input (the scene or test image), a stored template or filter (derived from the reference image), and an output (correlation).

Specifically, composite correlation filters were developed to handle the more general types of distortions that cannot be mathematically modeled by coordinate transforms that we presented in [14]. These kinds of filters are derived from several training images, which are representative views of the object or pattern to be recognized. These filters can be trained to recognize any object or type of distortion as long as the expected distortion can be captured by the training images.

In our research, 124 mammograms studies of Mexican women were labeled by the radiologists. From these studies, the research team analyzed the studies and defined 13 patterns that could allow the identification of findings with different features, such as vascular, ring-like, popcorn-like, round, suture, and rod-like (see Fig. 3). For each pattern, the clearer greyscales were kept, and the darker grayscales were ignored. To do this, first it was necessary to find a medium grey value between the maximum and minimum. Then, only grey values that were clearer than the medium were allowed in the final picture. To organize the patterns the following data was needed: minimum and maximum value for each pattern, size of the selected image in x and y coordinates, and the total size in pixels (see Fig. 3). This data was stored in an *npy* file (default extension for Python's numpy data) in order to keep all the original data from the mammography. These patterns were validated by the radiologists and used for building the composite correlation filters.

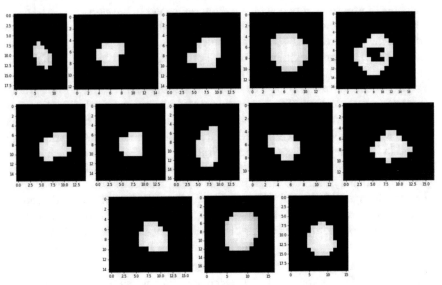

Fig. 3. Patterns used for building the composite correlation filters, where the axes indicate pixels.

3   Results

To test our detection algorithm and the 13 patterns defined above, a total of 2,015 images, obtained from three public datasets, were processed: INBreast [15] (410 images), CBIS-DDSM [16] (1,497 images), and BCDR-D01 [17] (108 images). The bit-depth per image from each dataset is of 14, 16 and 8 bits respectively (Fig. 4). The size of the images could be either of 2736x3580, 3328x4084, 2560x3328 or 4186x7111 pixels. The detection algorithm adjusted the image resolution of the images to 4100x4100 pixels, along with the filter employed for INBreast and BCDR-D01 datasets, while the resolution for the CBIS-DDSM dataset was adjusted 7200x7200 pixels.

Table 2. Percent of detection effectiveness for the INBreast dataset.

BI-RADS category	Percent of images	No. Images	*Labels*	*Correlation results*	Percent of detection
B1	16	67	0	16	–
B2	54	220	214	179	84
B3	6	23	12	8	67
B4A	3	13	11	7	64
B4B	2	8	6	5	83
B4C	5	22	16	14	88
B5	12	49	48	41	85
B6	2	8	6	5	83
Total	100	410	313	275	**88**

Table 3. Percent of detection effectiveness for the CBIS-DDSM dataset.

BI-RADS category	Cases	No. Images	*Labels*	*Correlation results*	Percent of detection
B0	37	75	75	65	87
B2	105	314	578	471	81
B3	52	105	115	87	76
B4	463	831	897	409	46
B5	93	172	191	140	73
Total	750	1497	1856	1172	**63**

As mentioned before, the first step for the creation of complete automatic classification system is the detection of findings. In this case, we only focused on the detection of calcifications using only the 13 patterns defined from the mammographic studies described in the previous section. Tables 2, 3 and 4 presents the results obtained for each dataset, where each table presents the results according to the BI-RADS classification value. For instance, Table 2 shows that most cases fall in the B2 category, since these are benign cases, and most patients does not have malign findings. This issue is known as unbalanced data, and in the case of cancer or diseases it is common to have this problem.

Even though the number of patterns used for detection is small, the correlation results for most BI-RADS categories are promising. The last column of each table shows the percentage of detected calcifications for each category, which is calculated by dividing the *"Correlation results"* column by the *"Labels"* column. The total percentage of detection for the INBreast (Table 2) and BCDR-D01 (Table 4) datasets resulted in 88% total, while for the CBIS-DDSM dataset (Table 3) was only 63%. Category B4 (which means that it is possible or highly possible to have cancer) was difficult to detect with our defined patterns. It is important to notice that our 13 patterns were defined using

Table 4. Percent of detection effectiveness for BCDR-D01 dataset.

Type	Cases	No. Images	*Labels*	*Correlation results*	Percent of detection
Macrocalcifications					
Benign 2	1	2	2	2	100
Benign 4	11	44	26	22	85
Malign 1	1	1	1	1	100
Malign 3	1	3	2	1	50
Malign 4	5	20	12	9	75
Malign 5	1	5	1	1	100
Subtotal	8	29	16	12	**75**
Microcalcifications					
Malign 3	1	3	2	2	100
Malign 4	6	24	12	12	100
Malign 6	1	6	2	2	100
Subtotal	8	33	16	16	**100**
Total	28	108	60	52	**87**

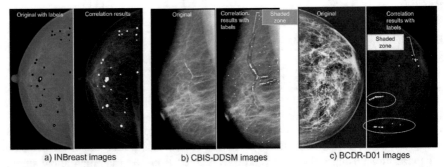

a) INBreast images   b) CBIS-DDSM images   c) BCDR-D01 images

Fig. 4. a) INBreast image with a depth of 14 bits, b) CBIS-DDSM image with a depth of 16 bits, c) BCDR-D01t image with a depth of 8 bits.

macrocalcifications, and these results show that they can detect microcalcifications as well. Thus, more patterns need to be defined to cover more cases and could increase the detection accuracy of the algorithm. Figure 4 presents three sample images from each dataset. Images on the left are original images and those on the right illustrates the application of the correlation filters and the detected findings.

4 Conclusions

Breast cancer is a disease that to affect to a large number of women, around the world. Having healthcare active systems that could expedicte the decision-making process of medical personnel could increase the number of people being treated on time, particularly for this kind of diseases where a timely detection of cancer is of great importance to save the patient's life. This paper presents the definition of a conceptual model for the creation of computational system that could automatically interpretate mammograms studies to help medical personnel in their own interpretations. This conceptual model was defined by analyzing the expertise of radiologists when interpreting mammograms using the BI-RADS medical taxonomy, instead of just using a binary value (cancer/not cancer). This model will help in the creation of the necessary context for healthcare active systems.

One of the first steps for the development of such automatic classification system is the detection of findings within mammograms. Once these findings are detected can be further analyzed and classified according to the BI-RADS system. In this paper we presented results for this detection process, specifically for macrocalcifications, a type of benign finding. By using only 13 patterns and correlation filters, we were able to obtain a promising detection accuracy for most BI-RADS categories applied to three datasets.

Further research needs to be done in terms of detecting calcifications (macro and micro), and other findings, such as tissue density and masses (and their different types). Since each type of finding also have several features to consider, the detection and classification problem escalates. Another important issue is the lack of datasets that consider details of findings (calcifications, masses, ganglions, among others), and their features, such as, asymmetries, density, measurements, relationships among each finding, etc.

References

1. Remagnino, P., Foresti, G.L., Ellis, T. (eds.): Ambient Intelligence: A Novel Paradigm. Springer, New York (2005). https://doi.org/10.1007/b100343
2. Ramos, C., Augusto, J.C., Shapiro, D.: Ambient Intelligence—the next step for artificial intelligence. IEEE Intell. Syst. **23**(2), 15–18 (2008). https://doi.org/10.1109/MIS.2008.19
3. ACR: ACR BI-RADS® Atlas Fifth Edition QUICK REFERENCE. https://www.acr.org/-/media/ACR/Files/RADS/BI-RADS/BIRADS-Reference-Card.pdf. Accessed 11 July 2023
4. Aghajan, H., Augusto, J.C., Delgado, R.L.C.: Human-centric interfaces for ambient intelligence. Academic Press (2010). http://portal.acm.org/citation.cfm?id=1816592. Accessed 11 July 2023
5. Mahmood, T., Li, J., Pei, Y., Akhtar, F., Imran, A., Yaqub, M.: An automatic detection and localization of mammographic microcalcifications roi with multi-scale features using the radiomics analysis approach. Cancers **13**(23), 5916 (2021). https://doi.org/10.3390/cancers13235916
6. Mordang, J.J., et al.: The importance of early detection of calcifications associated with breast cancer in screening. Breast Cancer Res. Treat. **167**(2), 451–458 (2018). https://doi.org/10.1007/s10549-017-4527-7
7. Alghamdi, M., Abdel-Mottaleb, M., Collado-Mesa, F.: DU-Net: convolutional network for the detection of arterial calcifications in mammograms. IEEE Trans. Med. Imaging **39**(10), 3240–3249 (2020). https://doi.org/10.1109/TMI.2020.2989737

8. Yu, X., Kang, C., Guttery, D.S., Kadry, S., Chen, Y., Zhang, Y.D.: ResNet-SCDA-50 for breast abnormality classification. IEEE/ACM Trans. Comput. Biol. Bioinforma. **18**(1), 94–102 (2021). https://doi.org/10.1109/TCBB.2020.2986544

9. Cai, H., et al.: Breast microcalcification diagnosis using deep convolutional neural network from digital mammograms. Comput. Math. Methods Med. **2019**, 1–10 (2019). https://doi.org/10.1155/2019/2717454

10. Khan, H.N., Shahid, A.R., Raza, B., Dar, A.H., Alquhayz, H.: Multi-view feature fusion based four views model for mammogram classification using convolutional neural network. IEEE Access **7**, 165724–165733 (2019). https://doi.org/10.1109/ACCESS.2019.2953318

11. Shu, X., Zhang, L., Wang, Z., Lv, Q., Yi, Z.: Deep neural networks with region-based pooling structures for mammographic image classification. IEEE Trans. Med. Imaging **39**(6), 2246–2255 (2020). https://doi.org/10.1109/TMI.2020.2968397

12. Understanding breast calcifications (2022). https://www.breastcancer.org/screening-testing/mammograms/what-mammograms-show/calcifications. Accessed 11 July 2023

13. Arancibia Hernández, P.L., Taub Estrada, T., López Pizarro, A., Díaz Cisternas, M.L. Sáez Tapia, C.: Breast calcifications: description and classification according to BI-RADS 5th edition. Rev. Chil. Radiol. **22**(2), 80–91 (2016). https://doi.org/10.1016/j.rchira.2016.06.004

14. Martinez-Perez, F.E., González-Fraga, J., Tentori, M.: Automatic activity estimation based on object behaviour signature. In: Proceedings of SPIE, San Diego CA, 2010, pp. 77980E. http://link.aip.org/link/?PSISDG/7798/77980E/1. Accessed 11 July 2023

15. Moreira, I.C., Amaral, I., Domingues, I., Cardoso, A., Cardoso, M.J., Cardoso, J.S.: INbreast: toward a full-field digital mammographic database. Acad. Radiol. **19**(2), 236–248 (2012). https://doi.org/10.1016/j.acra.2011.09.014

16. Heath, M., Bowyer, K., Kopans, D., Moore, R., Jr Kegelmeyer, P.: The digital database for screening mammography. In: Proceedings of the Fourth International Workshop on Digital Mammography (2000). https://doi.org/10.1007/978-94-011-5318-8_75

17. Ceta-ciemat: BCDR. https://www.ciemat.es/cargarAplicacionNoticias.do;jsessionid=70C792CCF8908DF0D7289D48D83DC7CB?identificador=391 Accessed 11 July 2023

Age-Inclusive Design of Video Streaming Services: Integrating Insights and Design Implications

Sruti Subramanian$^{(\boxtimes)}$ [ID] and Katrien De Moor [ID]

Department of Information Security and Communication Technology,
Norwegian University of Science and Technology, Trondheim, Norway
{sruti.subramanian,katrien.demoor}@ntnu.no

Abstract. With the rapid growth of video streaming services, under-standing user needs, preferences, and behaviors has become crucial. How-ever, older adults, are a significant yet often overlooked user segment, have received limited attention. This paper aims to bridge this research gap by integrating insights from HCI, QoE, and health sciences. By fos-tering an age-inclusive design process, three key design implications for video streaming services are highlighted. These implications have the potential to enhance the user experience for older adults, specifically in the context of video streaming services. Additionally, the study's impli-cations provide valuable guidance for future empirical research on video streaming services with older adults. By considering interdisciplinary knowledge, we can address the unique requirements of older adults, pro-mote engagement, and improve accessibility within video streaming ser-vices. Ultimately, these implications contribute to healthy aging and the promotion of a healthy lifestyle through the utilization of video streaming services.

Keywords: video streaming · Age-inclusive design · dynamic adaptation

1 Introduction

Video streaming services (such as Netflix, YouTube, Disney+, etc.) have wit-nessed a significant surge in recent years. Consequently, there is a growing inter-est in understanding users' needs, preferences, and behaviors related to the use of these multimedia platforms. Improved comprehension of various human fac-tors are known to be crucial not only for enhancing user experience but also for increased technological optimization [3,13]. Currently, video streaming ser-vices are widely used by diverse age groups and user segments globally [8,20]. However, when identifying the needs, preferences, and behaviors of users, the focus has predominantly been on younger and middle-aged individuals, who are considered more accessible [10]. Although they are prominent users of video-on-demand (VoD) and video streaming services, there is a lack of consideration for

J. Bravo and G. Urzáiz (Eds.): UCAmI 2023, LNNS 835, pp. 51–56, 2023.
https://doi.org/10.1007/978-3-031-48306-6_5

other user groups, particularly older adults, who are also significant but less represented users in this context. This poses a significant challenge as different user groups have distinct requirements and preferences when it comes to using the same technology [7]. While previous studies have explored aspects of designing interactive technology for older adults in the fields of HCI, health sciences, and little knowledge in quality of experience (QoE), there is a lack of integration between these disciplines to ensure an age-inclusive design of video streaming services. This paper aims to bridge this research gap by integrating insights from the different disciplines.

2 Background

2.1 Physiological Changes in Older Adults

Medical advancements over the past century have resulted in increased life expectancy than ever before, resulting in an increasing global population of older adults. The global numbers of older adults are rapidly increasing and expected to reach over 1.5 billion in the year 2050 [15]. Older adults are a distinctive demographic as the natural process of aging contributes to various impairments affecting ones' sensory (i.e., visual, auditory), physical (e.g., skeletal, and muscular system), cognitive (e.g., memory, attention) and perceptual capabilities [16,22]. Hence, though numerous older adults are currently living longer lives, several of those years are with disabilities and chronic conditions [11]. Such physiological changes largely contribute to the needs, preferences and behaviors of older adults using VoD and video streaming services.

2.2 Understanding Older Adults' Preferences and Needs with Interactive Technology

Numerous technological applications are widely being used by older adults for various purposes such as healthcare, social well-being, and entertainment to mention a few. Despite not having previous experience or sufficient knowledge about technology, older adults have shown to be tech savvy to adopt and use new technologies that are of value and relevance to them [6,18].

Despite the speculations of older adults and the use of modern technology, it is known however, that Televisions (TVs) are a well accepted form of technology and are widely used by this segment of the population[2,17]. TV viewing is one of the most popular leisure time activities which is particularly prominent among older adults. Studies [1,14] further suggest that as people get older, they watch more TV, with viewing time significantly increasing after retirement. In this regard, with the modern smart TV's functional spectrum encompassing not only traditional network channels but also VoD and video streaming services among others, the need for opting an age-inclusive design of such services is imperative.

Coelho et al., [4] suggest integrating multi-modal interaction, and various adaptation techniques in the design process of TV applications to adapt user interfaces to the individual needs and limitations of elderly users. Bobeth et al., [2] further explored the use of free hand gestures for TV menu control among older adults and identified that participants were positive towards gesture-based interaction. Furthermore, with respect to the UI design of applications for older adults in general, numerous studies have provided design guidelines to accommodate some of the common physiological limitations experienced by the user group providing recommendations such as increasing font size, increasing button size, increasing contrast, reducing content, etc., [19].

2.3 Optimized Video Streaming Services

Adaptive bitrate streaming (ABR) [21,23] dynamically adjusts the quality of a video stream based on the viewer's network conditions. It enables the video player to automatically switch between different bitrates (i.e., levels of video quality) during playback in response to changes in the available bandwidth and network stability. This adaptive approach ensures a smoother viewing experience by reducing buffering, minimizing startup delays, and optimizing video quality based on the viewer's specific device and network capabilities. With the use of mobile devices, due to the limited screen size, the increase in bitrate becomes unrecognizable for the user after a certain degree [21]. The authors further state that more energy-efficient decisions can be made by avoiding higher bitrates if there is no gain in the perceived quality.

3   Design Implications and Discussion

3.1   Dynamic Adaptation of Age-Inclusive Design

Considering that older adults' needs, and preferences are vastly different from other user segments, implementing a complete age-inclusive design may not be attractive to the other users such as the younger adults. However, incorporating a dynamically adaptive version of the application which is age-inclusive, and encompasses features such as multi modal interaction techniques [5,9,12], in addition to various guidelines and adaptation techniques specific to older adults [19] could be a potential implementation. Such that, based on the identified users' interaction with the application (e.g., ease/difficulty) the version (e.g., age-inclusive version, original version) of the application could be dynamically adapted to the individual user and displayed accordingly. This also has implications for potential QoE evaluation of such services, which should also be adapted to the abilities and preferences of the older adults user segment (e.g., increasing font size, increasing button size, increasing contrast, reduced complexity of the set-up, etc.).

3.2   Optimizing Video Bit Rate

A lot of research, and resources are being implemented towards increasing the Quality of Experience while simultaneously trying to achieve technological optimization [21,23]. In this regard, considering older adults' needs, preferences and behaviors and designing accordingly may not only have a positive influence on the user experience and QoE of this demographic but in turn may also have a positive influence in terms of technological optimization. For example, considering the natural visual and auditory decline among the user segment [16,22], the video and audio quality might be scaled down without necessarily compromising QoE and the overall user experience. As mentioned by [21], perceived quality does not increase beyond a specific point despite increasing the bit rate. For instance, a slight difference in video/audio quality may not even be distinguishable. So watching a video with bitrate of 750/450 kbps instead of 1500 kbps for e.g., may not make a significant difference in terms of the viewing experience among older adults. In such a case, while user experience may not necessarily be influenced by the difference in bit rate, from a technological perspective this difference can potentially contribute towards achieving better optimization. Such an implementation could further be integrated into the dynamically adaptive age-inclusive version of the application.

3.3   Integrating External Sensors for Multi-modal Interaction

Earlier studies [5,9,12] have emphasized the need to design for multi-modal interaction techniques when designing for older adults. Furthermore, various studies [2] have identified that movement/body-based interaction techniques (such as hand gestures/whole-body movements) for controlling interactive applications and Televisions have been well received by the demographic. In this regard, integrating additional sensors to the TV to allow for the implementation of such body-based interaction could potentially be an additional means of interacting with the application. With the option of being either the primary mode of interaction or a supplement, such a hardware integration could be of potential benefit. In addition to the purpose of facilitating body-based interaction, the integrated sensors can also be programmed for a more serious purpose such as detecting unusual movements among older adults (such as falls). Hence, such an implementation of external sensors to not only enable multi-modal interaction but to also detect falls could be highly beneficial particularly among the older adults who are living alone.

Conclusion and Future Work

Based on integration of existing studies, we provide the following three implications for the design and evaluation of age-inclusive VoD and video streaming services: 1. Dynamic adaptation of age-inclusive design 2. Optimizing video bit rate 3. Integrating external sensors for multi-modal interaction.

While the discussed implications largely reflect the physiological changes experienced by older adults, future studies should also consider the various other factors (age, sex, culture, etc.,) influencing this heterogeneous user groups' use of VoD. In this regard, further conducting extensive user studies focusing on older adults' use of VoD and video streaming would allow for gaining better insight and reflections for the design and evaluation of such services and actively contribute to expanding current knowledge as there are several potential areas to explore. For instance, field studies in addition to different research methods can be opted to better understand the various human factors (i.e., physical, cognitive, social, cultural, and emotional) influencing older adults use of VoD and video streaming services and contribute to the limited knowledge available. Furthermore, while the current work focuses more on TVs, further studies could explore older adults' use of VoD and streaming services on different devices (such as mobile phones, tablets). Studies focusing on older adults behavior in this regard could be further beneficial and contribute to the existing design knowledge pertaining to the demographic, as well as unravel important considerations for QoE evaluation involving this user segment.

Acknowledgments. The research leading to these results has received funding from the Norwegian Financial Mechanism 2014-2021 under project 2019/34/H/ST6/00599.

References

1. Barnett, I., van Sluijs, E., Ogilvie, D., Wareham, N.J.: Changes in household, transport and recreational physical activity and television viewing time across the transition to retirement: longitudinal evidence from the epic-norfolk cohort. J. Epidemiol. Community Health **68**(8), 747–753 (2014)
2. Bobeth, J., Schmehl, S., Kruijff, E., Deutsch, S., Tscheligi, M.: Evaluating performance and acceptance of older adults using freehand gestures for tv menu control. In: Proceedings of the 10th European Conference on Interactive tv and Video, pp. 35–44 (2012)
3. Choi, J., Reaz, A.S., Mukherjee, B.: A survey of user behavior in vod service and bandwidth-saving multicast streaming schemes. IEEE Commun. Surv. Tutorials **14**(1), 156–169 (2011)
4. Coelho, J., Duarte, C., Biswas, P., Langdon, P.: Developing accessible tv applications. In: The Proceedings of the 13th International ACM SIGACCESS Conference on Computers and Accessibility, pp. 131–138 (2011)
5. Coelho, J., Guerreiro, T., Duarte, C.: Designing tv interaction for the elderly–a case study of the design for all approach. In: A Multimodal End-2-End Approach to Accessible Computing, pp. 49–69. Springer, Cham (2013). https://doi.org/10.1007/978-1-4471-5082-4_3
6. Fausset, C.B., Harley, L., Farmer, S., Fain, B.: Older adults' perceptions and use of technology: a novel approach. In: Stephanidis, C., Antona, M. (eds.) UAHCI 2013. LNCS, vol. 8010, pp. 51–58. Springer, Heidelberg (2013). https://doi.org/10.1007/978-3-642-39191-0_6
7. Himmelsbach, J., Schwarz, S., Gerdenitsch, C., Wais-Zechmann, B., Bobeth, J., Tscheligi, M.: Do we care about diversity in human computer interaction: A comprehensive content analysis on diversity dimensions in research. In: Proceedings of the 2019 CHI Conference on Human Factors in Computing Systems. pp. 1–16 (2019)
8. Hussain, A., Abd Razak, M.N.F., Mkpojiogu, E.O., Hamdi, M.M.F.: Ux evaluation of video streaming application with teenage users. J. Telecommun. Electron. Comput. Eng. (JTEC) **9**(2-11), 129–131 (2017)

9. Jian, C., et al.: Towards effective, efficient and elderly-friendly multimodal interaction. In: Proceedings of the 4th International Conference on PErvasive Technologies Related to Assistive Environments, pp. 1–8 (2011)
10. Jumisko-Pyykkö, S., Weitzel, M., Strohmeier, D.: Designing for user experience: what to expect from mobile 3d tv and video? In: Proceedings of the 1st International Conference on Designing Interactive User Experiences for TV and Video, pp. 183–192 (2008)
11. Kespichayawattana, J., Jitapunkul, S.: Health and health care system for older persons. Ageing Int. **33**(1), 28–49 (2008)
12. Khosla, R., Chu, M.T., Kachouie, R., Yamada, K., Yoshihiro, F., Yamaguchi, T.: Interactive multimodal social robot for improving quality of care of elderly in australian nursing homes. In: Proceedings of the 20th ACM International Conference on Multimedia, pp. 1173–1176 (2012)
13. Ma, G., Wang, Z., Zhang, M., Ye, J., Chen, M., Zhu, W.: Understanding performance of edge content caching for mobile video streaming. IEEE J. Sel. Areas Commun. **35**(5), 1076–1089 (2017)
14. Mares, M.L., Woodard, E.H., IV.: In search of the older audience: adult age differences in television viewing. J. Broadcasting Electron. Media **50**(4), 595–614 (2006)
15. McNicoll, G.: World population ageing 1950–2050. Popul. Dev. Rev. **28**(4), 814–816 (2002)
16. Saunders, G.H., Echt, K.V.: An overview of dual sensory impairment in older adults: perspectives for rehabilitation. Trends Amplif. **11**(4), 243–258 (2007)
17. Skorupska, K., Núñez, M., Kopec, W., Nielek, R.: Older adults and crowdsourcing: Android tv app for evaluating tedx subtitle quality. In: Proceedings of the ACM on Human-Computer Interaction 2(CSCW), 1–23 (2018)
18. Subramanian, S., De Moor, K., Vereijken, B., Dahl, Y., Svanæs, D.: Exploring the influence of tech savviness and physical activity in older adults playing an exergame. In: 2022 IEEE 10th International Conference on Serious Games and Applications for Health (SeGAH), pp. 1–8. IEEE (2022)
19. Subramanian, S., Skjæret-Maroni, N., Dahl, Y.: Systematic review of design guidelines for full-body interactive games. Interact. Comput. **32**(4), 367–406 (2020)
20. Tahir, R., Ahmed, F., Saeed, H., Ali, S., Zaffar, F., Wilson, C.: Bringing the kid back into youtube kids: detecting inappropriate content on video streaming platforms. In: 2019 IEEE/ACM International Conference on Advances in Social Networks Analysis and Mining (ASONAM), pp. 464–469. IEEE (2019)
21. Turkkan, B.O., et al.: Greenabr: energy-aware adaptive bitrate streaming with deep reinforcement learning. In: Proceedings of the 13th ACM Multimedia Systems Conference, pp. 150–163 (2022)
22. Volkow, N.D., et al.: Association between decline in brain dopamine activity with age and cognitive and motor impairment in healthy individuals. Am. J. Psychiatry **155**(3), 344–349 (1998)
23. Yitong, L., Yun, S., Yinian, M., Jing, L., Qi, L., Dacheng, Y.: A study on quality of experience for adaptive streaming service. In: 2013 IEEE International Conference on Communications Workshops (ICC), pp. 682–686. IEEE (2013)

Review on Internet of Things for Innovation in Nursing Process- A PubMed-Based Search

Aurora Polo-Rodríguez[1,4], Jose Romero-Sanchez[2], Elena Fernández -García[3],
Olga Paloma-Castro[2], Ana-María Porcel-Gálvez[3],
and Javier Medina-Quero[4(✉)]

[1] Department of Computer Science, University of Jaén, 23071 Jaén, Spain
apolo@ujaen.es
[2] Faculty of Nursing and Physiotherapy, University of Cadiz, 11009 Cadiz, Spain
{josemanuel.romero,olga.paloma}@uca.es
[3] Department of Nursing, University of Sevilla, 41004 Sevilla, Spain
{efernandez23,aporcel}@us.es
[4] Department of Computer Engineering, Automation and Robotics,
University of Granada, 18071 Granada, Spain
javiermq@ugr.es

Abstract. The nursing process is a systematic framework that guides nurses in delivering patient-centred care involving assessment, diagnosis, planning, implementation, and evaluation, allowing nurses to assess human needs and formulate care plans. On the other hand, the Internet of Things has revolutionised nursing by enabling the sensing and monitoring of patients in clinical and home contexts and providing more efficient and personalised care employing ubiquitous computing. In this review, we evaluate and review five years of research works based on PubMed search, which are focused on innovation generated by nursing professionals.

Keywords: Nursing · Internet of Things (IoT) · Systematic review

1 Introduction

The Internet of Things (IoT) has significantly transformed nursing, providing new avenues for patient care and management [48]. By seamlessly integrating smart devices and sensors into healthcare environments, IoT enables continuous monitoring and real-time data collection. This interconnected network of devices allows healthcare professionals to sense and track patients' vital signs, activity levels, and other health-related metrics in clinical settings and the comfort of their homes [18].

The transformative impact of IoT in nursing is evident in providing more efficient and personalized care. Through ubiquitous computing, nurses can access valuable patient information instantly, facilitating prompt interventions and tailored treatment plans. IoT technology has also improved the early detection of medical conditions, allowing timely and proactive measures to be taken [10].

J. Bravo and G. Urzáiz (Eds.): UCAmI 2023, LNNS 835, pp. 57–70, 2023.
https://doi.org/10.1007/978-3-031-48306-6_6

This article delves into the various applications of IoT in nursing, including wearable sensors, mobile devices, and embedded systems for continuous health monitoring. We also explore the research objectives, health conditions addressed, and evaluation methodologies used to validate the effectiveness of IoT-based nursing solutions. Our focus is on highlighting emerging trends and gaps in the application of IoT in nursing, with the aim of pinpointing areas that require further attention and development in the field. We intend to provide an overview of future possibilities and challenges.

2   Methodology

This section details the methodology and eligibility criteria established a prior. According to the PRISMA 2020 (Preferred Reporting Items for Systematic Reviews and Meta-Analyses) [36], the work is conducted as a literature review.

2.1   Eligibility Criteria

This review focused on research on the Internet of Things (IoT) technology and its application in nursing. The objective was to evaluate and review research articles conducted in the last five years (from 2018 to 2023) that focused on innovation generated by nursing professionals and met specific inclusion criteria. The inclusion criteria were as follows:

- Studies with a direct relationship between IoT technology and its application in nursing.
- Studies involved case studies or evaluations in some participants, which implies that the proposed technology was applied and evaluated in real situations or with study subjects.
- Studies published between January 2018 and June 2023.
- Scientific article published in a journal or conference proceedings.

2.2   Information Sources and Search Strategy

In this scientific review, we performed a meticulous literature search using the extensive PubMed database, encompassing more than 28 million biomedical article citations from MEDLINE, life science journals and online books. Maintained by the esteemed National Center for Biotechnology Information (NCBI) at the National Library of Medicine (NLM), PubMed was an invaluable tool for focusing on pertinent publications investigating the integration of cutting-edge technologies in nursing.

Our search strategy adhered to pre-defined inclusion and exclusion criteria. The search spanned January 1, 2018, to December 31, 2023. Our primary search terms were "Internet of Things" and "nursing" included in the abstract or title of the studies. To ensure thoroughness, we also considered synonyms and related terms. Specifically, we include "IoT" as a synonym for "Internet of Things." However, no additional synonyms were necessary for the term "nursing" because of its specificity. We specifically selected papers published in English in reputable computer science or engineering journals or proceedings. Furthermore, we focused on papers that incorporate case studies or participant evaluations, demonstrating real-world applications of IoT technology in nursing. The resulting PubMed query is as follows: *(IoT OR "Internet of Things") AND ("nursing")*

2.3 Study Selection

Selection criteria consisted of articles describing the use of IoT in nursing for monitoring, care, diagnosis, etc., including articles with any of the following specifications, with a focus on the nursing domain:

- Developing systems or applications that facilitate the detection of medical symptoms.
- Development of hardware and software systems for monitoring or rehabilitation purposes.
- Evaluations or questionnaires on how the use of technology might affect the care process or the patients themselves.

Figure 1 shows the study's PRISMA diagram, showing the scoping review flow.

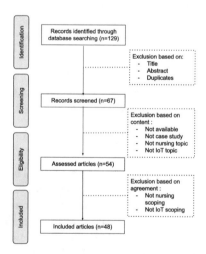

Fig. 1. Flowchart for the literature search based on PRISMA protocol.

All identified titles and abstracts of pre-selected articles (n = 129) were classified as related, unrelated, and unclear, and eligibility was checked by consensus of two researchers (A.P.R., J.M.Q.). Two researchers (A.P.R., J.M.Q.) reviewed the full text of unclear articles before they were labelled related or unrelated. The related articles (n = 54) were retrieved, and the full texts were checked independently by all researchers (J.M.Q., A.P.R., J.R., O.P., A.P., E.F.) to ensure the eligibility criteria were met. Through this process, 48 articles were included in the review.

To better understand how IoT approaches have permeated the nursing field, we focused on the studies that developed a case study with participants. We classified the journals into three categories: medical, informatics, or mixed.

All authors reviewed the included studies and extracted relevant information from each article by dividing it into four groups (bibliometric analysis, sensing technology, application of IoT to healthcare and study characteristics), from which various categories have been derived. The articles have been counted according to the following criteria: articles in more than one category have been counted as many times as the number of categories they fulfil, which is why some percentages are above 100%.

– Exclusive categories: journal type, context of participants (children, pregnancy, older, urgency, etc.), research goals, participant's health condition (disorder), participant's context (hospitalized, under treatment, etc.), number of participants of case study, role of participants of the case study (patients, professionals and students), place where the case study is developed (hospital, home), conditions of the case study (controlled, naturalistics), duration of case study and evaluation type (survey, sensor metrics).
– Non-exclusive categories: device name, type of device (wearable, mobile, vision, embedded, etc.), sensor metric (heart rate, spO2, etc.), monitored variable (sleep, activity, stress, weight, etc.)

3   Results

This section presents findings regarding bibliometric analysis, sensing technology, application of IoT to health care, and study characteristics.

3.1   Bibliometric Analysis

In the modern age of ubiquitous technology and AI's progress, engineering's impact on nursing has steadily risen. This interdisciplinary partnership has benefited nurses and significantly improved the patient recovery process. Notably, these advantages include tailored medication administration, early disease recognition, successful treatment plans, and the facilitation of remote rehabilitation activities, among other remarkable achievements. These remarkable accomplishments have been made possible through the combined efforts of various teams consisting of medical personnel and engineers.

The reviewed manuscripts are published in 25 distinct journals within this collection of articles. We have categorised these journals into three primary classifications: medical journals, accounting for 43.8% of the articles [3–7,9,13,17,19,21,25,32,34,37–40,42,45,51,52]; computer science journals, comprising 22. 9% of the articles [1,8,16,23,24,41,43,44,47,49,50]; and multidisciplinary journals, comprising 33. 3% of the articles [2,11,12,14,15,20,22,26–31,33,35,46], based on their main topics. Graph 2 displays the number of articles published annually by area. These data reveal that technology improvement has enabled the expansion of research that incorporates technology into nursing, not just in medical journals.

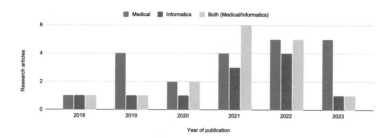

Fig. 2. Number of articles published annually by areas.

3.2 Sensing Technology and Approach

The continuous advancement of technology in research has seen remarkable progress in developing more user-friendly and less invasive devices. In this context, wearable sensors have emerged as a major tool in the latest scientific research, being present in 56.25% of the studies reviewed [3, 7, 9, 13, 15, 21, 23–27, 29–32, 34, 37–43, 46, 49–51], including activity bracelets, smart rings, cardiac and physiological sensors worn on different body parts (head and chest). The second most widely used type has been the embedded device (47.92% of the studies [1, 2, 4, 6, 7, 11, 12, 16, 19–21, 25–27, 29, 30, 35, 38, 39, 44, 45, 47, 49]), which includes primarily environmental sensors such as opening and closing sensors, temperature, light, and presence sensors. However, it also includes location or identification beacons with Bluetooth Low Energy, WiFi, or RFID protocols. Some studies classified as embedded systems were commercial systems that monitored specific parameters such as infusion time. The third most used type of device is the mobile device (33. 3% of studies [3, 5, 7, 9, 14, 17, 21, 25, 26, 29, 31, 33, 42, 43, 46, 49]), including smartphones, PDAs and tablets, as they allow for the monitoring of values obtained through other sensors or systems. Regarding audio and vision devices, such as microphones, cameras, and speakers, their use has decreased due to privacy concerns, being present in only 12.5% of the studies [8, 14, 17, 27, 28, 49]. It is worth noting that 29.17% of the total studies [7, 9, 21, 25–27, 30, 31, 38, 39, 42, 43, 46, 49] combine more than one technology.

Figure 3 shows the evolution over the years of the types of devices used in the studies, with the growing trend of employing diverse data sources from multimodal sensors becoming more common.

Some studies use specific commercial devices such as the MediHandTrace [11], Patient Smart Reader, Raspberry Pi 4 Model B [4], various Samsung devices (Gear Sport) [2, 23, 32, 34, 40, 42, 43], Garmin Vivosmart [11, 13, 41], Oura

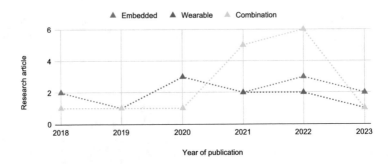

Fig. 3. Evolution of the use of different types of devices in scientific articles.

ring[5, 16, 19, 23, 32, 40], ActiGraph[8, 29, 32], Fitbit Charge 5 [1, 21, 38], Withings BPM Connect [38, 49], and InBody Dial h20b[25], among others. Some studies use devices like the Aeotec hub, Aeotec door and window sensor 6, Aeotec multisensory 6, Aeotec Smart Switch 6 (smart plugs), Lenovo smart tablet M8, Danalto positioning Wearable device with associated hubs [49], and Hexiwear biometric bracelets [9]. Almost half of the studies (47.72%) do not mention which device they use [1–3, 7, 12, 14, 16, 17, 19, 22, 24, 26, 27, 29–31, 33, 35, 39, 44, 46, 47, 50]. Figure 4 shows an overview of the technological devices used.

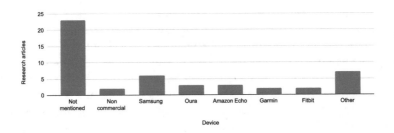

Fig. 4. Technological devices used by each study.

Thanks to these devices, it is possible to obtain metrics from the sensors that compose them, the most common being sleep monitoring (25%) [4, 7, 13, 19, 21, 23, 32, 34, 38, 40, 41, 43, 49], heart rate (22.92%) [3, 13, 32, 34, 38, 40, 41, 47, 51], body composition (18.75%) [21, 23, 25, 26, 37–39, 45, 47] and environmental parameters (light, temperature, opening and closing of doors, etc.) (16.67%) [1, 6, 16, 27, 35, 37, 39, 49]. Figure 4 shows in detail each of the metrics used in the studies.

3.3   Application of IoT to Healthcare

The technological application covers the detection or prevention of specific symptoms that can lead to illness, monitoring users, and optimising the nursing process.

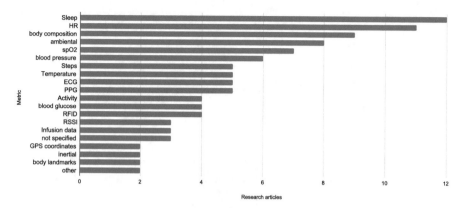

Fig. 5. Sensor metrics used by each study.

According to the research objectives of each study, 14.6% [1,21,28–31,34] of the studies focus on the detection or prevention of medical symptoms or specific conditions through the use of IoT technology or IoT-based systems. 14.6% [7,14,25,37,40,45,51] of the studies address the rehabilitation process, i.e. recovery and improvement of the physical or mental function of patients who have suffered an injury, illness or medical procedure. 4.2% [27,46] of studies focus on detection and rehabilitation. 25% [2,8,9,12,15,19,26,33,35,38,39,44] of the studies concentrate on enhancing and optimising the nursing process in various healthcare environments. The research objectives in these articles strive to utilise IoT technology or IoT-based systems to help and improve the duties and activities of nursing personnel, which could result in more efficient and effective patient care. 25% [4,6,11,13,16,17,20,32,41–43,49] of the studies focus on monitoring users or patients in different healthcare settings. 16.7% [3,5,22–24,47,50,52] of the studies investigate the acceptance and perception of IoT technology or IoT-based systems among nursing professionals, nursing students or other relevant users in healthcare.

The graph in Fig. 6 shows that most studies have concentrated on general healthcare (58.3%) [3–9,11–13,15,16,19,20,22–24,26,28,32,33,35,38,41,44, 47,49,52] and dementia (8%) [1,17,39,50]. Specifically, 25% of the studies [5,7–9,17,22,23,30,31,38,42,52] focus on older people, 16.7% on pregnant women and people hospitalized [4,6,12,15,19–21,25,26,28,33,35,39,41,46,51], 14.6% on patients under treatment for illness [1,14,27,32,37,47,50], 6.3% on nurses [24,34,43], 4.2% on people in emergency rooms, children, healthy adults and the general public [2,3,11,13,16,29,40,44,45,49], showed in Fig. 6.

Integration of technology into the nursing process has been approached in different ways. According to the monitored variable, we found that almost half of the studies (45.83%) [1,4,6,9,13–15,21,22,24,25,27,32–35,37–39,41,43,51] monitor vital signs, especially those focused on hospitalised, under treatment or pregnant women. Controlling and monitoring of sleep have been considered in 27.08% [4,7,13,19,21,29,38–41,43,45,49] studies, especially in pregnant

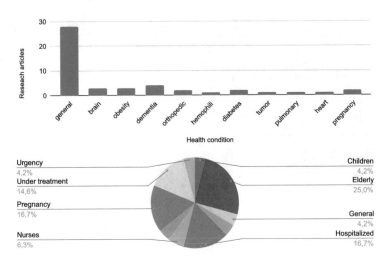

Fig. 6. Above) Number of studies per health condition, below) Scientific articles classified according to the participant's context.

women, being the only variable to be monitored in some of them. Monitoring the user's position or movement has also taken on great importance, with 20.83% [4,11,16,17,20,24,25,28,38,45,46,49,50] of the studies on this subject, primarily those centred on nurses. Other monitored variables are stress (12.5%) [1,3,21,43,45,50], activity (10.42%) [1,3,21,43,45,50], mostly in older people, specific parameters such as infusion time in emergency or hospitalised patients (10.42%) [2,12,26,30] and mental health (8.33%) [17,44,45,52] in older people above all. Only three studies (6.25%) [5,42,47] could not be classified in any group.

3.4   Study Characteristics

Evaluations play a key role in validating the outcomes resulting from the integration of technology in the nursing domain. Consequently, articles lacking a case study were excluded from the scope of this review. The mean number of participants across the studies is 94, with a minimum sample size of one subject [23] and a maximum of 568 [2]. In particular, 33.33% [1,5,8,9,13,15–17,19,23,38,41, 43,44,50] of the studies encompassed fewer than 30 participants, while a considerable 37.50% [4,6,24,25,27–34,37,40,42,45,46,49] involved a participant range of 30 to 90. Moreover, 22.92% of the studies [2,3,12,14,21,22,26,39,47,51,52] encompassed a more substantial number of participants, exceeding 100 subjects. The remaining articles (6.25%) do not specify the number of participants.

Regarding the duration of the case studies, it should be noted that 14.58% [7,8,12,32,38,40] of the included studies conducted evaluations lasting seven days or less. A significant proportion, 10.42% [20,23,25,27,28] of the studies, ranged from one week to one month. Furthermore, 14.58% [2,6,19,37,45,50] of the studies extended their evaluation period between one month and three months, while 18.75% [13, 16, 26, 33, 39, 49] lasted from three to six months. Additionally, 12.5% [3, 5, 43] of the studies spanned six months and one year, and an equal percentage (10.42%) [11, 31, 34, 44] of the studies persisted for more than one year. Notably, the shortest evaluation period observed was 55 min [15], whereas the most extensive study lasted three years [30]. 18.75% [4, 9, 14, 24, 35, 46, 47, 51, 52] of the studies did not specify the duration of their case studies.

The majority of study participants were patients (66.67% of the total studies [1, 2, 6–8, 12–14, 16, 17, 21, 23, 25–31, 34, 35, 37–45, 51, 52]), while 10.42% of the studies [5, 9, 11, 20, 47] were evaluated solely by professionals, and another 10.42% [19, 33, 46, 49, 50] involved both patients and professionals. Nursing students and healthy adults without pathology were represented in 6.25% of the studies [3, 4, 15, 22, 24, 32]. More than half of the evaluations were conducted in clinical or hospital settings (58.33%) [1, 2, 4, 5, 7–9, 11, 12, 17, 19–21, 25, 26, 28–31, 33–35, 38, 46, 47, 50–52], with 65.38% [2, 4, 7, 8, 11, 12, 17, 25, 26, 28–31, 33–35, 51] of them under controlled conditions and only 34.62% [1, 9, 19–21, 38, 46, 47, 50] under naturalistic conditions. On the other hand, in private homes (33.33%) [6, 13, 14, 16, 22, 27, 32, 37, 39–45, 49], 81.25% of the studies [6, 13, 14, 16, 22, 27, 32, 37, 39, 41–43, 49] were conducted under naturalistic conditions, while only 18.75% [40, 44, 45] were under controlled conditions. A small number of studies were carried out in laboratories or universities (8.33%) [3, 15, 23, 24], where there was uniformity in the study conditions.

Regarding the type of evaluation that has been carried out to show the study's validity, 58.33% [2, 4, 8, 12, 15, 17, 19–22, 24, 25, 28, 30, 32, 34, 35, 37–46, 50, 51] of the studies use sensor metrics, 20.83% [1, 3, 5, 9, 11, 14, 16, 33, 47, 52] conduct surveys, and 18.75% [6, 7, 13, 23, 26, 27, 29, 31, 49] use both techniques.

4 Discussion

Integrating technology in nursing has shown remarkable advances in improving patient care and healthcare outcomes. This review focuses on studies that present innovation in the nursing process through IoT and have demonstrated its validity through a case study.

Analysis of 48 reviewed articles revealed that technology had been integrated into various aspects of nursing, with significant contributions from medical, informatics, and multidisciplinary journals. This integration has led to advances in patient monitoring using wearable sensors such as activity wristbands, smart rings and physiological sensors. Embedded devices, such as environmental sensors and location beacons, have also been prevalent in this research. These devices have enabled the collection of essential metrics, such as sleep monitoring, heart rate, body composition and environmental parameters, which have proven valuable in detecting or preventing medical symptoms and optimising the nursing process. Above all, there has been substantial growth in using these devices in combination, as more variables can be monitored, providing helpful information (e.g. vital signs and location). However, privacy concerns have limited the use of audio and vision devices in patient monitoring, as reflected in the number of studies using such sensors over the years.

Studies have focused on various healthcare objectives, such as disease detection and prevention, rehabilitation, user or patient monitoring, and assessing the acceptance of IoT technology among healthcare professionals. General healthcare and dementia have been the main areas of concentration in these studies. It makes sense that most studies focus on optimising the nursing process, as it saves healthcare costs and increases patient care quality. Evaluation methods have evolved, with a significant number of studies presenting prototypes and proofs of concept for research purposes. The duration of case studies has varied, with evaluations lasting between 55 min and three years. Most study participants were patients, followed by professionals and a combination of patients and professionals. Evaluations have been conducted in various settings, with clinical or hospital settings being the most common, often under controlled conditions. Sensor metrics and surveys have been the main methods used for evaluation, and some studies have used both techniques to validate their results. Several studies have omitted crucial details such as the duration of their case studies, the technology used, and the metrics tracked, making it challenging to comprehend the breadth and significance of their research. This lack of transparency hinders the complete evaluation and limits the possibility of drawing meaningful conclusions. Henceforth, it is paramount that future studies adopt a transparent and comprehensive reporting approach to ensure the reproducibility and dependability of research outcomes.

Overall, integrating technology in nursing has opened up new possibilities for patient monitoring and disease prevention, improving the healthcare process and contributing to better patient outcomes. The results of this review provide valuable information for future research and development in this rapidly evolving field. For future systematic reviews, it would be essential to study the interventions' effectiveness, usefulness, or applicability. It would be interesting to see the participation of nurses or other health professionals in the studies, even if they are focused on patients, as many times technology is developed that may not respond to real problems or that the impact could be more significant if the opinion of experts had been obtained beforehand.

5 Conclusion

The healthcare industry is undergoing a change towards patient-centred approaches and less invasive methods for diagnosis and treatment. This shift has been made possible by the integration of wearable, mobile, and textile sensing technology, leading to significant advancements in continuous health monitoring. After analyzing 48 articles from PubMed, our review found a varied research landscape in nursing technology. The studies were published in 25 different journals, with medical journals making up 43.8%, computer science journals making up 22. 9%, and multidisciplinary journals comprising 33.3%. This result suggests that technology integration in nursing research expands beyond medical publications.

The popularity of wearable sensors (56.25% of studies) and embedded devices (47.92% of studies) is on the rise, as they offer user-friendly and less invasive data collection. Additionally, mobile devices (37.5% of studies) play a significant role in obtaining valuable data. However, privacy concerns limit the usage of audio and vision devices (12.5% of studies). The research objectives in this field cover a wide range, including medical symptom detection and prevention (14.6%), rehabilitation (14.6%), optimizing nursing processes (25%), continuous monitoring of users (25%), acceptance and perception of IoT technology among nursing professionals (16.7%), and others. Various health conditions have been studied, focusing on general healthcare (58.3%) and dementia (8%). The most monitored variables in the studies are vital signs (45.83%) and sleep (27.08%).

Acknowledgements. This contribution has been supported by the Spanish Institute of Health ISCIII through the project DTS21-00047. Moreover, this research has received funding from EU Horizon 2020 Pharaon Project 'Pilots for Healthy and Active Ageing', Grant agreement no. 857188.

References

1. Antón, M.Á., Ordieres-Meré, J., Saralegui, U., Sun, S.: Non-invasive ambient intelligence in real life: dealing with noisy patterns to help older people. Sensors **19**(14), 3113 (2019)
2. Bai, X., et al.: Application of infusion control system based on internet of things technology in joint orthopedics nursing work. J. Healthc. Eng. **2021** (2021)
3. Bodur, G., Gumus, S., Gursoy, N.G.: Perceptions of Turkish health professional students toward the effects of the internet of things (IoT) technology in the future. Nurse Educ. Today **79**, 98–104 (2019)
4. Chen, P.J., Hu, T.H., Wang, M.S.: Raspberry Pi-based sleep posture recognition system using AIoT technique. In: Healthcare, vol. 10, p. 513. MDPI (2022)
5. Chien, S.C., et al.: Investigating nurses' acceptance of patients' bring your own device implementation in a clinical setting: a pilot study. Asia Pac. J. Oncol. Nurs. **10**(3), 100195 (2023)
6. Choi, Y.K., Thompson, H.J., Demiris, G.: Use of an internet-of-things smart home system for healthy aging in older adults in residential settings: pilot feasibility study. JMIR Aging **3**(2), e21964 (2020)
7. Duan, N., Lin, G.: Effect of intelligent medical data technology in postoperative nursing care. BioMed Res. Int. **2022** (2022)
8. Dunn, M., et al.: Notes from the field: a voice-activated video communication system for nurses to communicate with inpatients with COVID-19. JMIR Formative Res. **6**(3), e31342 (2022)
9. Durán-Vega, L.A., et al.: An IoT system for remote health monitoring in elderly adults through a wearable device and mobile application. Geriatrics **4**(2), 34 (2019)
10. Farahani, B., Firouzi, F., Chakrabarty, K.: Healthcare IoT. In: Intelligent Internet of Things: From Device to Fog and Cloud, pp. 515–545 (2020)
11. Florea, O., Gonin, J., Tissot Dupont, H., Dufour, J.C., Brouqui, P., Boudjema, S.: Internet of things to explore moment 2 of who my five moments for hand hygiene. Front. Digital Health **3**, 684746 (2021)

12. Gao, Y., Kong, D., Fu, X.J., Pi, H.Y.: Application and effect evaluation of infusion management system based on internet of things technology in nursing work. In: Nursing Informatics 2018, pp. 111–114. IOS Press (2018)

13. Grym, K., et al.: Feasibility of smart wristbands for continuous monitoring during pregnancy and one month after birth. BMC Pregnancy Childbirth **19**(1), 1–9 (2019)

14. Gu, Y., et al.: Empirical analysis of the nursing effect of intelligent medical internet of things in postoperative osteoarthritis. Computat. Math. Methods Med. **2022** (2022)

15. Han, H.J., Labbaf, S., Borelli, J.L., Dutt, N., Rahmani, A.M.: Objective stress monitoring based on wearable sensors in everyday settings. J. Med. Eng. Technol. **44**(4), 177–189 (2020)

16. Hjelm, K., Hedlund, L.: Internet-of-things (IoT) in healthcare and social services-experiences of a sensor system for notifications of deviant behaviours in the home from the users perspective. Health Inf. J. **28**(1), 14604582221075562 (2022)

17. Hung, L.P., Huang, W., Shih, J.Y., Liu, C.L.: A novel IoT based positioning and shadowing system for dementia training. Int. J. Environ. Res. Public Health **18**(4), 1610 (2021)

18. Islam, M.M., Rahaman, A., Islam, M.R.: Development of smart healthcare monitoring system in IoT environment. SN Comput. Sci. **1**, 1–11 (2020)

19. Itoh, S., Tan, H.P., Kudo, K., Ogata, Y.: Comparison of the mental burden on nursing care providers with and without mat-type sleep state sensors at a nursing home in Tokyo, Japan: Quasi-experimental study. JMIR Aging **5**(1), e19641 (2022)

20. Karvonen, S., et al.: Key flow processes on wards. HERD: Health Environ. Res. Design J. **16**(2), 208–222 (2023)

21. Kawasaki, M., et al.: Protocol for an interventional study to reduce postpartum weight retention in obese mothers using the internet of things and a mobile application: a randomized controlled trial (SpringMom). BMC Pregnancy Childbirth **21**, 1–14 (2021)

22. Kivekas, E., Mikkonen, S., Koponen, S., Saranto, K.: Technology supporting nursing at homecare-seems to be lacking. In: Integrated Citizen Centered Digital Health and Social Care (2020)

23. Lai, J., et al.: Using multimodal assessments to capture personalized contexts of college student well-being in 2020: case study. JMIR Formative Res. **5**(5), e26186 (2021)

24. Laplante, N.L., Laplante, P.A., Voas, J.M.: Could the internet of things be used to enhance student nurses' experiences in a disaster simulation? On-line J. Nursing Inf. **22**(1) (2018)

25. Lee, Y.J., Hong, J.H., Hur, M.H., Seo, E.Y.: Effects of virtual reality exercise program on blood glucose, body composition, and exercise immersion in patients with type 2 diabetes. Int. J. Environ. Res. Public Health **20**(5), 4178 (2023)

26. Li, L., et al.: Design and implementation of hospital automatic nursing management information system based on computer information technology. Comput. Math. Methods Med. **2021** (2021)

27. Li, X., Ren, S., Gu, F.: Medical internet of things to realize elderly stroke prevention and nursing management. J. Healthc. Eng. **2021** (2021)

28. Li, Y., Zhang, P., Zhang, Y., Miyazaki, K.: Gait analysis using stereo camera in daily environment. In: 2019 41st Annual International Conference of the IEEE Engineering in Medicine and Biology Society (EMBC), pp. 1471–1475. IEEE (2019)

29. Liu, M., et al.: Research on the impact of home nursing based on intelligent medical internet of things on the quality of life of patients with hemophilia. Comput. Math. Methods Med. **2022** (2022)
30. Liu, S., Jiang, L., Wang, X., et al.: Intelligent internet of things medical technology in implantable intravenous infusion port in children with malignant tumors. J. Healthc. Eng. **2021** (2021)
31. Lu, L., Huang, T.: Effects of early nursing monitoring on pregnancy outcomes of pregnant women with gestational diabetes mellitus under internet of things. Comput. Math. Methods Med. **2022** (2022)
32. Mehrabadi, M.A.: Sleep tracking of a commercially available smart ring and smart-watch against medical-grade actigraphy in everyday settings: instrument validation study. JMIR Mhealth Uhealth **8**(11), e20465 (2020)
33. Nie, X., et al.: Construction and application of comprehensive nursing information service platform based on internet of things technology. J. Healthc. Eng. **2022** (2022)
34. Niela-Vilén, H., et al.: Pregnant women's daily patterns of well-being before and during the COVID-19 pandemic in Finland: longitudinal monitoring through smartwatch technology. PLoS ONE **16**(2), e0246494 (2021)
35. Ou, T., et al.: A novel method of clinical nursing under the medical internet of things technology. J. Healthc. Eng. **2021** (2021)
36. Page, M.J., et al.: Prisma 2020 explanation and elaboration: updated guidance and exemplars for reporting systematic reviews. BMJ **372** (2021)
37. Park, S., et al.: Mobile phone app-based pulmonary rehabilitation for chemotherapy-treated patients with advanced lung cancer: pilot study. JMIR Mhealth Uhealth **7**(2), e11094 (2019)
38. Paulauskaite-Taraseviciene, A., et al.: Geriatric care management system powered by the IoT and computer vision techniques. In: Healthcare, vol. 11, p. 1152. MDPI (2023)
39. Rostill, H., Nilforooshan, R., Morgan, A., Barnaghi, P., Ream, E., Chrysanthaki, T.: Technology integrated health management for dementia. Br. J. Community Nurs. **23**(10), 502–508 (2018)
40. Saarikko, J., et al.: Supporting lifestyle change in obese pregnant mothers through the wearable internet-of-things (slim)-intervention for overweight pregnant women: study protocol for a quasi-experimental trial. PLoS ONE **18**(1), e0279696 (2023)
41. Saarikko, J., et al.: Continuous 7-month internet of things-based monitoring of health parameters of pregnant and postpartum women: prospective observational feasibility study. JMIR Formative Res. **4**(7), e12417 (2020)
42. Sarhaddi, F., Azimi, I., Axelin, A., Niela-Vilen, H., Liljeberg, P., Rahmani, A.M., et al.: Trends in heart rate and heart rate variability during pregnancy and the 3-month postpartum period: continuous monitoring in a free-living context. JMIR Mhealth Uhealth **10**(6), e33458 (2022)
43. Sarhaddi, F., et al.: Long-term IoT-based maternal monitoring: system design and evaluation. Sensors **21**(7), 2281 (2021)
44. Sauzéon, H., Edjolo, A., Amieva, H., Consel, C., Pérès, K., et al.: Effectiveness of an ambient assisted living (HomeAssist) platform for supporting aging in place of older adults with frailty: protocol for a quasi-experimental study. JMIR Res. Protoc. **11**(10), e33351 (2022)
45. Seo, E.Y., Kim, Y.S., Lee, Y.J., Hur, M.H.: Virtual reality exercise program effects on body mass index, depression, exercise fun and exercise immersion in overweight middle-aged women: a randomized controlled trial. Int. J. Environ. Res. Public Health **20**(2), 900 (2023)

46. Song, Y., et al.: Medical data acquisition and internet of things technology-based cerebral stroke disease prevention and rehabilitation nursing mobile medical management system. Comput. Math. Methods Med. **2022** (2022)
47. Tak, S.H., Choi, H., Lee, D., Song, Y.A., Park, J.: Nurses' perceptions about smart beds in hospitals. Comput. Inform. Nurs. **41**(6), 394 (2023)
48. Thilakarathne, N.N., Kagita, M.K., Gadekallu, T.R.: The role of the internet of things in health care: a systematic and comprehensive study. Available at SSRN 3690815 (2020)
49. Timon, C.M., et al.: Development of an internet of things technology platform (the NEX system) to support older adults to live independently: protocol for a development and usability study. JMIR Res. Protoc. **11**(5), e35277 (2022)
50. Wang, G., Albayrak, A., Kortuem, G., van der Cammen, T.J., et al.: A digital platform for facilitating personalized dementia care in nursing homes: formative evaluation study. JMIR Formative Res. **5**(5), e25705 (2021)
51. Wen, J., et al.: Analysis of perioperative nursing intervention effect of cerebrovascular intervention patients based on intelligent internet of things. BioMed Res. Int. **2022** (2022)
52. Zhao, Y., Sazlina, S.G., Rokhani, F.Z., Su, J., Chew, B.H.: The expectations and acceptability of a smart nursing home model among Chinese elderly people: a mixed methods study protocol. PLoS ONE **16**(8), e0255865 (2021)

Detection of Sets and Repetitions in Strength Exercises Using IMU-Based Wristband Wearables

Aurora Polo-Rodriguez[1,3], David Diaz-Jimenez[1], Miguel Angel Carvajal[2], Oresti Baños[3], and Javier Medina-Quero[3(✉)]

[1] Department of Computer Science, University of Jaén, Jaén 23071, Spain
apolo@ujaen.es

[2] ECsens, CITIC-UGR, IMUDs, Department of Electronics and Computer Technology, University of Granada (UGR), Granada 18014, Spain

[3] Department of Computer Engineering, Automation and Robotics, University of Granada, Granada 18071, Spain
javiermq@ugr.es

Abstract. This study presents a model for classifying four strength exercises (chest, back, biceps, and triceps). The system is built on the data collected from an IMU sensor embedded in a wearable wristband device. We gathered data from multiple sessions involving three users while they performed these exercises on fitness machines in real-world scenarios. A Deep Learning model incorporating CNN and LSTM networks was developed to classify these exercises and estimate repetition progressions accurately. The performance results are highly promising, suggesting the feasibility of integrating such exercise recognition capabilities into both mobile and wearable devices.

Keywords: IMU-based wristband wearable · Deep Learning · Strength exercises

1 Introduction

In recent years, the growing popularity of wearable devices has revolutionised various aspects of our lives, including fitness and health monitoring [9]. Among the different types of wearable, wristbands based on an Inertial Measurement Unit (IMU) have emerged as a promising tool to track physical activities [3,4]. With their ability to capture motion data accurately and non-invasively, IMU-based wristbands offer new opportunities for analysing and quantifying strength exercises. This article focuses on detecting sets and repetitions in strength exercises using IMU-based wristband wearables to provide a reliable and convenient solution for monitoring and evaluating exercise performance.

Strength exercises, such as weightlifting and resistance training, are fundamental components of fitness routines to improve muscular strength and endurance at all stages of human rehabilitation [2,10]. Tracking and analyzing sets and repetitions during these exercises are crucial for assessing performance, ensuring proper form, and preventing injuries. Traditional methods for

© The Author(s), under exclusive license to Springer Nature Switzerland AG 2023
J. Bravo and G. Urzáiz (Eds.): UCAmI 2023, LNNS 835, pp. 71–80, 2023.
https://doi.org/10.1007/978-3-031-48306-6_7

monitoring strength exercises, such as manual counting or reliance on visual cues, are prone to human error and lack objectivity. By leveraging the power of IMU-based wristband wearables, we can overcome these limitations and provide a more accurate and automated solution for detecting sets and repetitions. IMU-based wristband wearables combine accelerometer and gyroscope sensors to capture three-dimensional motion data, enabling precise measurement of wrist movements during strength exercises [4].

In general, this work aims to comprehensively explore the detection of sets and repetitions in strength exercises using IMU-based wristband wearables using Deep Learning methods. By harnessing the capabilities of these devices, we can improve exercise monitoring, improve performance assessment, and contribute to the advancement of personalised fitness training. The remainder of this paper is organised as follows. Section 2 details some related work. Section 3 describes the preprocessing of the data and the CNN+LSTM baseline model used for classification and estimating the progression of exercises. Section 4 shows the results of a real-life case study. In the end, 5 concludes the article and discusses ongoing.

2   Related Works

In recent years, the application of Convolutional Neural Networks (CNN), Long Short-Term Memory ($LSTM$), and $CNN+LSTM$ models has gained significant attention in various domains, particularly in the field of activity recognition. These techniques have proven to be powerful tools for extracting meaningful information from the data, allowing accurate identification and classification of different activities.

The use of CNN models has been widely explored for activity recognition tasks. CNNs excel at capturing spatial features from visual data, making them suitable for analyzing image or video-based activities. In this sense, Polo-Rodríguez et al. [7] proposed using thermal vision sensors to estimate the landmarks of the frontal body of an inhabitant using a residual neural network. Their approach includes an auto-labelling system with dual visible-spectrum and thermal cameras, and they achieved encouraging results in a case study with four participants. Furthermore, Medina-Quero et al. [6], developed a methodology to detect falls from non-invasive thermal vision sensors using CNNs. The authors collected a dataset that includes several cases where data augmentation techniques were used to increase the learning capabilities of the classification. The results show an encouraging performance, with up to 92% accuracy.

The use of LSTMs in activity recognition is significant because of their ability to model the temporal dependencies in sequential data effectively. This makes them highly suitable for processing time series data, such as sensor activations. Medina-Quero et al. [5] propose an ensemble of activity-based classifiers for balanced training, where each classifier in the ensemble is a long-short-term memory (LSTM), used to classify activities based on binary sensor data.

The combination of CNNs and LSTMs has also been addressed in activity recognition, which presents several advantages. CNNs help extract spatial features from sensor data, while LSTMs effectively model temporal dependencies. When these two models are combined, the resulting model can capture both spatial and temporal features, leading to improved performance in activity recognition. Polo-Rodríguez et al. [8] propose a method to classify five human activities related to sport using a thermal vision sensor. The authors then used a deep learning model that combines a CNN and an LSTM network to extract relevant features from the spatial domain and model the sequence of images to compute the final classification. The results show promising performance and quick learning capabilities.

CNNs and LSTMs have also been treated in a wide range of fields. Almonacid-Olleros et al. [1] explore the use of machine learning models to predict the behaviour and energy production of a photovoltaic system using Internet of Things environmental data. The authors evaluated several machine learning models and compared them with the state-of-the-art analytical model with respect to error metrics and learning time. The results show that machine learning models offer improved results compared to the analytical model, with significant differences in learning time and performance. The use of multiple temporal windows is shown as a suitable tool for modelling temporal features to improve performance.

3   Materials and Methods

In this section, we describe the materials, pre-processing of data and the Deep Learning model for detecting sets and repetitions in strength exercises using IMU-based wristband wearables. First, IMU-based wristband wearable preprocessing of data for computing sequence features is described. Second, a DL model based on CNN + LSTM is detailed to classify and estimate the progression of the exercises.

3.1   IMU-Based Wristband Wearable

The Fitbit Versa 3 is a wearable that offers a range of sensors for comprehensive tracking, including an optical heart rate monitor for continuous heart rate measurement, an accelerometer + gyroscope for tracking movement, a barometer for measuring changes in atmospheric pressure, a built-in GPS for accurate outdoor activity tracking, and a SpO2 sensor for monitoring oxygen saturation levels in the blood. This device offers the ability to develop applications using JavaScript, allowing developers to harness its potential fully. The accelerometer sensor has been used for this work, with a sampling frequency of 50 samples per second. In Fig. 1, we show the Fitbit Versa 3 wearable device and its wearable application.

Fig. 1. IMU-based wristband wearable: Fitbit Versa 3 and wearable application for collecting data.

3.2 Segmentation and Sequence of Features

A data stream S obtained from a sensor s consists of a set of measurements $s = s_0, \ldots, s_i$. Each measurement s_i corresponds to a specific time-stamp t_i within the defined time interval of interest $[t_0, t_N]$. In our study, the input sensors include the 3-axis acceleration (X, Y, Z) of an IMU-based wristband wearable. Given the high and variable sample rate of the wearable (ranging from 100 Hz to 150 Hz), we initially established a method for signal segmentation to aggregate and resample the data. This segmentation is achieved through short-term temporal windows T whose size is 50ms, $T = 50ms$. An aggregation function denoted $\bigcup$ is applied to aggregate and reduce the sampled data within these windows. In our specific case, we use the mean value represented as $\bigcup = \mu$. A mean was computed because the variation in inertial data within a 50 ms time-frame is not crucial for activity recognition and there was no need to include other aggregation measures like maximum or minimum. After this preprocessing method, we note the sample size reduction in raw data from 100-150 to 20 per second.

Additionally, we introduce a time-step parameter defined as $\Delta = 250ms$ to identify specific time points $t_i = t_0, t_0 + \Delta + \ldots + t_N - \Delta + t_N$ within the time interval $[t_0, t_N]$ to collect a sequence of data and label it according to the exercise developed. This time step allows us to collect aligned and variable samples from the raw data. The aggregated value of the raw data for each time point t_i is represented as $S_i = \bigcup_{s_i}^{s} s_i, t_i \in [t_i - T, t_i]$.

Since real-time models require current and previous measurements as inputs, for each sample at time t_i, we define a sequence W configured by previous data S_i^*, whose length is 40 samples (2 s) $W = 40$, for each sensor source (axis X, Y and Z). The size of the sequence remains consistent across all sources S_s and is structured as follows:

$$S_i^* = S_i \rightarrow S_{i-1}, \ldots, S_{i-W}$$

It is important to note that the size of input features for computing data stream features is determined by the segmented temporal window T and the length of the sequence W, resulting in a total window size of $W \times T$.

3.3 Deep Learning Model for Classification and Estimation of Progression Level of Exercises

The same baseline model is proposed for the classification and estimation of the progression of exercises, composed by:

- Three layers of 1D CNN are firstly integrated as spatial feature extractors. Each 1D convolution layer is defined by Conv1D(K,S,ST), where K defines the number of kernels, S the size of the convolution, and T the stride, whose stride=2 value reduces the size from the input matrix to half size in the output matrix. The inner layers include a rectified linear unit activation function (ReLU) to overcome the problem of vanishing gradients and improve learning performance.
- A LSTM network models the temporal sequence of spatial features extracted by the CNN for each sample. A 2-layer LSTM of 32 units is integrated.
- Two dense multilayer perceptron-based layers configure the final output to relate the spatio-temporal features with the output.
- In the classification of exercises, a final dense layer of 4 units for the classes (chest, back, bicep, tricep) is included. The model learns from loss function as *categorical cross-entropy* with optimizer *adam* using the metrics of *accuracy*.
- In the regression of the progression level of exercises, a final dense layer of 1 unit is included to estimate the progression value. The model learns from the loss function as *mean squared error* with the optimiser *Adam* using the metrics of *mean absolute error*.

In Fig. 2, we describe the configuration and layers of the proposed model.

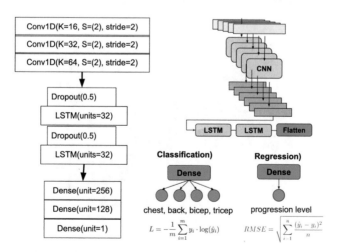

Fig. 2. CNN+LSTM model for classification and estimation of progression of body exercises

4   Results

In this section, we present the case study conducted to evaluate the performance of our proposed model in classifying exercises and estimating the progression of repetitions using IMU-based wristband wearables. We conducted multiple exercise sessions using the Fitbit Versa 3 wearable device in natural, real-world conditions. The wearable collected real-time acceleration data along the X, Y, and Z axes. Specifically, we selected four strength exercises - chest, back, biceps, and triceps - due to their relevance in strength training. During eight strength exercise sessions, a total of 132,408 data samples were collected. To facilitate labelling, each exercise was identified during the data collection process. Subsequently, we assigned progression-level labels to each set of exercises based on observations from an external observer using video recordings. The repetition labels were determined by identifying repetition peaks within each set of strengths. The progression level for each timestamp was calculated by linear interpolation between repetitions. Next, we aggregated the acceleration data into short temporal windows of $T = 50ms$. For each time step of $\Delta = 250ms$, the data served as input to the DL model for learning. We found that using a sequence of previous data $W = 40$, corresponding to a window size of 2 s for each axis, yielded optimal performance.

4.1   Results for Classification of Strength Exercises

A 10-cross-validation was conducted to evaluate the data. The metric results of the classification of chest, back, biceps, and triceps exercises are shown in Fig. 3 and detailed in Table 1, including precision, recall, and f1 score.

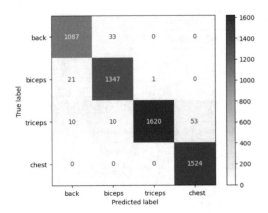

Fig. 3. Results for classification of strength exercises (chest, back, biceps, and triceps)

Table 1. Summary of classification metrics

Exercise	Precision	Recall	F1-score
back	0.97	0.97	0.97
biceps	0.97	0.98	0.98
triceps	1.00	0.96	0.98
chest	0.97	1.00	0.98
Accuracy		0.98	
Macro average		0.98	
Weighted average	0.98		

4.2 Results for Estimation of Progression Level in Repetitions

Before delving into the results, it is essential to clarify the primary objective of our model, which is to identify the initiation, progression, and ending of repetitions. We achieved this by implementing a labelling system based on peaks associated with the conclusion of each repetition. The progression levels within these intervals are calculated through linear interpolation using these peaks as reference points. Therefore, while we include the Mean Absolute Error (MAE) in Table 2, it serves as a supplementary performance metric rather than a critical one. Accurate identification of peaks aligning with the progression level and prediction labelled is of greater significance, depicted in Fig. 4 in the temporal dimension. In the back exercise graph, although the prediction scale differs from the labelling, the alignment of the beginning and end of the progression matches. We highlight the model's capability to effectively recognise the initiation and termination times and the duration of repetitions.

Table 2. Summary of MAE metrics for estimation of progression level

Exercise	MAE
back	0.1517
biceps	0.1518
triceps	0.2001
chest	0.1566

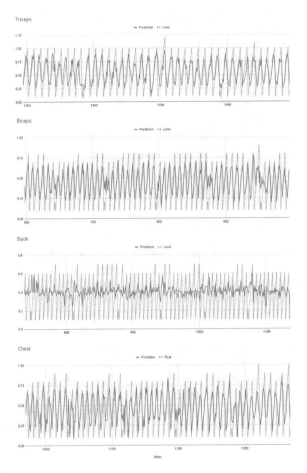

Fig. 4. Results for estimation of progression level in repetitions (chest, back, biceps, and triceps)

5   Conclusions and Ongoing Works

This study successfully classified four strength exercises (chest, back, biceps, triceps) using an IMU sensor embedded in a wearable wristband device. The research encompassed several pivotal phases. Initially, data was collected from multiple sessions involving three users performing exercises in real-world gym settings, ensuring the practicality and representativeness of the data collection process. Subsequently, a Deep Learning model was constructed, merging Convolutional Neural Networks (CNN) and Long Short-Term Memory (LSTM) architectures. This model was meticulously designed to classify exercises and accurately estimate the progression of repetitions. The fusion of CNN and LSTM enables the model to extract pertinent features from the IMU sensor data and capture temporal dependencies, facilitating precise exercise recognition. Ultimately, the study's results underscore the promising performance of the developed model, with encouraging accuracy in exercise classification and repetition progression estimation. These findings indicate the potential for integrating the model into mobile and wearable devices, facilitating real-time exercise movement recognition. In ongoing work, we plan to gather more diverse datasets to enhance the system's ability to generalise across various users, exercise variations, and environmental conditions. Furthermore, the integration of the recognition system into mobile and wearable devices will necessitate the evaluation of non-deep learning models for assessing metrics related to power consumption, real-time processing, and user-friendly interfaces. Additionally, we are actively exploring the utilisation of inertial units with higher sampling frequencies to further improve system performance and classification accuracy.

Data and Source Code The data collected, and the source code of the deep learning models are available in https://github.com/AuroraPR/Sports-UCAmI.

Acknowledgment. This contribution has been supported by the Spanish Institute of Health ISCIII through the project DTS21-00047. Furthermore, this research has received funding from the EU Horizon 2020 Pharaon Project 'Pilots for Healthy and Active Ageing', Grant agreement no. 857188; and Andamove project (ref. EXP74829) founded by the European Union by the NextGeneration EU programme and the SensorSportLab II ref. 11/UPB/23 (Redes de Investigación en Ciencias del Deporte 2023) project by Consejo Superior de Deportes (Ministerio de Cultura y Deporte).

References

1. Almonacid-Olleros, G., Almonacid, G., Fernandez-Carrasco, J.I., Espinilla-Estevez, M., Medina-Quero, J.: A new architecture based on iot and machine learning paradigms in photovoltaic systems to nowcast output energy. Sensors **20**(15), 4224 (2020)
2. Anderson-Hanley, C., Nimon, J.P., Westen, S.C.: Cognitive health benefits of strengthening exercise for community-dwelling older adults. J. Clin. Exp. Neuropsychol. **32**(9), 996–1001 (2010)
3. Hou, C.: A study on imu-based human activity recognition using deep learning and traditional machine learning. In: 2020 5th International Conference on Computer and Communication Systems (ICCCS), pp. 225–234. IEEE (2020)
4. Ishii, S., Nkurikiyeyezu, K., Luimula, M., Yokokubo, A., Lopez, G.: Exersense: real-time physical exercise segmentation, classification, and counting algorithm using an imu sensor. Activity Behav. Comput., 239–255 (2021)
5. Medina-Quero, J., Zhang, S., Nugent, C., Espinilla, M.: Ensemble classifier of long short-term memory with fuzzy temporal windows on binary sensors for activity recognition. Expert Syst. Appl. **114**, 441–453 (2018)
6. Medina-Quero, J.M., Burns, M., Razzaq, M.A., Nugent, C., Espinilla, M.: Detection of falls from non-invasive thermal vision sensors using convolutional neural networks. In: Proceedings, p. 1236. MDPI (2018)
7. Polo-Rodríguez, A., Lupión, M., Ortigosa, P.M., Medina-Quero, J.: Estimating frontal body landmarks from thermal sensors using residual neural networks. In:

International Work-Conference on Bioinformatics and Biomedical Engineering, pp. 330–342. Springer (2022). https://doi.org/10.1007/978-3-031-07704-3_27

8. Polo-Rodriguez, A., Montoro-Lendinez, A., Espinilla, M., Medina-Quero, J.: Classifying sport-related human activity from thermal vision sensors using cnn and lstm. In: International Conference on Image Analysis and Processing, pp. 38–48. Springer (2022). https://doi.org/10.1007/978-3-031-13321-3_4

9. Seneviratne, S., et al.: A survey of wearable devices and challenges. IEEE Commun. Surv. Tutorials **19**(4), 2573–2620 (2017)

10. Van Der Heijden, G.J., et al.: Strength exercise improves muscle mass and hepatic insulin sensitivity in obese youth. Med. Sci. Sports Exercise **42**(11), 1973 (2010)

Enhancing Unobtrusive Home Technology Systems with a Virtual Assistant for Mood and Social Monitoring

Sara Comai[4]([✉]) [iD], Giovanna Viana Mundstock Freitas[1] [iD], Kelly Xu[1,2],
Marta Conte[1], Anita Colombo[1], Senja Pöyhönen[1,3], Marco Ajovalasit[1] [iD],
and Fabio Salice[4] [iD]

[1] Department of Design, Politecnico di Milano, Milan, Italy
{giovanna.viana,marta.conte,anita.colombo}@mail.polimi.it,
marco.ajovalasit@polimi.it
[2] TU Delft, Delft, The Netherlands
K.Xu-2@student.tudelft.nl
[3] South-Eastern Finland University of Applied Sciences - Xamk, Tampere, Finland
csepo002@edu.xamk.fi
[4] Department of Electronics Information and Bioengineering, Politenico di Milano,
Milan, Italy
{sara.comai,fabio.salice}@polimi.it

Abstract. As the global population ages, there is a growing need for support in daily activities among older individuals. Information and Communication Technologies (ICT) have the potential to ease caregiver responsibilities and worries and enhance the independence of older individuals. The objective of this study is to enrich traditional indoor monitoring systems, which mainly focus on safety and functional aspects, with features that consider both the needs of the caregivers and those of the monitored person. A triangulation of methods approach is employed, utilizing personas, surveys, and interviews to identify both parties' specific needs and preferences and guide the selection of suitable technologies. Results recognize the importance of addressing the mood and social needs of the monitored persons and consider the barriers that hinder the installation of such systems due to privacy and independence concerns. A general framework is presented, which extends traditional monitoring systems to incorporate these additional needs.

Keywords: Elderly care · Independent living · Mental health · Ageing in place · Caregivers · Smart homes · Remote Monitoring · Well-being

1 Introduction

According to the United Nations, the global population of people aged 65 years or older was 771 million in 2022. This number is projected to increase to 994 million by 2030 and 1.6 billion by 2050 [1]. Moreover, historical data show that older individuals have been increasingly living alone instead of co-residing with

J. Bravo and G. Urzáiz (Eds.): UCAmI 2023, LNNS 835, pp. 81–93, 2023.
https://doi.org/10.1007/978-3-031-48306-6_8

others, and the majority of older persons in the more developed countries prefer to live independently [2].

Elderly individuals living alone face challenges like social isolation, cognitive decline, and multiple health issues, increasing the risk of depression [3].

Many older people are supervised by informal caregivers, who devote a significant amount of time to assisting the older person in their daily activities, such as personal care, medication management, meal preparation, and overall surveillance. Information and Communication Technologies (ICT) can offer opportunities to alleviate the caregiving burden and promote the autonomy of the older person but older individuals often perceive such systems as a threat to their independence due to the sensation of constant surveillance. Indeed, several studies have identified the potential risks associated with the improper use of monitoring technologies, including emotional abuse or excessive monitoring of older people [4–6].

The acceptance of technology as a supportive tool for fostering mutual tranquillity is a relatively recent development [7]. Caregivers see it as a compensatory tool that provides information on the person's well-being and alerts them to any potential issues. For the individual, it can serve as a means of requesting assistance when needed and providing a range of tools for managing daily activities [8].

Several indoor monitoring systems have been proposed in the literature [9, 10], but their adoption is still limited.

The goal of this paper is to better understand older individuals' needs with respect to such systems and to suggest improvements and changes to an existing system to meet unaddressed user needs.

In particular, the study builds upon an existing non-intrusive monitoring system called BRIDGe [7], which collects information about the status of both the household and the individual residing in it. BRIGDe detects anomalies in daily habits or agreed-upon rules, enables the person to send predefined notifications, monitors sleep patterns and quality, tracks environmental parameters and space utilization over time, and analyzes behavioural trends to identify deviations.

Using a *triangulation of methods* [11], based on personas, the BRIDGe system, surveys and interviews with caregivers and older people, we propose a general framework to incorporate additional needs to improve the usefulness and acceptance of indoor monitoring systems.

The rest of the paper is organized as follows: Sect. 2 provides an overview of monitoring systems characteristics and of their acceptance; Sect. 3 introduces the method, while Sect. 4 reports the results collected with interviews and surveys. Finally, Sects. 5 and 6 discuss the results and present the general framework; Sect. 7 draws the conclusions.

2   Related Work

A review of smart home solutions for older people introduces six categories that define different types of health-related smart home monitoring types and corresponding technologies [10]. The different category types with their main goals are summarized in Table 1.

Table 1. Categories of monitoring

Monitoring type	Goals
Physiological Monitoring	Monitoring vital signals and body metrics, such as heart rate, blood pressure, and respiratory rate
Functional Monitoring	Detecting emergencies; focusing on Activities of Daily Living (ADL) and detecting significant events like falls
Safety Monitoring and Assistance	Detecting environmental hazards, such as gas leaks or fires
Security Monitoring and Assistance	Detecting and responding to potential intruders or security breaches
Social Interaction Monitoring and Assistance	Collecting data on social interactions or enabling communication support, such as making phone calls or video calls
Cognitive and Sensory Assistance	Provide reminders (e.g., for medication or instructions), as well as aids that compensate for sensory deficits, such as impaired vision, hearing, or touch

In relation to these aspects, the literature highlights that most of the available monitoring solutions focus on safety and functional monitoring [9,12], and physiological monitoring [13], while cognitive solutions are scarce, and social and security monitoring options are extremely rare [9].

When it comes to technology acceptance, several papers are available in the healthcare domain, such as [14,15]. An interesting study, although focused on Dubai, is [16], which provides a detailed analysis of technology acceptance for smart home technologies (SHT) in relation to gender, age, income, and diseases. The study found that the most requested SHT by the elderly are environmental control and health monitoring and that individuals in their sixties show less demand compared to those in their forties. Motion and voice recognition sensors are favoured, while cameras are less preferred for privacy concerns. Anxiety towards technology and age-related preferences impact the acceptance of smart home solutions, with differences observed in the technology preferences of individuals in their forties compared to previous generations.

Motivations, barriers, and risks of adoption from a consumer perspective are analysed in [17]. The primary motivations for adoption include efficient energy management, improved home health services, and a better quality of life. The major barriers to adoption include distrust, resistance, and anxiety towards technology, limited perceptions of smart homes, concerns about financial issues, as well as privacy and security concerns.

3   Method

To better understand the factors and requirements of older persons and caregivers that could enhance the acceptance of monitoring solutions in home technology systems, we applied a "triangulation of methods" [11], i.e. we combined various methods, data resources, and perspectives. In particular, we developed personas to represent the main target audience and keep a human-centered perspective; we considered an existing technological solution as a reference to understand the current landscape and identify potential enhancements; finally, we conducted interviews with the elderly and administered surveys to caregivers to better explore their perspectives, needs, attitudes and preferences.

Personas. We created three personas based on real older individuals, using an approach known as extreme personas [18]. Persona 1 represents users who are initially resistant to the idea of monitoring and assistance. They may have concerns about privacy, loss of independence, or other barriers that prevent them from embracing new technologies. Persona 3, on the other hand, represents users who are enthusiastic about monitoring systems. They are open to the idea of assistance and are actively interested in exploring new technologies to enhance their well-being. Persona 2 represents an average user who falls between the extremes of resistance and enthusiasm. This user is somewhat open to receiving assistance but still relies on family members for decision-making; this persona finds comfort in what is well-known and familiar.

Monitoring System. BRIDGe (Behavioral dRift compensation for autonomous and InDependent livinG) is a monitoring system currently under development by the *Assistive Technology Group (ATG)* of Politecnico di Milano. Its primary objective is to monitor the behaviour of fragile subjects at home in a non-intrusive way to preserve autonomy and correct behavioural drifts [7]. In the long-term BRIDGe aims at anticipating and preventing problems in order to provide timely care; in the short-term it aims to foster *mutual tranquillity* for families, friends, and individuals living alone. To achieve these objectives, BRIDGe utilizes various kinds of sensors that constantly transmit information about the status of both the home and the individual to the family; in case of problems, notifications are sent to the caregiver. In this way, BRIDGe enhances families' sense of security and increases the perceived level of safety for individuals living alone preserving their independence.

Surveys and Interviews. We conducted a survey involving family members/caregivers (n = 50) of older persons to determine the features they would like to have in monitoring systems. Moreover, we conducted interviews with older individuals (n = 4) living alone using semi-structured scripts to identify needs, concerns and also engaging activities. Table 2 provides a comprehensive overview of our research objectives and the planned investigation methods. In particular, the table reports the question to be validated; what we aim to understand; if the question focuses on capturing observable actions/behaviours (behavioural) or if it explores opinions/attitudes (attitudinal); if answers are collected with interviews or surveys; to whom the question is addressed.

4   Results

4.1   Survey Results

The survey results are summarized in Tables 3, 4, and 5. The "other" category includes all responses that accounted for less than 5% of the total.

Table 2. Research plan. B/A corresponds to Behavioural/Attitudinal.

Assumption/question that needs to be validated	What we want to know	B/A	Technique	With whom
Which ADLs do older individuals require more support with?	Activities that should be enabled or monitored	B	Interviews	Older people
Which activities (or ADLs) cause the greatest concerns?	What types of sensors are required	B	Interviews	Older people
What are the most common hobbies among older individuals?	What are the key features that enhance engagement?	B	Interviews	Older people
Why do older people often avoid interacting with technological devices?	What makes technological devices not usable to older people	B	Survey	Caregivers
Do you/do you not use a monitoring system? Why?	What are the motivating factors driving individuals to install a monitoring system, and what are the barriers that deter them from doing so	A	Survey	Caregivers
Which kind of behaviours would you like to be informed about?	What kind of data/behaviour is important to monitor	A	Survey	Caregivers not using a monitoring system yet
Which kinds of activities do you monitor? Which are important?	What kind of data/behaviour is important to collect/monitor	A	Survey	Caregivers already using a monitoring system

Use of Devices and Barriers to Technology Usage. (Table 3) The majority of individuals are able to use a TV autonomously, 25% can use a smartphone. The main barriers reported by these individuals when using technology include memory difficulties (struggling to remember how to interact with devices), usability challenges (difficulties in understanding device interaction), and physical limitations (difficulties in seeing, hearing, or handling devices).

Interest in Indoor Monitoring Systems. (Table 3) 52% of participants expressed a lack of interest in indoor monitoring systems. However, it is noteworthy that a total of 68% of participants held a positive view regarding the concept of such systems.

Reasons for not Being Interested in Indoor Monitoring Systems. (Table 4) The primary reason, accounting for 31%, was that the receivers would be opposed to it. However, further details regarding the nature of this resistance were not specified. Other barriers mentioned include cost (27%), the presence of an alternative system (12%), and privacy concerns (8%).

Reasons for using Indoor Monitoring Systems. (Table 4) The most interesting areas to monitor include emergencies (21%), emotional distress (17%), medical data (14%), and sleep-related information (10%).

Characteristics of Personal Indoor Monitoring Systems in Use.
(Table 5) Out of all the participants, 11 individuals use some indoor monitoring systems. They mainly use motion sensors (36%), video recognition devices (36%), voice assistants (21%). Regarding the types of information collected by their systems, the most common categories include emergencies (18%), unusual body parameters (12%), entrance door openings (12%), shower usage (10%), and emotional distress (10%).

Table 3. Results of all participants of the survey (n=50)

1. Care receivers are able to use autonomously		2.Tech. barriers for elderly		3. Interest in indoor monitoring	
Object	%	Barriers	%	Interest level	%
TV	54%	Memory	36%	No, I am not interested	32%
Smartphone	25%	Usability	34%	Yes, but I currently do not use any	30%
Voice Assistance	10%	Physical	23%	No, but it is a good idea	20%
Monitoring Sys.	7%	Other	6%	Yes, I already use one or mor	18%
Other	4%				

Table 4. Results of participants who do not have an indoor monitoring system but are interested in one/the idea (n=25)

4. Barriers against installing indoor monitoring systems		5. Most important information to gather withan indoor monitoring system	
Barriers	%	Kind of information	%
Resistance from the individual	31%	Emergencies and critical situations	21%
Costs	27%	When the person is in emotional distress	17%
Other system already in use	12%	Whether the person took their medications	14%
Privacy	8%	When the person is sleeping	10%
Other	23%	Unusual body parameters	8%
		Whether and how much ate during the day	7%
		When the entrance door is opening	7%
		Other	16%

4.2   Interview Results

We conducted qualitative semi-structured interviews with four older women living alone (P1: 88 y.o. Brazilian; P2: 85 y.o. Brazilian; P3: 75 y.o. Italian; P4: 80 y.o. Finnish). The objective was to explore their preferences for activities that would enhance their independence, identify areas where they require more support, and uncover opportunities for providing engaging experiences. We specifically focused on women as they represent the largest demographic of older individuals living alone worldwide [2].

Activities They Used to do Independently, now Relied on Others for.
Currently, all the participants rely on their families to handle grocery shopping.

Table 5. Results of participants who already have an indoor monitoring system (n=11)

6. Current systems in place		7. Collected Information	
Systems	%	Kind of information	%
Motion sensors	36%	Emergencies and critical situations	18%
Video recognition	36%	Unusual body parameters	12%
Sound recognition	21%	When the entrance door is opening	12%
Identification bracelet	7%	When/how long the shower is being used	10%
		When the elder is in emotional distress	10%
		Whether the elder took their medications	8%
		When the elder is sleeping	8%
		Whether and how much ate during the day	6%
		When the elder access a room	6%
		Other	8%

Two participants (P1 and P4) mentioned depending on their families to purchase materials for their crafts, while another participant expressed a desire to independently buy gifts for her family (P2). All participants expressed a longing for going out on the streets unaccompanied but cited insecurity about falling as a reason for avoiding it. Three participants (P1, P2, P4) miss their involvement in religious activities due to depending on someone to attend. Two participants (P3, P4) mentioned missing doing gardening activities but acknowledged the physical demands involved, which would necessitate support to perform.

Fears and Concerns about ADLs. All the participants mentioned the fear of falling as their primary concern. Specifically, some participants expressed worries about falling on the stairs within their homes (P1, P2, P3), while others cited concerns about falling in the shower (P1) or while attempting to clean furniture (P3). There were mentions of fears related to leaving the iron on (P2) or forgetting to close the sink (P3). One participant (P3) also mentioned a social risk, i.e. the fear that arises from receiving calls from unknown numbers due to past experiences of being a victim of a scam.

Routines and Frustrations. Participants reported various frustrations and sources of joy. In particular, they worry about family and friends during difficult situations, such as when relatives are sick (P1 and P2). They feel bored on rainy days, as they are confined to their homes with nothing to do (P3). One participant (P4) mentioned encountering difficulties with household chores, such as hanging clothes to dry or reaching high kitchen cabinets. On the other hand, the participants also shared activities that brought them joy, which primarily centred around social interactions, for example, when they receive visits and calls from family (all) and friends or when they go out with friends and family (P1,

P2, P3), experiencing nature and encountering different people. In this context, relatives are good companions but mostly offer safety support.

Activities That Bring Joy and Mental Engagement. For their spare time, participants commonly mentioned activities such as solving puzzles (e.g., jigsaw puzzles and crossword puzzles), watching TV, participating in crafts, and gardening. However, what emerged as the most significant factor for maintaining mental proactivity for all the participants was the opportunity to meet and interact with other people.

5   Discussion

5.1   Limits, Barriers, Interesting Aspects of Indoor Monitoring Systems

One of the primary barriers mentioned by family members against installing an indoor monitoring system was the resistance from their older family members. Although our study did not specifically investigate the nature of this resistance, other researchers have delved into this topic [4,8]. In fact, the prospect of having their activities monitored can evoke a sense of loss of *privacy* and *control*. Older individuals may fear that their every move and action will be scrutinized, leading to a degradation of their peace of mind and a perceived violation of their *personal boundaries*. For these reasons, the introduction of monitoring systems can trigger feelings of distrust, suspicion, anxiety, or even paranoia among some older individuals. However, some older individuals have systems in place for emergency monitoring. This suggests that there is a willingness to embrace technology for specific purposes.

5.2   Data/Behavior That Should Be Monitored

The existing technology solution aims to provide mutual tranquillity by monitoring deviations in routines and sending notifications. While this serves the purpose of ensuring safety and security, it falls in addressing the emotional and social needs that older individuals highlighted in the interviews and surveys.

Emotional Monitoring. Family members expressed their interest in monitoring both physical and emotional emergencies, as well as activities such as medication intake and sleep patterns. However, surveys conducted with individuals who already have an indoor monitoring system indicate that emotional distress is rarely monitored. This finding is consistent with a study [9] and validates the need for an evolution in the current technological solution to focus on a more comprehensive emotional monitoring spectrum.

Importance of Social Engagement. Based on the interviews, the fears expressed by participants primarily involved physical activities, while their sources of joy were predominantly rooted in social interactions. The correlation between emotional engagement, social activities, and overall well-being is well-supported in the literature [19]. An indoor monitoring system could facilitate communication between individuals and their family members, encourage participation in events or engagements in hobbies, help caregivers in scheduling visits and plan social activities for the older person on the basis of possible deviations from the person's regular routines at home (e.g., in case of prolonged periods of inactivity).

5.3   Which Technology Should Be Used?

According to the survey, some respondents indicated that their systems already incorporate the usage of voice assistants in their monitoring system. This suggests that their integration could be generally well-received to support emotional monitoring. Notice that this approach is supported by research in the field of speech emotion recognition (SER), which has demonstrated its effectiveness in measuring emotions [20, 21]. Emotion recognition can significantly enhance indoor monitoring systems designed to track activities and behavior, enabling the detection of potential emotional problems. By integrating emotional indicators alongside changes in sleeping patterns, eating habits, and physical activity levels, the system can provide more comprehensive insights into an individual's well-being.

However, according to the survey findings, only 10% of the care receivers reported being able to take advantage of a Virtual Assistant (VA) autonomously. This result should not be surprising considering that the generation surveyed is not used to technology and, indeed, distrusts it (intergenerational digital divide) [4, 6]. Moreover, it indicates that there is room for improvement in the design and usability of VAs to make them more accessible and user-friendly for older people, and also for future generations. The advantage of VAs is manifold; besides providing a convenient interface for a whole range of already established services (e.g., music, games, weather forecasts, news, alarms, etc.) it is useful for preserving ties between persons through communication and images. All these stimuli and opportunities go in the direction of counteracting moments of loneliness. A recent survey found that "those who start using technology at home then want to know what else can be done" [22]. This, albeit in a more limited form, also applies to the elderly.

6   A New Scenario

Based on the interview and survey results, we extend the typical architecture of systems supporting safety and functional monitoring (e.g., systems like BRIDGe [7]), with emotion monitoring and social support (see Fig. 1). In particular, a voice assistant (e.g., through Alexa or Google home) plays the role of analyzing and monitoring cognitive and social aspects. For example, in addition to encouraging older users to actively participate in an activity, pursue hobbies and engage in social interactions, the response to these stimuli is stored and analyzed allowing for the identification of a socio-cognitive drift with respect to a baseline.

The voice assistant can facilitate mood monitoring [20] by measuring emotions using speech emotion recognition (SER). Emotions and moods are distinct concepts, with emotions being short-lived responses to specific stimuli while moods are more enduring, subtle, and background-oriented, influencing an individual's affective and social state. Researchers commonly focus on basic emotions such as anger, fear, sadness, and happiness, with some including surprise and boredom (e.g., [23,24]). By detecting emotions, it becomes possible to identify the underlying mood, which involves considering the time course of emotional events [25]. While the identification of mood through emotions over time remains an under-explored area, it holds relevance as a non-intrusive technological monitoring method to support autonomous living in the home context.

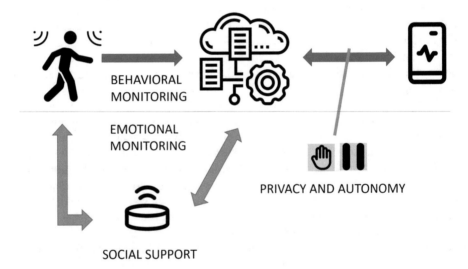

Fig. 1. Architecture. Existing services and proposal of the new personal service.

The voice assistant acts as the initial point of contact, interacting with the older user to gather data, useful for trend analysis, or intercept atypical situations. The system sends alerts to the caregiver only when unexpected activities or events or when a drift in behaviour (social or mood) are detected. In this way, the system minimizes unnecessary notifications, reducing caregiver burden and improving overall system efficiency. Moreover, the vocal assistant can have a further role in the context of the continuous investigation of cognitive impairment through the dispensing of questionnaires, in vocal form, and derived from those commonly used; for this activity, the voice system proposes a question from time to time, changing the way and format and collecting the answers. In this way, it is possible to have a continuous trace of the person's status and also to identify variations in responses over time.

Finally, the voice assistant can play an intermediary role from the person to the system itself. While it is fundamental that emergencies are always supported, this new architecture can introduce a verification process that allows older users to have control over the information collected by sensors before it is shared with caregivers. This empowers older individuals to maintain their privacy while still benefiting from the monitoring system. For example, for some types of notifications, the person may decide to pause the transmission or may avoid sending false alarms.

The proposed architecture builds upon insights from various studies in the field, ensuring that the new scenario aligns with users' needs and incorporates best practices. In particular, we considered studies like [4, 26–28].

7   Conclusions and Future Work

This work has addressed the question of the relationship between technology and autonomous life in the home environment, also taking into account the role and the emotional involvement of caregivers. The goal was to identify the real needs of both people and caregivers and to propose further technological solutions to be coupled with existing non-intrusive monitoring systems. The current state of monitoring systems mainly focuses on safety and functional aspects. The results obtained underline that, for those who accept technology as a compensatory system, technologies must focus on the identification and monitoring of sociality and mood aspects, taking into due consideration that there are natural barriers to installation caused by privacy and independence. Another result of the survey is that some respondents use voice assistants. This aspect suggests that the integration of voice assistants could be generally well received. The voice assistant, therefore, represents a useful tool for the collection and analysis of data on emotions, cognitive impairment, and sociality. Future work will test the tool with other user groups and relevant scenarios.

References

1. World population prospects: Summary of results, p. 2022. United Nations Department of Economic and Social Affairs, Population Division (2022)
2. World Population Ageing 2020: Highlights: Living Arrangements of Older Persons. United Nations Department of Economic and Social Affairs (2021)
3. Perini, G., Ramusino, M.C., Sinforiani, E., Bernini, S., Petrachi, R., Costa, A.: Cognitive impairment in depression: recent advances and novel treatments. Neuropsych. Disease Treatm. **15**, 1249–1258 (2019).pMID: 31190831
4. Berridge, C., Zhou, Y., et al.: Control matters in elder care technology: evidence and direction for designing it in. In: Designing Interactive Systems Conference, ser. DIS 2022. Association for Computing Machinery, New York (2022). https://doi.org/10.1145/3532106.3533471
5. Knowles, B., Hanson, V.L., et al.: The harm in conflating aging with accessibility. Commun. ACM **7**, 66–71 (2021). https://doi.org/10.1145/3431280

6. Berridge, C., Demiris, G., Kaye, J.: Domain experts on dementia-care technologies: mitigating risk in design and implementation. Sci. Eng. Ethics **27**(14) (2021)
7. Mangano, S., Saidinejad, H., Veronese, F., Comai, S., Matteucci, M., Salice, F.: Bridge: mutual reassurance for autonomous and independent living. IEEE Intell. Syst. **30**(4), 31–38 (2015)
8. Yusif, S., Soar, J., Hafeez-Baig, A.: Older people, assistive technologies, and the barriers to adoption: a systematic review. Int. J. Med. Informatics **94**, 112–116 (2016)
9. Maswadi, K., Ghani, N.B.A., Hamid, S.B.: Systematic literature review of smart home monitoring technologies based on iot for the elderly. IEEE Access **8**, 92244–92261 (2020)
10. Demiris, G., Hensel, B.K.: Technologies for an aging society: a systematic review of "smart home" applications. Yearb. Med. Inform. **17**(01), 33–40 (2008)
11. Wang, W., Duffy, A.: A triangulation approach for design research. In: DS 58–2: Proceedings of ICED 2009, the 17th International Conference on Engineering Design, vol. 2, pp. 275–286 (2009)
12. Sokullu, R., Akkaş, M.A., Demir, E.: Iot supported smart home for the elderly. Internet of Things **11**, 100239 (2020)
13. Pandia Rajan, J., Edward Rajan, S.: An internet of things based physiological signal monitoring and receiving system for virtual enhanced health care network,". Technol. Health Care **26**(2), 379–385 (2018)
14. Tseng, K., Hsu, C., Chuang, Y.: Designing an intelligent health monitoring system and exploring user acceptance for the elderly. J. Med. Syst. **37**(9967), 2013 (2013)
15. Mann, W.C., Marchant, T., Tomita, M., Fraas, L., Stanton, K.: Elder acceptance of health monitoring devices in the home. Care Manag. J. **3**(2), 91–98 (2002)
16. Arar, M., Jung, C., Awad, J., Chohan, A.: Analysis of smart home technology acceptance and preference for elderly in Dubai, UAE. Designs **5**(4), 70 (2021). https://doi.org/10.3390/designs5040070
17. Li, W., Yigitcanlar, T., Erol, I., Liu, A.: Motivations, barriers and risks of smart home adoption: from systematic literature review to conceptual framework. Energy Res. Soc. Sci. **80**, 102211 (2021)
18. Djajadiningrat, J.P., Gaver, W.W., Fres, J.: Interaction relabelling and extreme characters: methods for exploring aesthetic interactions. In: Proceedings of the 3rd Conference on Designing Interactive Systems: Processes, Practices, Methods, and Techniques, pp. 66–71 (2000)
19. Rimé, B., Philippot, P., Boca, S., Mesquita, B.: Long-lasting cognitive and social consequences of emotion: social sharing and rumination. Eur. Rev. Soc. Psychol. **3**(1), 225–258 (1992)
20. Alonso, J.B., Cabrera, J., Travieso, C.M., de Ipiña, K.L., Sánchez-Medina, A.: Continuous tracking of the emotion temperature. Neurocomputing **255**, 17–25 (2017). bioinspired Intelligence for machine learning. https://www.sciencedirect.com/science/article/pii/S0925231217305490
21. Fahad, M.S., Ranjan, A., Yadav, J., Deepak, A.: A survey of speech emotion recognition in natural environment. Digital Signal Process. **110**, 102951 (2021). https://www.sciencedirect.com/science/article/pii/S1051200420302967
22. Kim, S., Choudhury, A.: Exploring older adults' perception and use of smart speaker-based voice assistants: a longitudinal study. Comput. Hum. Behav. **124**, 106914 (2021). https://www.sciencedirect.com/science/article/pii/S0747563221002375

23. Kimmatkar, N.V., Babu, B.V.: Novel approach for emotion detection and stabilizing mental state by using machine learning techniques. Computers **10**(3), 37 (2021)
24. Tariq, Z., Shah, S.K., Lee, Y.: Speech emotion detection using iot based deep learning for health care In: IEEE International Conference on Big Data (Big Data), vol. 2019, pp. 4191–4196 (2019)
25. Khorram, S., Jaiswal, M., Gideon, J., McInnis, M.G., Provost, E.M.: The PRIORI emotion dataset: Linking mood to emotion detected in-the-wild, CoRR, vol. abs/ arXiv: 1806.10658 (2018)
26. Dixon, E., Michaels, R., et al.: Mobile phone use by people with mild to moderate dementia: uncovering challenges and identifying opportunities: Mobile phone use by people with mild to moderate dementia, ASSETS 2022. ACM (2022)
27. Mehta, V., Gooch, D., Bandara, A., Price, B., Nuseibeh, B.: privacy care: a tangible interaction framework for privacy management. ACM Trans. Internet Technol. **21**(1) (2021)
28. McKay, D., Miller, C.: Standing in the way of control: a call to action to prevent abuse through better design of smart technologies. In: Proceedings of the 2021 CHI Conference on Human Factors in Computing Systems, CHI 2021. Association for Computing Machinery, New York (2021)

New Trends in Machine Learning Techniques for Human Activity Recognition Using Multimodal Sensors

Jesús González-Lama[1], Alicia Montoro[2], Macarena Espinilla[2],
Juan Carlos Valera[3], David Gil[3(✉)] iD, Jesús Peral[4] iD, and Magnus Johnsson[5]

[1] Maimonides Biomedical Research Institute of Cordoba (IMIBIC),
Reina Sofia University, Madrid, Spain
[2] University of Jaén, Jaén, Spain
[3] Department of Computer Technology and Computation, University of Alicante,
03690 Alicante, Spain
david.gil@ua.es
[4] LUCENTIA Research Group, Department of Software and Computing Systems,
University of Alicante, 03690 Alicante, Spain
[5] University of Malmö, Malmö, Sweden

Abstract. The ageing of today's society, according to demographic and epidemiological data, presents a significant increase in the elderly population. Following the experienced pandemic, research in telemedicine to improve the lives of elderly people through a comprehensive program developed by multidisciplinary teams has become a top priority. This enables the provision of remote healthcare services, facilitating access to specialists, disease monitoring, medication management, and health indicator tracking to address the medical, social, and emotional needs of elderly individuals. This study proposes a sensor-based approach to identify activity patterns without prior labels. The system architecture responsible for collecting data from the monitored user in the assisted living facility consists of a beacon, multiple anchors, and various sensors for motion, opening and closing, temperature, and humidity. The experimentation was carried out with distinct activities such as sleeping, eating, taking medication, walking, showering, and brushing teeth, inferred from the identified patterns. This approach offers an automatic and objective way to understand the routines and behaviours of older individuals, thereby improving their care and attention through personalized interventions tailored to their individual needs. Furthermore, it lays the groundwork for future research on the detection and monitoring of changes in activities over time, identifying possible signs of impairment or changes in the health of elderly people.

Keywords: Machine Learning · HAR · multimodal sensors · smart environment

J. Bravo and G. Urzáiz (Eds.): UCAmI 2023, LNNS 835, pp. 94–99, 2023.
https://doi.org/10.1007/978-3-031-48306-6_9

1 Introduction

In recent decades, technological advances have played a fundamental role in improving people's quality of life. These advances have enabled experiments and the development of new methods that aim to improve different aspects of our daily lives. One of these advances is human activity recognition (HAR), which in short is a technique that uses sensors to identify and classify the actions that a person performs on a daily basis.

HAR is based on the use of sensors that can distinguish between different everyday activities, such as walking, running, showering, cooking, etc. These sensors can be found in mobile devices such as smartphones and smartwatches.

In addition to mobile sensors, there are also fixed sensors that are placed in specific locations to constantly collect information. Examples of these sensors are video cameras, acoustic sensors, proximity sensors, etc.

Our focus of work is in the field of healthcare, specifically in the monitoring and care of elderly people with diabetes living in sheltered housing. A study is currently being conducted in ten supervised flats, the primary goal is to develop a comprehensive system of care and support that will improve the health and wellbeing of these people, giving them greater independence, confidence and security in their daily lives.

The remainder of this paper is structured as follows: in Sect. 2, the related work is reviewed. Thereafter, in Sect. 3, we propose an architecture for HAR using multimodal sensors. Section 4 summarizes the experimentation carried out. The paper finalizes with the conclusions (Sect. 5).

2 Background

Accurate identification of the person performing an action is crucial to avoid errors in HAR area. Indoor location tracking (indoor localization) plays an essential role in this aspect. However, current devices often exhibit false or incorrect activations, leading to misleading results [1]. To address this challenge, the Received Signal Strength Indicator (RSSI) value is utilized, which measures the power of a radio signal. Although RSSI offers advantages such as low cost and good performance, signal variability among devices can affect localization accuracy [2,3].

The most commonly used method worldwide for indoor localization is the utilization of Bluetooth Low Energy (BLE) in conjunction with smartphones or wearable devices that implement this technology. This approach employs beacons that collect RSSI data. Although it has limitations such as signal variability and the need to install multiple beacons in large indoor spaces, it remains popular due to its ease of implementation and low cost [4]. In addition to using RSSI, this work employs the Message Queing Telemetry Transport (MQTT) protocol for sensor communication. MQTT is a lightweight and energy-efficient messaging protocol designed for efficient communication between Internet-connected devices [5,6].

3   Architecture

This section will provide a description of the architecture used and a detailed explanation of the devices employed.

Firstly, multiple Raspberry Pi are used as anchors in the architecture. These Raspberry Pi are responsible for scanning BLE beacons, i.e., searching for and detecting nearby BLE beacons, then collect the RSSI of each detected beacon and send it to the fog node.

The anchors are deployed evenly in the bedroom, bathroom, kitchen, and dining room. Additionally, an activity wristband is utilized as a beacon, it is a small device that emits a signal or broadcast messages at regular intervals to nearby electronic devices or receivers. This wristband has BLE connectivity and is used to locate the inhabitant using the RSSI signal.

To detect the performance of activities, different types of sensors have been installed throughout the rooms. These sensors include motion sensors, interruption sensors to detect the opening and closing of drawers and doors, and temperature and humidity sensors. The sensors utilize the ZigBee protocol to facilitate communication between them, enabling the detection of patient activity and monitoring of relevant environmental conditions. Figure 1 presents the plans of the house with the distribution of the devices used to capture data.

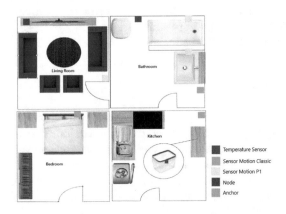

Fig. 1. House Plans.

4   Case Study

The focus of this work is to examine the effectiveness of various sensors, anchors, and beacons installed in a floor to monitor the routines of users suffering from a disease, specifically diabetes.

The main idea is to track the daily routines and activities of users by installing sensors. These activities are prescribed by a family physician and encompass actions such as brushing teeth three times a day, going for a walk, showering, sleeping, eating, and taking medication.

The physician conducts regular visits to assess the patients' health status. In order to enhance monitoring and facilitate the physician's work, the collected data is utilized to generate graphs that provide the physician with a quick overview of the patients' routines over the past days, weeks, or months. This will enable the physician to promptly identify any potential anomalies, non-compliance, discrepancies, or deception by the users, which would not be beneficial for their health condition.

After studying the different data captured by the devices about the state of the sensors, the RSSI, and the humidity and combining them, we can create a dataset in which we label the activities performed by the monitored user by day and hour. The purpose is to create highly visual and easy-to-understand graphs and statistics to facilitate the physician's task when it comes to monitoring and ensuring compliance with the proposed routines.

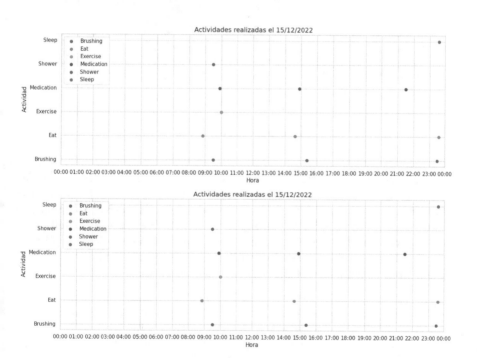

Fig. 2. Analysis of the Routine of Activities.

From the Fig. 2, we can observe how, during the 2 d analyzed, the user has performed all the activities required by the physician. When examining the graph, we can identify patterns in the user's daily routine. Generally, the user starts the day between 8:00 and 10:00 and goes to sleep around 23:00, indicating compliance with the recommended minimum of 8 h of sleep. It is also observed that most of the time the user brushes their teeth after meals, which usually occur shortly after waking up, at 14:00, and at 21:00. Additionally, the user typically showers in the morning before going for a walk to exercise.

By analyzing this graph, we obtain valuable information about the established routine of the monitored user. This is useful for assessing if the routine is maintained over time, and in the case of treating another user with a less established routine, it could be helpful in providing advice on how to correct certain habits and determining if they are being improved.

Additionally, it is also possible to study the duration of activities, but this information is only recorded for exercise and sleep. For this purpose, a time series spanning the entire month has been created, as shown in Fig. 3.

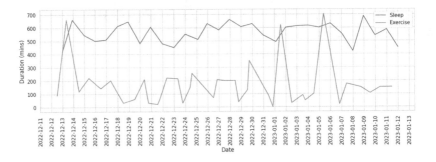

Fig. 3. Duration of Activities.

Upon examining the graph, it is observed that there are three data points representing exercise durations exceeding 10 h, which clearly must be attributed to measurement errors and do not reflect real values. To correct these outliers caused by measurement errors, data points that exceed the upper limit of 2.5 times the standard deviation above the mean have been removed. By performing this procedure, the three identified outliers have been eliminated, as shown in Table 1, resulting in a significant change in the statistics.

Table 1. Comparison of Duration Statistics With and Without Outliers

Activity	count	mean	std	min	25%	50%	75%	max
Exercise	38.0	166.81	167.02	0.0	60.0	136.0	206.5	702.0
Exercise without outliers	35.0	124.48	83.04	0.0	55.0	120.0	200.0	350.0
Difference	3.0	42.33	83.98	0.0	5.0	16.0	6.5	352.0

5 Conclusions

In this study, we propose a sensor-based approach to identify activity patterns. The architecture of the system in charge of collecting data from the monitored user in the assisted living facility consists of a beacon, multiple anchors, and several sensors for motion, opening and closing, temperature, and humidity. Experimentation was carried out with different activities such as sleeping, eating, taking medication, walking, showering and brushing teeth, inferred from the identified patterns. This approach offers an automatic and objective way of understanding elderly people's routines and behaviours, enhancing their care and attention through personalised interventions customised to their individual needs. Future lines are quite challenging although they are intended to be carried out in the long term. Clustering is intended to correlate the groups obtained with this unsupervised learning technique and the activities proposed by the doctors. Uncorrelated elements should be explored to try to deduce their meaning.

Acknowledgements. This research has been partially funded by the BALLADEER project (PROMETEO/2021/088) from the Consellería Valenciana, by the AETHER-UA (PID2020-112540RB-C43) project from the Spanish Ministry of Science and Innovation, and by the Spanish Government project PID2021-127275OB-I00, FEDER.

References

1. ElHady, N.E., Provost, J.: A systematic survey on sensor failure detection and fault-tolerance in ambient assisted living. Sensors **18**(7), 1991 (2018)
2. Ismail, M.I.M., et al.: An RSSI-based wireless sensor node localisation using trilateration and multilateration methods for outdoor environment. arXiv preprint arXiv:1912.07801 (2019)
3. Singh, N., Choe, S., Punmiya, R.: Machine learning based indoor localization using Wi-Fi RSSI fingerprints: an overview. IEEE Access **9**, 127150–127174 (2021)
4. Ometov, A., et al.: A survey on wearable technology: history, state-of-the-art and current challenges. Comput. Netw. **193**, 108074 (2021)
5. Dinculeană, D., Cheng, X.: Vulnerabilities and limitations of MQTT protocol used between IoT devices. Appl. Sci. **9**(5), 848 (2019)
6. Soni, D., Makwana, A.: A survey on MQTT: a protocol of internet of things (IoT). In: International Conference on Telecommunication, Power Analysis and Computing Techniques (ICTPACT-2017). vol. 20, pp. 173–177 (2017)

Exploring the Potential of an AI-Integrated Cloud-Based mHealth Platform for Enhanced Type 2 Diabetes Mellitus Management

Michel Bergoeing[1]([⊠]) [ORCID], Andres Neyem[2,3] [ORCID], Paulina Jofré[1,4], Camila Hernández[4], Juan Chacón[2], Richard Morales[2], and Matías Giddings[2]

[1] Escuela de Medicina, Pontificia Universidad Católica de Chile, Santiago, Chile
`{mbergoei,pejofre}@uc.cl`
[2] Computer Science Department, Pontificia Universidad Católica de Chile, Santiago, Chile
`{aneyem,jpchacon,rgmorales1,matias.giddings}@uc.cl`
[3] National Center for Artificial Intelligence (CENIA), Santiago, Chile
[4] Hospital Clínico Eloísa Díaz La Florida, Santiago, Chile
`camila.hernandez.s@hospitallaflorida.cl`

Abstract. Type 2 Diabetes Mellitus (T2DM) is a globally escalating health issue, with its complications significantly reducing life expectancy and imposing considerable economic and social burdens. Particularly in Chile, prevalence soars to 25–30% among individuals aged 65 and over from mid to low socioeconomic backgrounds. However, only 34% of T2DM patients achieve a target glycated hemoglobin (HbA1c) level of less than 7%. Our proposed project aims to address this challenge by developing and validating an Artificial Intelligence (AI)-integrated, cloud-based mobile health (mHealth) platform designed to enhance T2DM management. This platform facilitates lifestyle modifications and medication adherence, reducing treatment burden, while fostering patient education and self-management. It leverages biometric sensors in smartphones and wearable devices to generate metabolic control indicators to improve glycemic control beyond HbA1c and utilizes AI image processing techniques for early detection of feet ulcers. The mHealth concept, bolstered by increasing mobile connectivity, integrates patients and health providers into digital platforms, thereby creating a collaborative care model. Evidence has shown that multidimensional strategies have superior outcomes to unidimensional approaches, emphasizing the importance of incorporating various determinants into T2DM management. This novel AI-integrated cloud-based mHealth platform has the potential to revolutionize T2DM care by providing comprehensive, personalized, and efficient management solutions that not only address medical needs but also empower patients. By integrating cutting-edge AI and mHealth technology, our project is set to transform the landscape of T2DM management and improve the quality of life for millions of patients.

Keywords: Artificial Intelligence in Healthcare · mHealth platforms · Type 2 Diabetes Mellitus Management

© The Author(s), under exclusive license to Springer Nature Switzerland AG 2023
J. Bravo and G. Urzáiz (Eds.): UCAmI 2023, LNNS 835, pp. 100–111, 2023.
https://doi.org/10.1007/978-3-031-48306-6_10

1 Introduction

Type 2 Diabetes Mellitus (T2DM) is a significant global health issue, especially critical in middle to low socioeconomic groups [1]. These demographics bear the brunt of the disease due to the increased prevalence and serious complications associated with T2DM. Given the disease's considerable impact on life expectancy and overall health, there is an urgent need for a comprehensive, multidisciplinary approach to manage T2DM, involving lifestyle modifications, medication adherence, and patient education.

In Chile, the situation is quite alarming. The prevalence of T2DM is notably high, particularly in individuals over 65 years of age and those belonging to mid to low socioeconomic backgrounds [2]. Moreover, only a third of these patients achieve the recommended glycated hemoglobin (HbA1c) level of 7% or less, highlighting an urgent need for improved strategies to effectively address this escalating health crisis.

The advent of mobile technology presents an array of promising solutions to enhance T2DM management [3]. Known as Mobile Health (mHealth), this technological development, augmented by the surge in mobile connectivity, connects patients and healthcare providers within a shared digital platform. This connectivity fosters a patient-centric, data-driven collaborative care model that goes beyond traditional HbA1c measures. Moreover, the integration of biometric sensors in mobile and wearable devices allows for continuous monitoring of metabolic control, thereby opening new avenues for patient self-management [4, 5].

This paper introduces an Artificial Intelligence (AI) integrated, cloud-based mHealth platform (Platform) designed to optimize T2DM management with the overarching goal of improving patient education, self-management, and metabolic control. The platform utilizes AI image processing techniques for the early detection of diabetic foot ulcers (DFU), a common and severe complication of T2DM. This feature can facilitate early intervention, potentially reducing the risk of amputation and enhancing overall patient outcomes.

In conclusion, we present this platform as a beacon of hope in leveraging technology to address a global health challenge. Our proposed Platform offers a viable, effective, and scalable strategy to tackle the growing challenge of global T2DM management. We believe that this innovative, technology-driven approach will empower patients to actively participate in their care, representing a significant stride in using technology to confront T2DM and other similar health issues.

2 Related Work

The exploration of mobile technology in healthcare, known as mHealth, has unveiled a wealth of potential in managing chronic diseases such as T2DM. An extensive body of research has detailed the numerous benefits of mHealth platforms in disease management, including but not limited to enhanced patient adherence to treatment plans, increased accessibility to medical care, and improved patient-provider communication [6–9]. It has been noted that many existing mHealth applications primarily concentrate on monitoring physiological parameters and supplying feedback. While these features undoubtedly serve a purpose, they may fall short when dealing with complex chronic conditions like T2DM.

One area of keen interest is the development and impact of mHealth applications geared toward diabetes management [10–14]. These applications typically provide a range of services such as glucose logging, automated feedback provision, and communication facilitation between patients and their healthcare providers. Empirical evidence suggests that these functionalities have contributed to improved glycemic control among users. However, these technologies have their limitations [9, 15]. Many of these applications have not yet harnessed the potential of advanced analytics or personalized care strategies. The integration of these elements into the design and functionality of these applications could significantly enhance the management of T2DM.

Simultaneously, another branch of innovation in this field has been the advent of cloud-based mHealth systems specifically designed for diabetes management. These systems incorporate features like nutrition management, exercise logging, and medication reminders, thus proving beneficial in increasing patient adherence to treatment regimens [16]. However, there exist gaps in these solutions. Notably, these systems often lack real-time monitoring capabilities and personalized care recommendations. They also tend to underutilize AI in data analysis and decision-making processes [4, 17].

AI has emerged as a promising tool in the domain of diabetes management. AI-based algorithms have demonstrated their value in predicting blood glucose levels and designing tailor-made treatment plans [18, 19]. However, current research primarily focuses on creating standalone predictive models, with little emphasis on integrating these models into comprehensive mHealth platforms [20].

To summarize, while mHealth and AI have made significant strides in the field of T2DM management, notable limitations persist. Current solutions often lack personalization, real-time monitoring, and comprehensive AI integration. Additionally, they may lack sufficient resources for patient education and self-management. The present study aims to rectify these deficiencies by proposing a Platform for T2DM management, striving not only to offer physiological parameter tracking but also to provide personalized feedback, foster patient education and self-management, and employ AI for advanced data analysis and decision-making.

3   The Proposed AI-Integrated Cloud-Based mHealth Platform

The prior sections indicate the potential of mHealth and AI solutions to augment the management of chronic illnesses like T2DM, and the need to address certain shortcomings in the current landscape. To meet these challenges, we present the design of a Platform to enhance the care for T2DM patients by leveraging the power of mobile and cloud technologies and AI-driven image analysis. The platform's foundation is Mobile Cloud Computing, a strategy where data storage and processing can occur off the mobile device when necessary, enhancing mobile hardware capabilities and compensating for its limitations [5, 21].

The platform's general architecture, as illustrated in Fig. 1, is segregated into three core areas:

a) AI and Cloud-based backend services: They form the backbone of the platform, providing computational resources and data storage capacities needed for efficient

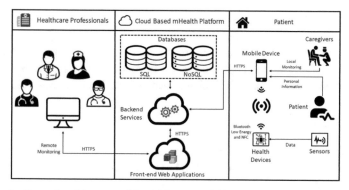

Fig. 1. General architecture of the AI-Integrated Cloud-Based mHealth Platform.

operation. The cloud services guarantee data availability, scalability, and security, providing a smooth experience for users.

b) Web Application: It serves healthcare professionals with a tool for continuous monitoring and management of patient-related data. This user-friendly application displays vital information such as blood glucose levels, medication adherence, and lifestyle patterns in real-time, assisting healthcare providers in timely decision-making.

c) Mobile Application: Tailored for patients and caregivers (family members), it delivers the same core data points as the web application in a patient-oriented, mobile format. Features include real-time glucose tracking, dietary management tools, medication reminders, and exercise logging, encouraging active participation in managing their condition.

3.1 AI and Cloud-Based Backend Services

The backbone of the Platform is rooted in its cloud-based backend services, which is hosted on the Amazon Web Services (AWS). This advanced infrastructure enables high availability, extensive storage capabilities, system flexibility, efficient processing power, and seamless integration with multiple service providers. Crucially, AWS provides robust mechanisms for security and privacy, employing digital certificates and password-based user authentication.

The backend services collect and synthesize data from a range of sensors (podometer, continuous glucose monitor device – CGM), leading to the creation of a comprehensive Personal Health Record for each patient [22]. These services are underpinned by a microservices architecture, thereby ensuring independent scalability of each component, including authentication, main service, time series data, and logger. Each of these components is developed leveraging the versatile Django framework.

For storage, our Platform employs PostgreSQL for most of the services due to its reliability, scalability, and compatibility with SQL language. However, the time series data service specifically utilizes TimescaleDB, an optimized extension of PostgreSQL designed for efficient management of time series data.

The system is equipped with an "Image Recognition Engine AI", an AI tool that classifies images of diabetic feet based on the Wagner Classification System (WCS)

[23]. It consists of a scale from zero to five (W0 to W5), the higher the value, the greater the degree of injury and amputation, and it effectively predicts amputation risk. Finally, to ensure the effective distribution of data across various platforms, we incorporate a mix of technologies. For web applications, the Web-Socket protocol is employed, while for mobile applications, Google's Firebase push notification service is used.

3.2 Web Application

The Web Application of our Platform is principally aimed at providing healthcare teams and medical experts with a tool for continuous monitoring of patients with T2DM. This application focuses on presenting three key pieces of data in real-time: the patient's most recent blood glucose levels, relevant medication details, and the patient's connection status to the application.

This application has been designed using JavaScript, HTML, and CSS, and it utilizes the React library. React's component-based architecture facilitates the creation of efficient and complex user interfaces. Its ability to self-manage individual components allows a clear separation of presentation, data, and logic components, thereby ensuring faster rendering and enhanced performance. Moreover, the application integrates Redux, a state management tool for JavaScript applications, to manage the application's global state transparently and predictably. Redux consolidates the state of the application into a single, consistent source, which simplifies the flow of data and tracks changes over time—an essential feature in a real-time monitoring application like ours.

The application also provides role-based access control, modifying the interface according to the user's role—health professional, health facility administrator, or super administrator. Each role has unique capabilities and responsibilities, ensuring the application serves its purpose efficiently across all user levels. To provide a more comprehensive understanding, Figs. 2 and 3 are referenced for further elaboration.

Figure 2 presents the patient profile view from a super admin perspective. The top section contains a primary navigation bar with various features, supplemented by a secondary bar for displaying key medical information. The screen is divided into two primary sections: the left side shows the patient's image and personal details, while the right side presents medical data. A bottom bar provides access to five critical aspects of patient care: Next Goals, Medical Appointments, Allergies, Carers, and Health Facilities. This layout aligns with the application's design principles of clarity, user-friendliness, and immediate visibility of critical data.

Figure 3 illustrates the patient's glycemia report within the application. The top features a primary navigation bar, accompanied by the secondary bar for key medical information. The interface is divided into three areas: on the left, various glycemia measures with their target values are listed; in the middle, a graph displays the percentage of time spent in each glycemia range; and on the right, the corresponding time in minutes is clearly shown. This layout provides an accessible and comprehensive view of the patient's glycemic control.

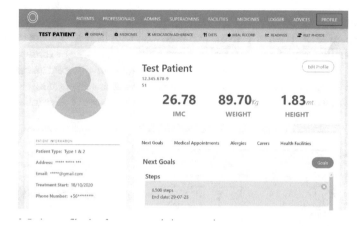

Fig. 2. Patient profile view from a super admin perspective.

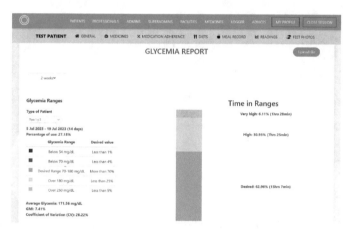

Fig. 3. Patient's glycemia report within the application.

3.3 Mobile Application

The Mobile Application of our Platform is specifically designed for patients and their caregivers. Available on both iOS and Android operating systems, it focuses on the delivery of vital, individual data points to users, rather than the continuous monitoring feature central to the web application. Employing the open-source React Native framework, this application provides a native user experience whilst ensuring cross-platform consistency and efficiency. React Native's robust framework is instrumental in providing our users with a seamless, intuitive interface that is both user-friendly and highly functional.

This application is programmed to alert users of any anomalies detected in the patient's health data, thereby ensuring timely and accurate updates regarding their health condition. It provides real-time updates on blood glucose levels and medication details and informs patients of their connection status to the application. This feature equips patients and their caregivers with the information necessary to manage their condition effectively and to maintain effective communication with their healthcare team.

To better illustrate these features, we can refer to Fig. 4. (a) exhibits the patient diet logger, a crucial feature that facilitates the search for different meals and individual food items, providing essential nutritional feedback such as nutrients, calories, and carbohydrates. (b) and (c) present comprehensive reports of glycemia values and compliance.

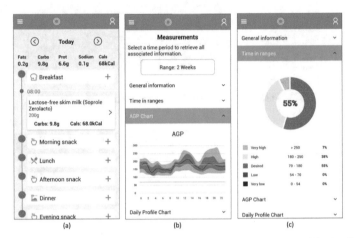

Fig. 4. Nutritional Logging and Glycemic Reporting Features in the Mobile Application.

4   Preliminary Results

This section presents the preliminary findings from our study, which comprises two parts. The first subsection describes the process and results of using AI for image recognition and classification of DFU. The second subsection explores the actual usage of our platform in a real-world context with patients diagnosed with T2DM.

4.1   AI Image Recognition and Classification - Wagnernet

The compilation of DFU and healthy feet was made through three sources: patients receiving care at the UC Christus Health Network Santiago, Chile with prior consent; Google and Instagram search (healthy feet - HF). Unlike other studies we did not set a standard on image acquisition (lighting, distance, angle, etc.) because we wanted to generalize as best as possible. DFU pictures were taken at the time of diagnosis and prior to receiving any wound care (90% via a smartphone). Once the compilation was made, physicians categorized each of the photographs. To balance the data set, images from classes with a big number of images were randomly eliminated. Due to the small sample set of W5 they were included in one category along with W4 lesions, since immediate clinical care is required, the purpose of the program remains intact. The resulting dataset had a similar number of images in all categories (Group I; 1,213 images). Then the dataset was randomly divided between training set (80%) and testing set (20%) maintaining proportions between categories. Unlike previous reports, we did not trace the ulcer contour. The following data augmentation techniques were used: rotation, translation, turning, and contrast adjustment. EfficientNetB4 was the chosen architecture. We transformed each image to a resolution of 380 pixels according to the input shape of the model. An AveragePooling2D layer was added to the end of the EfficientNet to convert the 4D tensor to a 2D one, dropout to decrease overfitting, and finally, a dense layer with 6 outputs to perform the categorization (W0, W1, W2, W3, W4orW5, HF) of an image. To obtain the best hyperparameters for training, Hyperband technique was used. The loss function used was crossentropy, the optimizer SGD, learning rate 0.01, moment 0.9, and decay 0.01. The training process was separated into two stages: 1) Transfer Learning (TL): So that the model could generalize better and not have to learn from scratch, the weights acquired in training with the ImageNet set were used. Regarding training, all the EfficientNet layers were frozen, and the model was trained for 10 iterations to converge more quickly. 2) Fine Tuning: After TL the model was completely unfrozen and then trained for an additional 25 epochs. Finally, the weights with the best accuracy in the test set were stored. After training and testing, to validate that the trained model did not significantly lose its classification capacity with a different data set than the training one, newly acquired images (DFU and HF) were analyzed with the model (Group II; 126 images). Finally, to be able to corroborate that the algorithm is performing as expected, Gradient-weight Class Activation Mapping (Grad-CAM) technique which analyzes the reaction of the model neurons was used to confirm correct detection of the DFU.

The results yielded varying levels of accuracy across the different sets. For the training set, we observed an accuracy of 90%, while the testing and validation sets recorded 72% and 65% respectively. A detailed summary of these results can be found in Table 1. We noted the most successful results in classifying healthy feet, whereas the least accurate results were associated with the recognition of W3 DFU.

An analysis using the Receiver Operating Characteristic (ROC) curve and Area Under the ROC curve (AUROC) reflected an area of 0.89, indicative of robust discriminative capabilities of our model. In conclusion, the results from our AI Image Recognition and Classification have shown promising results. Further optimization is required to improve recognition of intermediate ulcer stages.

4.2 Feasibility Study

In this preliminary study, we enlisted a cohort of four patients (2 female), each of whom utilized the platform over a one-month period. Within this timeframe, the patients actively engaged with the platform's features, logging their daily blood sugar levels through CGM, keeping track of medication adherence, and recording lifestyle activities via our mobile app. Table 2 shows patient demographics.

Table 1. Summary of results obtained from the validation set.

	Precision	Recall	F1-score	n
W0	0.82	0.67	0.74	21
W1	0.59	0.59	0.59	32
W2	0.64	0.50	0.56	14
W3	0.25	0.67	0.36	9
W4orW5	0.80	0.53	0.64	30
Healthy	0.91	1.00	0.95	20
Accuracy			0.65	126
Macro average	0.67	0.66	0.64	126
Weighted average	0.71	0.65	0.67	126

Table 2. Patient demographics, clinical characteristics and HbA1c pre and post intervention values.

	Average (SD)
Age	50.5 (9.4)
Diabetes duration (months)	89.1 (91.3)
Body Mass Index	31.1 (5.9)
HbA1c (pre)	8.8 (0.6)
HbA1c (post)	7.0 (0.9)

Three out of 4 patients wore the CGM for the whole duration of the study. These 3 patients showed a significant reduction in HbA1c during the study period. However, all patients showed a trend in decreasing glycemia values over the study period. Figure 6 shows daily glycemia values for each patient. Non-standardized patient feedback was overall positive after using the application.

In conclusion, these initial findings are promising, showcasing our platform's potential to notably enhance the management of T2DM from both the patient and healthcare provider perspectives. However, these results, while encouraging, are based on a relatively small sample size. Therefore, further exploration involving a larger cohort is necessary to confirm and expand upon these initial results (Fig. 5).

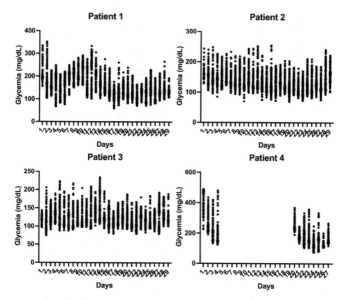

Fig. 5. Scatter dot plot showing daily glycemia values for each patient.

5   Conclusions and Further Work

This paper has presented an innovative approach to managing T2DM through an AI-integrated cloud-based mHealth platform. This proposed platform transcends being merely a medical tool, becoming an integral part of patients' lives and enabling a more effective disease management approach.

Our preliminary findings from the case study with four patients demonstrated enhanced disease monitoring and management, alongside an improved user experience. In the next phase of our research, we aim to conduct a larger-scale study involving a more diverse patient group from a public hospital. This will enable us to garner broader data and insights, facilitating the further refinement and optimization of the platform. We aim to investigate the platform's performance and impact on patient outcomes in a real-world healthcare setting.

In terms of future enhancements, we aim to enrich our platform by integrating sophisticated Generative AI like Large Language Models (LLMs), including those comparable to GPT or LLaMA, enhancing our platform's analytical prowess and responsiveness. The integration of such Foundation Models (FM) promises not only nuanced insights and recommendations but a transformative improvement in personalized, context-aware interactions for patients, caregivers, and providers in managing T2DM. However, to harness the full potential of these models, detailed fine-tuning is crucial to align their extensive knowledge precisely with the complex domain of T2DM, enabling the creation of an AI Assistant. This assistant, fueled by the platform's accumulated knowledge, will optimize treatment strategies and elevate patient-centric care, significantly impacting patient experiences and outcomes in managing T2DM.

Finally, our platform, primarily focused on T2DM, possesses the versatility to extend its services to manage other chronic conditions and promote preventive health measures, aiming to tackle the increasing prevalence of chronic diseases. We are resolutely aligned with public health strategies, seeking to integrate our innovations into the existing health-care system to revolutionize chronic disease management and enhance the overall quality of life for patients.

Acknowledgements. This work was partially supported by a grant from Red de Salud UCChristus and Glik.

Conflict of Interest. MB is stockholder of Glik.

References

1. OECD, Bank TW. Health at a glance: latin america and the caribbean 2023. OECD Publishing, Paris (2023). https://doi.org/10.1787/532b0e2d-en
2. Ministerio de Salud GdC: Encuesta Nacional De Salud 2016–2017. Primeros Resultados. https://www.minsal.cl/wp-content/uploads/2017/11/ENS-2016-17_PRIMEROS-RES ULTADOS.pdf. Accessed 18 Jan 2020
3. Mao, Y., Lin, W., Wen, J., Chen, G.: Impact and efficacy of mobile health intervention in the management of diabetes and hypertension: a systematic review and meta-analysis. BMJ Open Diabetes Res. Care 8(1), e001225 (2020). https://doi.org/10.1136/bmjdrc-2020-001225
4. Adu, M.D., Malabu, U.H., Malau-Aduli, A.E.O., Malau-Aduli, B.S.: The development of my care hub mobile-phone app to support self-management in Australians with type 1 or type 2 diabetes. Sci. Rep. 10(1), 7 (2020). https://doi.org/10.1038/s41598-019-56411-0
5. Al-Janabi, S., Al-Shourbaji, I., Shojafar, M., Abdelhag, M.: Mobile cloud computing: challenges and future research directions. In: 2017 10th International Conference on Developments in eSystems Engineering (DeSE), pp. 62–7 (2017)
6. Cruz-Ramos, N.A., Alor-Hernández, G., Colombo-Mendoza, L.O., Sánchez-Cervantes, J.L., Rodríguez-Mazahua, L., Guarneros-Nolasco, L.R.: MHealth apps for self-management of cardiovascular diseases: a scoping review. Healthcare 10(2), 322 (2022). https://doi.org/10. 3390/healthcare10020322
7. Hamine, S., Gerth-Guyette, E., Faulx, D., Green, B.B., Ginsburg, A.S.: Impact of mHealth chronic disease management on treatment adherence and patient outcomes: a systematic review. J. Med. Internet Res. 17(2), e52 (2015). https://doi.org/10.2196/jmir.3951
8. Song, T., Yu, P., Bliokas, V., Probst, Y., Peoples, G.E., Qian, S., et al.: A clinician-led, experience-based co-design approach for developing mhealth services to support the patient self-management of chronic conditions: development study and design case. JMIR Mhealth Uhealth 9(7), e20650 (2021). https://doi.org/10.2196/20650
9. Triantafyllidis, A., Kondylakis, H., Votis, K., Tzovaras, D., Maglaveras, N., Rahimi, K.: Features, outcomes, and challenges in mobile health interventions for patients living with chronic diseases: a review of systematic reviews. Int. J. Med. Inform. 132, 103984 (2019). https://doi.org/10.1016/j.ijmedinf.2019.103984
10. Liu, K., Xie, Z., Or, C.K.: Effectiveness of mobile app-assisted self-care interventions for improving patient outcomes in type 2 diabetes and/or hypertension: systematic review and meta-analysis of randomized controlled trials. JMIR Mhealth Uhealth 8(8), e15779 (2020). https://doi.org/10.2196/15779

11. Salas-Groves, E., Galyean, S., Alcorn, M., Childress, A.: Behavior change effectiveness using nutrition apps in people with chronic diseases: scoping review. JMIR Mhealth Uhealth **11**, e41235 (2023). https://doi.org/10.2196/41235

12. Shan, R., Sarkar, S., Martin, S.S.: Digital health technology and mobile devices for the management of diabetes mellitus: state of the art. Diabetologia **62**(6), 877–887 (2019). https://doi.org/10.1007/s00125-019-4864-7

13. Stephen, D.A., Nordin, A., Nilsson, J., Persenius, M.: Using mHealth applications for self-care - An integrative review on perceptions among adults with type 1 diabetes. BMC Endocr. Disord. **22**(1), 138 (2022). https://doi.org/10.1186/s12902-022-01039-x

14. Wang, Y., Min, J., Khuri, J., Xue, H., Xie, B., AK, L., et al.: Effectiveness of mobile health interventions on diabetes and obesity treatment and management: systematic review of systematic reviews. JMIR Mhealth Uhealth **8**(4), e15400 (2020). https://doi.org/10.2196/15400

15. Eberle, C., Löhnert, M., Stichling, S.: Effectiveness of disease-specific mHealth apps in patients with diabetes mellitus: scoping review. JMIR Mhealth Uhealth **9**(2), e23477 (2021). https://doi.org/10.2196/23477

16. Salari, R., Niakan, S.R., Kalhori, M.G., Jeddi, M., Nazari, M., Fatehi, F.: Mobile-based and cloud-based system for self-management of people with type 2 diabetes: development and usability evaluation. J. Med. Internet Res. **23**(6), e18167 (2021). https://doi.org/10.2196/18167

17. Lee, E.Y., Cha, S.-A., Yun, J.-S., Lim, S.-Y., Lee, J.-H., Ahn, Y.-B., et al.: Efficacy of personalized diabetes self-care using an electronic medical record-integrated mobile app in patients with type 2 diabetes: 6-month randomized controlled trial. J. Med. Internet Res. **24**(7), e37430 (2022). https://doi.org/10.2196/37430

18. Butayeva, J., Ratan, Z.A., Downie, S., Hosseinzadeh, H.: The impact of health literacy interventions on glycemic control and self-management outcomes among type 2 diabetes mellitus: a systematic review. J. Diabetes (2023). https://doi.org/10.1111/1753-0407.13436

19. Nomura, A., Noguchi, M., Kometani, M., Furukawa, K., Yoneda, T.: Artificial intelligence in current diabetes management and prediction. Curr. Diab. Rep. **21**(12), 61 (2021). https://doi.org/10.1007/s11892-021-01423-2

20. Ellahham, S.: Artificial intelligence: the future for diabetes care. Am. J. Med. **133**(8), 895–900 (2020). https://doi.org/10.1016/j.amjmed.2020.03.033

21. Abolfazli, S., Sanaei, Z., Ahmed, E., Gani, A., Buyya, R.: Cloud-based augmentation for mobile devices: motivation, taxonomies, and open challenges. IEEE Commun. Surv. Tutorials. **16**(1), 337–368 (2014). https://doi.org/10.1109/SURV.2013.070813.00285

22. Kim, J.W., Ryu, B., Cho, S., Heo, E., Kim, Y., Lee, J., et al.: Impact of personal health records and wearables on health outcomes and patient response: three-arm randomized controlled trial. JMIR Mhealth Uhealth **7**(1), e12070 (2019). https://doi.org/10.2196/12070

23. Wagner, F.: A classification and treatment program for diabetic, neuropathic, and dysvascular foot problems. Instr. Course Lect. **28**(1), 143–165 (1979)

Towards a Digital and Ubiquitous Ecosystem of Mobile Technology-Based Solutions to Facilitate Data Management Based on Sustainable Development Goals

Vladimir Villarreal[1,2]([✉]) [iD], Lilia Muñoz[1,2] [iD], Mel Nielsen[1,2] [iD],
Joseph Gonzalez[1,2] [iD], Dimas Concepcion[1,2] [iD], and Marco Rodriguez[1,2] [iD]

[1] Universidad Tecnológica de Panama, Panama City, Panama
{vladimir.villarreal,lilia.munoz,mel.nielsen,joseph.gonzalez3,
dimas.concepcion,marco.rodriguez1}@utp.ac.pa
[2] Grupo de Investigación en Tecnologías Computacionales Emergentes, Panama City, Chiriqui,
Panama

Abstract. The primary focus of research groups and centers is to develop sustainable solutions that address the problems faced by society. This article introduces a digital ecosystem platform that encompasses crucial areas like education, health, industry 4.0, and care environments, leveraging mobile and ubiquitous technology. The project aims to establish a digital and pervasive ecosystem through mobile technology, streamlining data management and the implementation of ubiquitous solutions in priority areas of the country. It's intended to provide accessible solutions in education, health, industry 4.0, and care environments for both civil society and research initiatives centered around open science. End-users will be able to access the developed platform and retrieve datasets and other resources generated by different research projects, all stored within the platform's infrastructure. Furthermore, a comprehensive dashboard will be provided to offer an insightful overview of the data's dynamics within the ecosystem.

Keywords: Digital Ecosystem · Mobile application · Open Data · Open Science · Sustainability · Ubiquitous technology

1 Introduction

During the last few years, the concepts Open Data and sustainability have been strongly promoted by governmental structures. At the same time, university research groups produce a lot of data and technological developments that can bet on these objectives, allowing society, academia, and the state in general, to take advantage of all technological solutions generated. For this to take place, the products developed, and data collected must be available and easily accessible, facilitating the reuse and sharing of data and technological developments, and encouraging the participation of all interested parts. Following an Open Science vision, we are developing a project that seeks the

J. Bravo and G. Urzáiz (Eds.): UCAmI 2023, LNNS 835, pp. 112–117, 2023.
https://doi.org/10.1007/978-3-031-48306-6_11

implementation of a digital and ubiquitous ecosystem through mobile technology, to facilitate data management and use of digital solutions in priority areas of the country, while developing solutions for education, health, industry 4.0 and care environments. This project provides a space to share all the technological developments and data that we are working on at universities, so they can be used by civil society, researchers, and organizations. We want to continue betting on Open Science and even more so when our projects receive public funding. In this article we present the conceptual design of a digital ecosystem to be implemented and validated with data and resources generated from our research group, previously formatted and structured to be open to end users.

2   Related Work

The development of digital ecosystems offers valuable potential in addressing societal challenges like the attainment of Sustainable Development Goals [1], by providing a comprehensive view of technology solutions adoption in priority areas and the effects experienced by their users. By emphasizing design principles, key components and interactions, these ecosystems help to generate actionable knowledge and identify effective interventions towards driving sustainable digital transformation [2]. Below, a series of implementations of digital ecosystems based on mobile and ubiquitous technologies are described according to the reports presented in the scientific literature:

Urban Digital Transformation. Digital transformation contributes to the development of smart cities, using digital solutions to satisfy the needs of their inhabitants. Thus, the authors of [3] presented a framework for the adoption of urban digital technologies like mobile applications, to benefit the quality of life among citizens, by optimizing resource management and district sustainability. Another example is the Digital Cities Challenge launched by the European Commission, which empowers European cities to design effective digital transformation strategies. Participating cities receive in-depth analysis of their digital situation, identifying strengths in areas like public services, innovation, and citizen engagement [4]. Thus, the inclusion of digital services and smart growth was addressed as part of a study conducted in three participating cities (Sofia, Granada, and Kavala). The process encompassed aspects like public infrastructure, smart services, intelligent mobility, environmental sustainability, innovation, and entrepreneurship [5].

Sustainable Development. Digitalization and sustainability are closely related in terms of information and data processes, involving interaction with digital tools to optimize productivity and operational quality, thereby facilitating data-driven decision-making. In study [6] authors identified the perspective of digital ecosystem architects to enhance strategic planning and the implementation of digital technologies and solutions that foster interconnectedness and collaboration within a collaborative organization. While this transition poses challenges, it also brings significant benefits to microenterprises and small and medium-sized enterprises (MSME). Thus, the authors of study [7] analyzed the development of a digital-based MSME system in Indonesia's economy, focusing on the Special Region of Yogyakarta (DIY) and its digitalization. The researchers explored how the integration and synergy of digital ecosystem are necessary to respond to changing environments and ensure productive developments for the future.

Industry 4.0. Digital transformation also extends to projects related to Industry 4.0 from a socio-technical perspective, emphasizing enterprise technology adoption to enhance efficiency and decision-making. Despite the value of Internet of Things (IoT)-generated data, platform adoption remains low, as Data governance in IoT platform adoption was examined in the Dutch horticultural industry [8]. Ecosystem data governance is pivotal in IoT platform adoption. Insights guide practitioners and policymakers in facilitating effective platform adoption, ensuring data security within ecosystems.

E-Government: In the end, innovation and digital transformation fundamentally has a significant impact on government digital ecosystems and public services [9]. It has brought profound changes in how governments interact with citizens, deliver services, and manage information [10]. For instance, projects in Uzbekistan have contributed to the digital economy with the development of a digital ecosystem of public platforms [11]. This transition enhances convenience and improves business transactions, emphasizing the importance of removing barriers to digitalization and digital commerce. Additionally, a comparative study examines Open Data platforms in Buenos Aires, Mexico City, and Montevideo [12]. The objective was to investigate how government data platform ecosystems can be governed and nurtured to stimulate innovation with the use of Open Data, highlighting the importance of establishing governance tools and rules to guide actors' participation within the ecosystem. In this sense, open information ecosystems, such as New Jersey's [13], play a crucial role in promoting transparency, accessibility, and the use of public information. They empower individuals by enabling them to utilize large datasets for research, consultation, and decision-making support.

3 Digital Ecosystem Architecture and Design

In a digital ecosystem, information can be discovered, collected, classified, secured, prepared, released, accessed, utilized, archived, purged, restored, and recycled. Following the theory proposed in [14] for homologation of digital and biological ecosystems, information can be grouped into nine elemental types: Organisms, Place, Time, Thing, Event, Mechanism, State, Intuition and Consequence. Digital ecosystems respond to a singular principle: information is relative and depends on the context in which it is found. It has a life cycle and can be discovered through interactions. Information can be transformed from one form to another, but it is never completely lost [15].

3.1 Conceptual Architecture Model

With the goal of addressing the objective of our study, the proposed digital ecosystem focuses on four priority areas: health, education, industry 4.0, and care environments with visual assistance technologies, operating in the realms of information generation, storage, and management for the specific users of each underlying application. Following the conceptual architecture model proposed in [16], the digital ecosystem structure is reflected in a set of four levels presented in Fig. 1, where the outer layer correspond to the user integration and the inner layers to the technological implementations.

The conceptual architecture model for elements of the digital ecosystem is structured from bottom up as follows: the inner level or **Semantic Core** represents data modeling

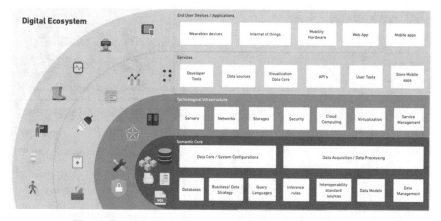

Fig. 1. Conceptual architecture model of the digital ecosystem.

technologies, databases, query languages, inference rules, and interoperability technologies that will ensure project scalability. The second level concentrates the ecosystem's **Technological Infrastructure**, including hardware, software, communication networks, information storage, security, and protection. The third level is reserved for user-facing **Services,** like search elements, visual interfaces, resources sharing and open data tools. Lastly, the fourth level encompasses **End User Channels** of interaction like wearables and IoT devices, as well as web and mobile application used as data sources.

3.2 Proposed Software Architecture for the Semantic Core

Considering software architecture design as a highly important element for successful implementations of digital ecosystems, we propose a modular development approach for the Semantic Core distributed across three domain layers: Presentation layer, Application layer, and Data layer (see Fig. 2). This architecture facilitates internal communication between components and integration with external services from the health, education, industry 4.0, and visual assistance technology applications to be developed.

The responsibilities of the different domain layers included in the software architecture are described as follows: the **Presentation layer** is based on an open-source Open Data management system with the implementation of the tools provided by the Comprehensive Knowledge Archive Network (CKAN) platform as common coordination core, enabling the open sharing of information like datasets, documents and dashboards among public and private entities, civil society organizations, and stakeholders. The **Application layer** includes the information transaction rules between the external Services from the applications and the Presentation layer, following business logic and Open Data strategies framed within the guidelines established by bioethics committees and data usage laws. Finally, the **Data layer** represents the databases and information resources where the data from each priority area of application is stored and managed.

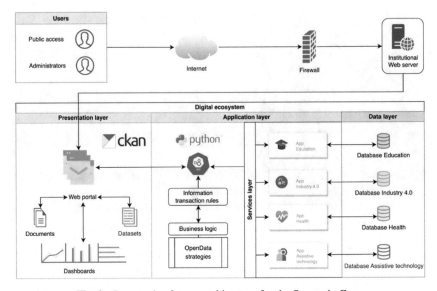

Fig. 2. Proposed software architecture for the Semantic Core.

4 Conclusions

The proposed ecosystem will allow end users to access datasets, resources, applications, and technologies produced from research projects over time. The data generated will be obtained both from external sources and from the digital applications that are able to interact with the ecosystem as needed. All services and resources that are generated by the applications can be shared, stored, and transferred to the semantic core to facilitate its acquisition. The implementation of this platform will facilitate the integration of new digital solutions, new data and above all the generation of new knowledge that can be shared and accessed making it open, adaptable, and distributed to the public.

Acknowledgments. Project funded by the National Secretariat for Science, Technology, and Innovation (SENACYT in Spanish), under the economic grant contract No. 006–2023, product of the Research Mobility Program - 2022. V. Villarreal y L. Muñoz are members of the National Research System (SNI in Spanish) of SENACYT.

References

1. United Nations: THE 17 GOALS - Sustainable Development (2023). https://sdgs.un.org/goals
2. König, P.: digital ecosystems and development. potentials and limitations of the digital ecosystem concept. SSRN Electr. J. (2023). https://doi.org/10.2139/SSRN.4382089
3. Elberzhager, F., Mennig, P., Polst, S., Scherr, S., Stüpfert, P.: Towards a digital ecosystem for a smart city district: procedure, results, and lessons learned. Smart Cities **4**, 686–716 (2021). https://doi.org/10.3390/smartcities4020035
4. European Commission: The digital cities challenge – Designing digital transformation strategies for EU Cities in the 21st century : final report (2019)

5. Komninos, N., Kakderi, C., Collado, A., Papadaki, I., Panori, A.: Digital transformation of city ecosystems: platforms shaping engagement and externalities across vertical markets. J. Urban Technol. **28**, 93–114 (2020). https://doi.org/10.1080/10630732.2020.1805712

6. Chen, Y., Wang, Z., Ortiz, J.: A sustainable digital ecosystem: digital servitization transformation and digital infrastructure support. Sustainability **15**, 1530 (2023). https://doi.org/10.3390/su15021530

7. Widyastuti, D.A.R., Wahyuni, H.I., Wastutiningsih, S.P.: Creating a digital ecosystem for sustainable development: insights from Indonesian micro, small and medium enterprises. Kasetsart J. Soc. Sci. **44**(1), 27–38 (2023). https://doi.org/10.34044/j.kjss.2023.44.1.04

8. de Prieelle, F., de Reuver, M., Rezaei, J.: The role of ecosystem data governance in adoption of data platforms by internet-of-things data providers: case of dutch Horti-culture industry. IEEE Trans. Eng. Manag. **69**, 940–950 (2022). https://doi.org/10.1109/TEM.2020.2966024

9. Ebert, C., Duarte, C.H.C.: Digital transformation. IEEE Softw. **35**, 16–21 (2018). https://doi.org/10.1109/MS.2018.2801537

10. Mas, J.M., Gómez, A.: Social partners in the digital ecosystem: will business organizations, trade unions and government organizations survive the digital revolution? Technol. Forecast. Soc. Change. **162**, 120349 (2021). https://doi.org/10.1016/j.techfore.2020.120349

11. Ashurov, Dr.Z., Makhmudova, G., Razakova, B.: Development of Digital Ecosystem and Formation of Digital Platforms in Uzbekistan. π-Economy. (2022)

12. Bonina, C., Eaton, B.: Cultivating open government data platform ecosystems through governance: lessons from Buenos Aires, Mexico City and Montevideo. Gov. Inf. Q. **37**, 101479 (2020). https://doi.org/10.1016/j.giq.2020.101479

13. Samuel, J., et al.: Garden state open data index for public informatics (GSODI): an integrated view of new Jersey's open information ecosystem (2023). https://rutgers.app.box.com/s/q5jg3xvr7ll3p8owzxp17mjt6ijd2npj

14. Gill, A.Q.: The digital ecosystem information framework: insights from action design research. J. Inf. Sci. 016555152210865 (2022). https://doi.org/10.1177/01655515221086593

15. Gill, A.Q.: A theory of information trilogy: digital ecosystem information exchange architecture. Information **12**, 283 (2021). https://doi.org/10.3390/info12070283

16. Akatkin, Y., Karpov, O., Konyavskiy, V., Yasinovskaya, E.: Digital economy: conceptual architecture of a digital economic sector ecosystem. Bus. Inform. **2017**(4), 17–28 (2017). https://doi.org/10.17323/1998-0663.2017.4.17.28

Towards Abnormal Behaviour Detection on Elderly People at Home Through Smart Plugs and Its Relationship with Activities of Daily Living

Adrián Sánchez-Miguel Ortega[1]([✉])[ID], Jesús Fontecha Diezma[1][ID], Iván González Díaz[1][ID], Luis Cabañero Gómez[1][ID], and Christopher Nugent[2][ID]

[1] MAmI Research Lab at Univeridad de Castilla-La Mancha, Ciudad Real 13071, Spain
{adrian.sortega,jesus.fontecha,ivan.gdiaz,luis.cabanero}@uclm.es
[2] Ulster University, School of Computing, Jordanstown, UK
cd.nugent@ulster.ac.uk

Abstract. Elderly population is growing, and it contributes significantly to the saturation of public healthcare services due to multiple reasons. Nowadays, an increasing number of elderly people are living alone at home, and they expect to live independently for a longer period without help from caregivers or others, or at least with a minimum of it. The performance of activities of daily living can be measured through several tests and techniques according to the literature, providing us a useful tool to detect health problems related to independence levels, which could lead to major issues. The integration of new technologies at home could support caregivers and experts to detect abnormal behaviour when people perform certain activities, listed as basic and/or instrumental activities of daily living. Connected devices together with data analysis can be used to perform this task. This paper describes a proposal to monitor load measurements using smart-plugs connected to appliances distributed in a home environment, in order to detect specific activities of daily living. A recurrent neural network has been trained based on previously acquired time series, from such monitoring, framed at different time periods. The preliminary results of this work present how it would be possible to detect and associate some tasks at home with activities of daily living, generating activity patterns and, therefore being able to detect abnormal behaviours to act conveniently with the aim of increasing the life quality of elderly people at home.

Keywords: Abnormal Behaviour · Ambient Assisted Living · Energy Consumption · Smart plugs · Recurrent Neural Network · Activities of Daily Living

J. Bravo and G. Urzáiz (Eds.): UCAmI 2023, LNNS 835, pp. 118–123, 2023.
https://doi.org/10.1007/978-3-031-48306-6_12

1 Introduction

Quality of life of elderly people who live alone at home depends mostly on the caregiving processes that they and their caregivers can provide, as well as the level of independence they have to manage daily situations. Nowadays, older adults wish to live independently in their homes and to perform daily activities without relying on or asking for external assistance. For some, provision of support for their activities of daily living (ADLs) may become a necessity to enable them to adequately cope with everyday life [1]. Thus, ADL measurement is a standard to determine the level of independence of an elderly person [2].

In general, ADLs are divided into three groups depending on their performance and autonomy levels as follows: basic activities of daily living (BADL), instrumental activities of daily living (IADL), and advanced activities of daily living (AADL); existing standardised scales and adaptations to measure the independence level per group of activities or as a whole [1]. In this sense, there are several activities from BADL and IADL that are performed at home, and can be analysed based on the usage of home devices and appliances, also to determine their impact on the dependence level of elderly people who mainly live alone at home.

This work presents a new approach to study activities of daily living patterns and detect abnormal behaviours through smart plugs connected to home appliances, distributed in the environment. Thus, it is intended to proficiently detect inconsistencies, such as the discontinuation of particular activities or the transposition of these activities to different time slots within a day. It is arguable that such deviations warrant due attention as well as a correct identification and analysis.

2 Background

2.1 ADL-Based Caregiving at Home

Since the beginning of the century, the increase in ubiquitous environments has allowed the rise of multiple studies which have covered the positive impact of caregiving based on ADLs for both the elderly or people with special needs, and their caregivers. This positive impact can be translated into both cost savings and the alleviation of healthcare systems [3]. The detection of patterns in activities using the information retrieved from patients can be performed using intrusive and non-intrusive digital methods, both having advantages and disadvantages.

2.2 Digital Intervention Approaches

In the Ambient Assisted Living (AAL) domain, non-intrusive monitoring methods aim to maintain privacy while still providing useful data. Ambient sensors installed around the home can unobtrusively monitor environmental factors such as movement and temperature. Smart home devices such as smart thermostats and bulbs provide usage pattern data, offering indirect insight into daily routines. Radio Frequency Identification (RFID) tags can be placed on frequently used objects in order to provide data on daily activities. Non-Intrusive Load Monitoring (NILM) is particularly notable for its ability to monitor energy usage without requiring equipment installation on each appliance, which minimizes disruption and favours scalability across different environments [4,5].

On the other hand, intrusive monitoring methods potentially offer more precise data, but they raise ethical and privacy concerns due to the intimate nature of the information collected. Examples include Video Monitoring Systems, which, while providing detailed visual data, are considered invasive because of significant privacy issues. Wearable devices like smartphones offer valuable insights into daily activities and health metrics, but may also be seen as intrusive. Additionally, Intrusive Load Monitoring (ILM), a less studied method due to requiring sensors to be installed on each appliance, also considered disruptive by the users, gives more detailed data and accuracy than NILM approaches [6,7].

3   Monitoring and Detection Approach

3.1   Data Acquisition, Communication and Storage

In the course of the study, a Rust-based Application Programming Interface (API) was used to acquire data from Tapo P110 smart plugs[1]. This information includes power status (on/off) and detailed energy consumption data. Each smart plug can supply an energy profile including the power consumption for the current day, week, and month. Furthermore, this smart plug allows tracking and report the real-time power flow, making it very useful for monitoring energy usage and efficiency. Figure 1 shows the necessary stages of this approach, from the data acquisition to the identification of ADLs.

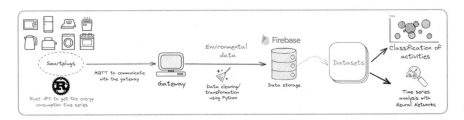

Fig. 1. General stages of the monitoring and detection approach

Data extraction was conducted by connecting to the smart plugs via their IP addresses. This method ensured direct and secure access to each individual smart plug, thereby enabling a seamless extraction process. A key factor that facilitated this process was the prior registration of each smart plug through the

---

[1] https://www.tapo.com/es/product/smart-plug/tapo-p110/.

Tapo application, which resulted in each plug having a unique identifier, making possible their recognition and access.

In the next phase of the data flow, a database was created on the Firebase platform[2] to store the encoded information from the smart plugs. To facilitate the data transfer from the Rust API to the Firebase database, we used the Message Queuing Telemetry Transport (MQTT) protocol[3].

This data, which was sent through MQTT, was then interfaced with a Python gateway. The role of this gateway was to act as a bridge between the Rust API and the Firebase database, allowing secure and efficient transmission of data from the smart plugs to the database.

Once the data was stored in datasets, it could be processed for their future analysis or presented to the user through a front-end application.

3.2 Preprocessing and Data Cleaning

The next phase involves the necessary processes to clean and prepare the dataset for our purposes. In this proposal, due to the unavailability of actual data on the environment, we used an existing dataset for the detection and analysis of ADLs, specifically the REFIT Electrical Load Measurements[4], a rich and public dataset of energy consumption records across 20 unique households in the United Kingdom.

The first steps in the preprocessing of the dataset included column renaming for interpretability, converting timestamp records into datetime objects for easier manipulation, and creating new 'activity' features in a separated dataframe. These features, based on the domain knowledge, added energy readings from appliances commonly used when doing a certain activity, into single columns. These columns were "Cooking", with appliances like the microwave or a toaster added together, "Working" with appliances like computers, "Entertainment" with appliances like televisions or game consoles, "Cleaning" with appliances like washing machines or dishwashers, and "Housekeeping" with appliances like electric heaters. Finally, before applying any machine learning algorithm, the activity dataframe was scaled and subjected to dimension reduction using Principal Component Analysis (PCA).

3.3 Categorization and Activity Detection

Application of K-Means Clustering. After preprocessing stage, K-Means clustering was employed to categorize the data into distinct groups representing different ADLs. Given the unsupervised nature of the dataset, K-Means serves as an ideal tool for exploratory analysis. After clustering, the categorical labels were transformed into a numerical format suitable for an upcoming Long Short-Term Memory (LSTM) model by using one-hot encoding.

---

[2] https://firebase.google.com/.

[3] https://mqtt.org/.

[4] https://pureportal.strath.ac.uk/en/datasets/refit-electrical-load-measurements-cleaned.

Detecting Patterns Using a Recurrent Neural Network. To detect patterns in time-series data, the recurrent neural network (RNN) architecture known as LSTM was employed. LSTMs are particularly adept at handling time-series data, due to their inherent design to capture long-term dependencies and deal with the vanishing gradient problem, a common issue in traditional RNNs [8].

The LSTM model consisted of three layers with 128, 64 and 32 units each, and a dense output layer to predict the probability distribution across the different ADL categories. The model was compiled using the 'Adam' optimizer and 'categorical crossentropy' as the loss function, given the multi-class nature of the problem.

4   Preliminary Results

The application of the K-Means algorithm shows that the inspection of the most representative points of each cluster suggested reasonable correlation with activities such as "Cooking", "Cleaning", "Entertainment" or "Working from Home", as can be seen in Fig. 2. The clusters extracted from the algorithm are represented on the y-axis of the confusion matrix in this figure, labelled from *Activity_0 to Activity_6*. An in-depth analysis of these preliminary results shows, for example, how the *Activity_0 and Activity_3* clusters have no significant correlation with none of the ADLs considered, while *Activity_1* cluster shows correlation with cooking ADLs, *Activity_2* cluster with entertainment ADLs (e.g. watching TV) and *Activity_4* with cleaning activities such as using the dishwasher or a washing machine. The confusion matrix also suggests that more than one ADL are represented by certain clusters, like cooking and cleaning in the *Activity_5* cluster or working and entertainment in the *Activity_6* cluster.

After training the LSTM RNN, the first model yielded a 98 percentage accuracy on the validation data across all 20 houses, which is a promising result showing the model's ability to detect patterns corresponding to ADLs predicted by the K-Means algorithm.

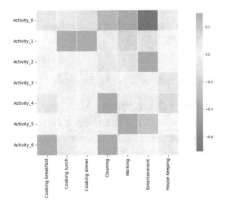

Fig. 2. Correlation between activities and K-Means clusters

5   Conclusions

The presented approach shows that, by employing a combination of data pre-processing techniques, K-Means clustering, and LSTM recurrent neural network models, different home activities were discerned, based on smart-plugs energy usage patterns associated with Activities of Daily Living (ADL).

The difficulties inherent in setting Intrusive Load Monitoring (ILM) projects have limited the breadth of research in this area, making Non-Intrusive Load Monitoring (NILM) a preferable choice for Activities of Daily Living (ADL) detection. Consequently, this paper introduces a novel methodology in ADL-based home caregiving by integrating ILM with various machine learning techniques.

The results show that the model generalizes well to different homes and users. However, it could be interesting to consider developing a house-to-house model in the future stages of the project to compare results with the generalised model. It would be relevant in case of concrete analyses of elderly people who live alone at home, who change their ADL execution patterns over time, what could be a sign of health issues.

Acknowledgements. This research is funded by the National Green and Digital Transition programme (MCIN/AEI/10.13039/501100011033) and the European Union NextGenerationEU/PRTR, with grant number TED2021-130296A-100.

References

1. Katz, S.: Assessing self-maintenance: activities of daily living, mobility, and instrumental activities of daily living. J. Am. Geriatr. Soc. **31**(12), 721–727 (1983)
2. Wallace, M., et al.: Katz index of independence in activities of daily living (ADL). Urol. Nurs. **27**(1), 93–94 (2007)
3. Bennett, J., Rokas, O., Chen, L.: Healthcare in the smart home: a study of past, present and future. Sustainability **9**(5), 840 (2017)
4. Alcalá, J.M., Ureña, J., Hernández, Á., Gualda, D.: Assessing human activity in elderly people using non-intrusive load monitoring. Sensors **17**(2), 351 (2017)
5. Devlin, M.A., Hayes, B.P.: Non-intrusive load monitoring and classification of activities of daily living using residential smart meter data. IEEE Trans. Consum. Electron. **65**(3), 339–348 (2019)
6. Franco, P., Martinez, J.M., Kim, Y.-C., Ahmed, M.A.: IoT based approach for load monitoring and activity recognition in smart homes. IEEE Access **9**, 45325–45339 (2021)
7. Fahim, M., Ahsan Kazmi, S.M., Khattak, A.M.: Appliancenet: a neural network based framework to recognize daily life activities and behavior in smart home using smart plugs. Neural Comput. Appl. **34**(15), 12749–12763 (2022)
8. Malhotra, P., et al.: Long short term memory networks for anomaly detection in time series. In: ESANN, vol. 2015, p. 89 (2015)

A Framework for Managing the Experimental Evaluation of Ambient Assisted Living Systems

Javier Jiménez-Ruescas[1]([✉])(ID), Roberto Sánchez[1](ID), Yuxa Maya[3],
Antonio Fernández-Caballero[1,2,4](ID), Arturo S. García[1,2](ID),
and Pascual González[1,2,4](ID)

[1] LoUISE Research Group, Instituto de Investigación en Informática (I3A),
Albacete, Spain
{javier.jruescas, roberto.sanchez, antonio.fdez, arturosimon.garcia,
pascual.gonzalez}@uclm.es
[2] Departamento de Sistemas Informáticos, Universidad de Castilla-La Mancha,
02071 Albacete, Spain
[3] Departmento de Psicología, Facultad de Medicina,
Universidad de Castilla-La Mancha, 02006 Albacete, Spain
yuxa.maya@uclm.es
[4] CIBERSAM-ISCIII (Biomedical Research Networking Center in Mental Health),
28016 Madrid, Spain

Abstract. Designing, setting up and running experiments is a key phases of studies, but it can be quite tedious due to the need to synchronise signals from data-collecting sensors, change variables between experiments, and manage volunteers. This article presents the architecture of a framework for managing and running experiments. It allows to easily manage the parameters of each of the tests assigned to the users and the information about them. It also enables the remote control of experiments via TCP communication, the integration and synchronisation of different devices via OpenViBE, and storage of data and signals in time series databases via InfluxDB. To illustrate the capabilities of the framework more clearly, we will also present a case study in an augmented reality environment using the HoloLens2 device. It describes the use of the framework in an experiment involving the display of different appearances of assistive drones to study participants to analyse their acceptance. After viewing the different drones in an augmented environment, participants are asked to provide subjective feedback through questionnaires and objective feedback through heart rate and eye-tracking sensors. This case study shows how the use of this framework streamlines testing by allowing test configuration, uniquely identifying the information provided by each participant and synchronising the information obtained from the different sensors.

Keywords: Experiment evaluation · Human-drone interaction · Augmented reality · Assistive drone

J. Bravo and G. Urzáiz (Eds.): UCAmI 2023, LNNS 835, pp. 124–135, 2023.
https://doi.org/10.1007/978-3-031-48306-6_13

1 Introduction

The field of Ambient Assisted Living (AAL) provides services that help users in their daily life, trying to be as less invasive as possible. Therefore, this type of applications can have a great impact on the life of potential users and in its development it is very relevant to analyse, through experiments with these users, how they perceive these applications in order to adapt their features to this perception. Thus, as recently pointed out, if AAL applications are designed with end-user input in mind, researchers will be able to accurately assess the acceptance of new AAL systems [6]. Furthermore, this assessment needs to be done not only at the end of development, but also during design.

One of the main areas of application for AAL applications is the elderly, as these technologies have great potential to improve the lives of this group. There are many applications that aim to provide solutions to improve the lives of the elderly by facilitating independent living. However, this type of technology also raises certain concerns that need to be addressed, having been identified eight perceived barriers [11]: lack of perceived need and usefulness; privacy, intrusiveness and control; lack of experience, technology anxiety and self-efficacy; fear of social stigma; reliability; lack of human interaction; cost; and health concerns. All of these barriers, along with communication, social skills and anthropocentrism [12], affect the successful adoption of AAL technologies and should be considered by the designer of new AAL applications. In general, these barriers are related to the novelty of these technologies and the lack of experience of potential users. Therefore, it is important to conduct experiments that can analyse the behaviour of these users when faced with different designs. This type of user evaluation is even more relevant to study the reaction of older people, as they have not grown up with technology and their relationship with these new technologies is uncertain.

To carry out these experiments, the use of simulated environments allows designers to bring the user into contact with their designs. The use of virtual reality (VR) or augmented reality (AR) applications allows the user to interact with these virtual objects and give an opinion on their design, usability and usefulness. However, the evaluation of the experimental design using these environments is difficult because they use different equipment, sometimes to be operated by inexperienced users. Moreover, the relationship between objective measurements, such as physiological signals obtained by sensors, and the action performed by the user is not easy to synchronise, a process whose importance has already been discussed on previous occasions [16].

Therefore, in this paper we present a holistic framework to facilitate the management of experimental evaluation in VR and AR environments. This framework allows the control of the participants in the experiment and synchronises all the information to ensure that it is properly managed. It integrates subjective information, obtained by questionnaires, with objective one, obtained by sensors and those derived from the task performed, by recording images that allow us to know what the user was doing at any given moment.

The remainder of the article is structured as follows. The state of the art of experiment management systems is briefly described in Sect. 2. The architecture of the proposed framework is then presented in Sect. 3. A case study is described in Sect. 4 to better understand how the framework is used. Finally, conclusions and future work are presented in Sect. 5.

2   State of the Art

The use of experiments to analyse human behaviour is a common activity in various fields such as psychology or medicine. As a result, there are currently several commercial and open source tools for managing these experiments. The first attempts to address the problem of experiment management were made at the end of the 80 s of the last century, with the appearance of systems such as B/C Power Lab, PsychLab, MEL or PsyScope among others [15]. The joint efforts of MEL and PsyScope resulted in E-Prime [18], the first commercial tool. This commercial tool is now one of the most widely used and provides, among other functions, the ability to design experiments, access a wide range of EEG, fNIRS, eye trackers and other psychological devices, and synchronise the different signals obtained from the sensors used.

The advent of VR and AR devices has opened the door to the development of fully customised virtual environments for conducting experiments in all kinds of fields, and new tools, methodologies and frameworks have emerged to conduct these experiments. The use of head-mounted display (HMD) devices for conducting experiments and the implementation of different sensors in the same devices has led to the creation of immersive questionnaires that are administered from the environment itself, usually VR environments, and the collection of data in the HMD device. For this reason, it has been necessary to create new frameworks or modify existing ones to enable the ability to define experiments and their configuration, control the workflow during the experiment, design questionnaires and store the results obtained in remote servers for later access [1].

One of the challenges faced since the beginning of experiments in virtual environments has been the adaptation of frameworks to facilitate their use by researchers. The use of virtual reality environments implies that they are developed in engines such as Unity or Unreal Engine, which are not only quite complex, but also oriented towards the design of video games. This has led to the emergence of frameworks and tools such as the Unity Experiment Framework (UXF) [4], which provides researchers with an easy and simple implementation to collect data and modify experiment variables in the Unity engine [5], or others, such as Vizard [17], which provides a toolkit for performing basic experimental tasks, such as creation, randomization, execution, and result storage, using Python to simplify development. Another very important issue is to facilitate the flow of the experiment for the participant working on it. For this reason, methods were also explored to allow instructors to remotely control the experiments [13].

For all these reasons, it has been seen that most of the articles related to VR focus on one part of the design of the experiment and, as it has been suggested on some occasions, a standardisation should be created to cover all the basic aspects of the whole experimental process (design, experiment, analysis and reproduction) [7]. With this goal in mind, this paper aims to propose a framework that meets these needs and is able to cover the whole process, freeing the user from unnecessary interactions with the device, focusing only on the experiment itself, and allowing the instructor to have full control over the experiment in a simple way. In addition, it has been observed that when introducing virtual or augmented reality into the experiment, in most cases the experiment focuses on taking in information provided by the device, making it difficult to synchronise this information with other information provided by other devices and sensors. This can be a limiting factor in conducting studies and is one of the objectives of the framework presented below.

3 Architecture

In this section we will analyse the proposed architecture to holistically control the design and execution of an experiment. Figure 1 shows the diagram of the proposed architecture of the framework. At least two actors are involved in the process of running an experiment: the instructors and the participants. Thus, the framework can be divided into two parts, one corresponding to the participants involved in the experiment and the elements associated with them (grey area with solid line) and the other corresponding to the elements running on the experimenter's computer (blue area with dashed line).

As we have seen, there are two actors involved in the process of the experiment. The first, the participant of the experiment, will use several devices. The first device is a computer running a VR or AR application (in our proposal Unity) that will have integrated UXF. This framework is an open source package of the Unity engine for developing experiments in virtual reality. It will be used to run a series of questionnaires, such as a demographic questionnaire to collect information about the participant, or other questionnaires to provide an evaluation of the experiment. It was decided to use UXF because it is easy to integrate into VR experiments, but at the same time gives the flexibility to use it in desktop, browser, etc. Another positive aspect of this framework is that it has an automated data collection process, systems to configure the different sessions or parts of our experiments, and an easily customisable user interface.

Participants will also use an HMD to conduct a VR or AR experiment. Although any type of VR or AR device can be used, in our case, since we are conducting an experiment using AR in our case study, we have used the Microsoft HoloLens 2 HMD, which allows us to replicate experiments previously conducted in VR in a much more realistic way by using the real environment. In addition, this device allows us to use functionalities such as eye-tracking to analyse the user's gaze behaviour during the experiment. In addition, thanks to the camera built into the device and the possibility of recording the HMD video, we will be able to analyse the different frames at specific moments of the experiment, allowing us to know what task the user is performing at a given time. Finally,

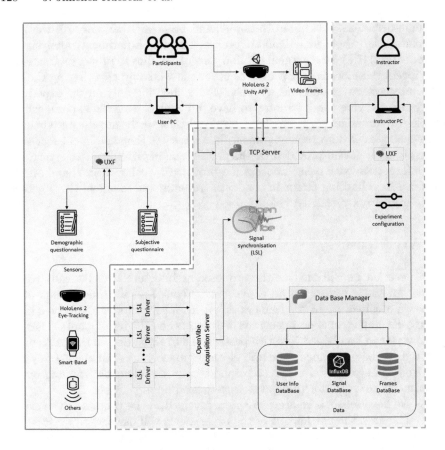

Fig. 1. Architecture chart.

the use of wearables allows additional sensors to be attached to the participant to collect other information (heart rate, perspiration levels, etc.). The nature of these sensors is very diverse, often using different connection methods (WiFi, Bluetooth), different refresh rates and data transmissions, etc., and yet they must be able to synchronise with each other.

On the other hand, the other actor involved in the experimental process is the instructor who will guide the participant during the experiment. The instructor will have a computer from which he/she can configure the variables involved in the experiment and subsequently control the process to be followed, also using UXF. In addition, the instructor's computer will have a TCP server, developed in Python, which will allow us to communicate with the other devices (the participant's computer and the HMD) and thus control the sequence of steps to be followed during the experiment, launch specific tests, etc. In this way, the user does not have to carry out any kind of configuration from the HMD, but can only concentrate on carrying out the experiment. The TCP protocol allows us to establish a connection between devices that work as a client-server. At the same time, it avoids saturation of the network, allows the simultaneous transmission of information from different devices and ensures that the data arrive at their destination in the order in which they were sent. On the other hand, the choice of Python as the programming language to create the server is justified, as we will see later, by OpenViBE's ability to include Python scripts that can be executed while the application is running.

Also available on the instructor's computer is the OpenViBE [10] software, a brain-computer interface (BCI) application development environment originally designed to enable real-time acquisition, processing and analysis of brain signals, although its ease of customisation means that it can now be used with any type of signal. Among the various tools available in the software, OpenViBE Designer was used in our architecture. Designer is a graphical application that allows us to define, configure and execute a part of the workflow of our experiment, in particular the one related to the acquisition and signal processing part.

We use this tool because it allows the implementation of modules for the acquisition of signal data, the use of the Lab Streaming Layer (LSL) for their synchronisation and the possibility of running scripts simultaneously, which are used to create a TCP server capable of connecting to the other devices used in the experiment, to control the phases of the process and to send the data received through the LSL to the appropriate databases. The use of LSL is a key factor, as it is a real-time data transmission protocol between devices that allows the integration of different signal sources in a synchronised manner.

On the other hand, we also have the OpenViBE Acquisition Server, an OpenViBE module that will allow us to connect to the devices to collect their signals. Since we will be using LSL, it is necessary to program an LSL driver for each device, but this allows us to connect any type of device, not just brain signal receivers as the software was originally intended. The signal data collected by the acquisition server is sent through a connection port assigned to each device to OpenViBE Designer, where a receiver module intercepts the information and can be used for processing.

Finally, there is a Python script for database management that runs concurrently in OpenViBE. This script takes as input the signals coming from the acquisition module and stores them in InfluxDB [9], a time series database commonly used for sensor monitoring. One of the main reasons for choosing this database is that it is optimised for time series data, i.e. it is able to handle large amounts of data in chronological order, each with a timestamp. The same script is used to store the participants' answers to the questionnaires provided during the experiment and to store specific frames captured by the HMD camera in order to analyse what happened at a given moment.

4   Case Study

The integration of drones in the field of active assistance or AAL offers many benefits to improve people's quality of life, but it also presents challenges that still need to be addressed, such as invasion of privacy by the device, discomfort with human-machine interaction, feeling of intrusion or lack of familiarity with the technologies.

A case study is presented in which, through VR and AR environments, the user is exposed to different appearances of a drone (shown in Fig. 2) that will perform several interactions with the user to study its acceptance and integration into their daily life. The main idea is to perform these tests in a VR or AR environment and finally in a real environment and with physical devices.

The choice of whether to use VR or AR will depend on the conditions and needs of the experimenters at the time. VR allows us to create much more controlled scenarios, in which the user is more isolated. However, the option of using AR will provide greater realism, but we will depend on external factors that may distract the user during the experiment, unfavourable lighting conditions that may impair the experience, etc. There may also be participants who are uncomfortable with any of the devices, especially older people. The possibility of using both technologies opens up a wide range of possibilities to carry out the experiments with the hardware that best suits the participant.

Using VR and AR technologies in the early stages of experimentation is a valuable way to test drone performance safely before using the actual physical device. In addition, introducing an intrusive element such as a drone into the daily lives of elderly people can have a significant impact and can lead to rejection. Therefore, conducting these tests in a simulated environment can help us to increase the level of acceptance of the drone before using the real device.

The following is an example of the process we recommend from the framework presented, through a process of experimentation in an AR environment.

Fig. 2. Appearances of the drones.

Figure 3a shows a diagram to guide the process to be followed. It shows the tasks to be performed, the tools and technologies involved in each of them, and the actors who develop these tasks. In this figure, we have marked the task modules developed by the experimenter with a solid line and those developed by the participants with a dashed line.

The session starts with an initial phase, developed only by the instructor In this phase, a pre-configuration of the experiment is carried out, in which the instructor enters data such as the number of participants and the conditions of the experiment. Specifically, in this case, the appearance of the drones and the distances at which they interact with the user with respect to their position are

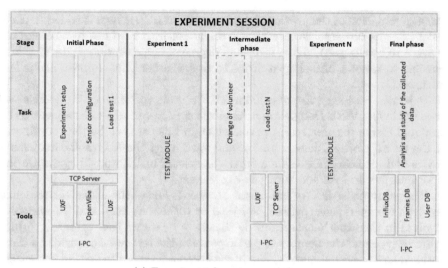

(a) Experimental session procedure.

(b) Procedure during a test module.

Fig. 3. Procedure and tools involved during the session

considered as independent variables [2]. All this configuration is done through a graphical interface developed with UXF, the Unit Experiment Framework seen in the previous section. In this way, the system automatically prepares as many combinations of tests as there are combinations of interaction distances and forms for each user. The shapes of the drones can vary between a quad-copter drone, a variation of the previous one, to which an LCD screen is added to show different emotions associated with the drone (such as tiredness indicating that it needs to be charged), giving it a small degree of humanisation [8,19]. It is also possible to select a bird-shaped drone to test whether a live appearance is less intrusive.

It is also necessary to pre-configure the sensors to be used. To do this, we have used OpenViBE Designer and Acquisition Server to select the sensors to be used and assign the connection ports, which in this case will be the coordinates of the user's position, head rotation and gaze using the Microsoft HoloLens 2 device and his heart rate using a smart bracelet. Finally, we will again use the UXF interface to load and start the experiment.

The setup phase is very important because it allows the instructor to define the type of experiment (between subjects or within subjects) and the variables involved in the experiment. Knowing that in our case we have chosen a within-subjects type and the number of participants in the test, we can decide to use a Latin square (REF), a matrix used in experimental design to control the effects of confounding factors or variables in a study. This will allow us to achieve a more balanced distribution of combinations of trials across the variables, improve the precision of the estimates, and introduce a small degree of randomness into the allocation of trials to participants. After this first phase of setting up the experiment, the experiment begins, as shown in Fig. 3b.

The instructor first explains the experiment to the participant and clarifies any doubts. Then, using the UXF interface and from his computer via TCP, he will send the demographic data questionnaire to the participant's computer (U-PC), where the participants will also fill it in using a UXF graphical interface and store the data in the DB relative to the users. Contrary to several articles in which the questionnaires and assessments were completed from the HMD [3,14], it was decided to complete the questionnaire on a computer rather than on the HMD device. This is because filling out questionnaires is a task that can be performed more easily on a conventional computer than on a virtual reality device. In addition, the experiments may be aimed at older people who may find it much more comfortable to complete these questionnaires on a traditional screen. Completing questionnaires outside the virtual or augmented reality environment is by no means mandatory. Depending on the age group targeted in your experiment, the platform used, the length of the questionnaire, etc., the instructor can adapt the questionnaire tasks presented in Fig. 3b to be completed from the HMD device.

Once the questionnaire has been completed, the instructor receives an automatic notification and has the option to launch the test generated for that user on their computer interface. The user, using the HoloLens 2 device, will observe different scenes in which different drones interact with the user in terms of their position, changing different parameters such as proximity to the user and appearance. In our case, all these tests will be performed consecutively, one after the other, until the stack of scenarios associated with the user is completed after the initial setup of the experiment. During the execution, the signals obtained, both those collected by the sensors and the images recorded by the HoloLens 2, are synchronously recorded for subsequent analysis. Once all the tests have been completed, the instructor receives a new notification and can send the command via the TCP server to the participant's computer to display the final evaluation questionnaire on its interface. The user will complete the questionnaire in the same way as in the first case and his answers will be recorded in the corresponding database. This questionnaire covers aspects such as the preference for the appearance of the drone, the evaluation of the usefulness of the incorporation of a screen on the drone, the feeling of intrusiveness of each of the drones, etc.

This is followed by an intermediate phase in which the participant changes and the instructor loads the test for the new participant from his/her interface. Since the number of participants in the experiment has been previously specified and the test scenarios have been assigned to each participant, no new configuration is required, but only the workflow displayed on the instructor's screen, which makes the session an automated and simple process for all users. The test module and the intermediate phase are repeated as many times as there are participants in the experimental session. Once all the tests have been completed, a final phase is carried out in which the instructor can access the databases from his or her computer to retrieve the information and data collected for analysis and processing.

5   Conclusions

As we have seen in the article, the field of AAL provides services that help users in their daily lives, trying to be as minimally invasive as possible. The analysis of the user's perception of these applications is fundamental in order to adapt their features and to evaluate the acceptance of new AAL systems. It is therefore of utmost importance to conduct experiments to overcome barriers such as lack of perceived need and usefulness, privacy and control concerns, lack of experience, technological anxiety, etc. Managing these types of experiments, both in virtual and augmented environments, can be challenging, especially when the number of devices involved in the process is significant, we have a large number of variables and participants, and we need to synchronise both objective measurements and actions performed by the user.

This paper has presented a framework to facilitate the management of evaluation and experimentation in AAL. Currently, commercial and open source tools are available for managing experiments in different scientific fields. However, due to their current popularity and the advantages offered by virtual and augmented environments, there is still a need to adapt and develop new tools and methodologies for experimentation focused on these types of environments. The proposal of a holistic framework that encompasses the whole process of experimentation in virtual and augmented environments, from design to analysis and reproduction, can facilitate the execution of experiments and free both researchers and participants from unnecessary interactions with the equipment, thus reducing the time spent on these tasks and making them easier for all users. The presented framework further facilitates the integration of subjective user input with objective sensor data and recorded process images, providing additional insights to researchers.

Experiment design frameworks should also consider that the people responsible for conducting experiments often come from different research fields and may not have a background in computer science. Therefore, the development of a framework that facilitates the modification of experiments and automates part of the experiment management process is essential to facilitate the task. Although psychologists with expertise in the design of experiments have collaborated in the design of this framework, we are conducting a study to validate with users with low computer skills how easy the proposed framework is to use and to identify potential ways to improve it to make it accessible to as many users as possible.

Acknowledgements. This paper has been funded by Junta de Comunidades de Castilla-La Mancha and by European Regional Development Fund (SBPLY/ 21/180501/000030), by Universidad de Castilla-La Mancha (2022-GRIN-34436), and by MCIN/AEI/ 10.13039/501100011033 and by "ERDF A way to make Europe" (PID2020-115220RB-C21). This work was also partially supported by CIBERSAM of the Instituto de Salud Carlos III.

References

1. Bebko, A.O., Troje, N.F.: bmlTUX: design and control of experiments in virtual reality and beyond. i-Perception **11**(4) (2020). https://doi.org/10.1177/2041669520938400
2. Belmonte, L.M., Garcia, A.S., Segura, E., Novais, P., Morales, R., Fernández-Caballero, A.: Virtual reality simulation of a quadrotor to monitor dependent people at home. IEEE Trans. Emerg. Top. Comput. **9**(3), 1301–1315 (2021). https://doi.org/10.1109/TETC.2020.3000352
3. Bovo, R., Giunchi, D., Steed, A., Heinis, T.: MR-RIEW: an MR toolkit for designing remote immersive experiment workflows. In: 2022 IEEE Conference on Virtual Reality and 3D User Interfaces Abstracts and Workshops, pp. 766–767. IEEE (2022). https://doi.org/10.1109/VRW55335.2022.00234
4. Brookes, J.: UXF - Unity Experiment Framework. https://immersivecognition.com/unity-experiment-framework/. Accessed 04 July 2023
5. Brookes, J., Warburton, M., Alghadier, M., Mon-Williams, M., Mushtaq, F.: Studying human behavior with virtual reality: the unity experiment framework. Behav. Res. Methods **52**(2), 455–463 (2020). https://doi.org/10.3758/s13428-019-01242-0
6. Choukou, M.A., et al.: Evaluating the acceptance of ambient assisted living technology (AALT) in rehabilitation: a scoping review. Int. J. Med. Informatics **150**, 104461 (2021). https://doi.org/10.1016/j.ijmedinf.2021.104461
7. Grübel, J.: The design, experiment, analyse, and reproduce principle for experimentation in virtual reality. Front. Virt. Realit. **4** (2023). https://doi.org/10.3389/frvir.2023.1069423
8. Herdel, V., Kuzminykh, A., Hildebrandt, A., Cauchard, J.R.: Drone in love: emotional perception of facial expressions on flying robots. In: Proceedings of the 2021

CHI Conference on Human Factors in Computing Systems, p. 716. ACM (2021). https://doi.org/10.1145/3411764.3445495

9. influxdata: InfluxDB — Real-time insights at any scale. https://www.influxdata. com/. Accessed 05 July 2023

10. INRIA Rennes: OpenViBE — Software for Brain Computer Interfaces and Real Time Neurosciences. http://openvibe.inria.fr/. Accessed 04 July 2023

11. Jaschinski, C.: Ambient assisted living: towards a model of technology adoption and use among elderly users. In: Adjunct Proceedings of the 2014 ACM International Joint Conference on Pervasive and Ubiquitous Computing, pp. 319–324 (2014). https://doi.org/10.1145/2638728.2638838

12. Jost, C., (eds.): Human-Robot Interaction. SSBN, vol. 12. Springer, Cham (2020). https://doi.org/10.1007/978-3-030-42307-0

13. Lee, J., Natarrajan, R., Rodriguez, S.S., Panda, P., Ofek, E.: RemoteLab: a VR remote study toolkit. In: Proceedings of the 35th Annual ACM Symposium on User Interface Software and Technology, p. 51. ACM (2022). https://doi.org/10. 1145/3526113.3545679

14. Liu, Z.M., Chen, Y.H.: A modularity design approach to behavioral research with immersive virtual reality: a SkyrimVR-based behavioral experimental framework. Behav. Res. Methods (2022). https://doi.org/10.3758/s13428-022-01990-6

15. Macwhinney, B., James, J., Schunn, C., Li, P., Schneider, W.: STEP - a system for teaching experimental psychology using E-Prime. Behav. Res. Methods **33**, 287–296 (2001). https://doi.org/10.3758/BF03195379

16. Sánchez-Reolid, R., Sánchez-Reolid, D., Pereira, A., Fernández-Caballero, A.: Acquisition and synchronisation of multi-source physiological data using microservices and event-driven architecture. In: Ambient Intelligence-Software and Applications–13th International Symposium on Ambient Intelligence, pp. 13–23. Springer, Cham (2023). https://doi.org/10.1007/978-3-031-22356-3_2

17. Schuetz, I., Karimpur, H., Fiehler, K.: vexptoolbox: a software toolbox for human behavior studies using the vizard virtual reality platform. Behav. Res. Methods (2022). https://doi.org/10.3758/s13428-022-01831-6

18. Tools, P.S.: E-Prime̋—Psychology Software Tools. https://pstnet.com/products/ e-prime/. Accessed 04 July 2023

19. Yeh, A., et al.: Exploring proxemics for human-drone interaction. In: Proceedings of the 5th International Conference on Human Agent Interaction, pp. 81–88. ACM (2017). https://doi.org/10.1145/3125739.3125773

Towards a Virtual Reality Visualization of Hand-Object Interactions to Support Remote Physical Therapy

Trudi Di Qi$^{(\boxtimes)}$![ORCID], LouAnne Boyd ![ORCID], Scott Fitzpatrick, Meghna Raswan, and Franceli L. Cibrian$^{(\boxtimes)}$![ORCID]

Fowler School of Engineering, Chapman University, Orange, CA 92866, USA
{dqi,lboyd,sfitzpatrick,raswan,cibrian}@chapman.edu

Abstract. Improving object manipulation skills through hand-object interaction exercises is crucial for rehabilitation. Despite limited healthcare resources, physical therapists propose remote exercise routines followed up by remote monitoring. However, remote motor skills assessment remains challenging due to the lack of effective motion visualizations. Therefore, exploring innovative ways of visualization is crucial, and virtual reality (VR) has shown the potential to address this limitation. However, it is unclear how VR visualization can represent understandable hand-object interactions. To address this gap, in this paper, we present VRMoVi, a VR visualization system that incorporates multiple levels of 3D visualization layers to depict movements. In a 2-stage study, we showed VRMoVi's potential in representing hand-object interactions, with its visualization outperforming traditional representations, and detailed features improved the hand-object interactions understanding. This study takes the initial step in developing VR visualization of hand-object interaction to support remote physical therapy.

Keywords: Virtual Reality · Data Visualization · Object manipulation skills · Remote Physical Therapy · Health Monitoring

1 Introduction

Upper limb rehabilitation typically entails executing specific functional activities like reaching, grasping, moving, and handling objects (i.e., hand-object interaction) [17]. As a key motor skill in physical therapy, improving object manipulation skills is crucial to enhance patients' dexterity and coordination, thus facilitating their daily activities with increased ease and assurance [12].

Limited healthcare resources result in a long waiting time and delayed feedback for in-person physical therapy. To address this, physical therapists suggest patients perform remote exercise routines followed by phone, video calls, or even chat [24]. Unfortunately, with this communication, therapists can only view one angle of the movements, based on the patient's camera position, which may

J. Bravo and G. Urzáiz (Eds.): UCAmI 2023, LNNS 835, pp. 136–147, 2023.
https://doi.org/10.1007/978-3-031-48306-6_14

result in inaccurate assessments [13]. Therefore, 3-dimensional (3D) visualizations are needed to effectively present hand-object interactions to improve the motion skill assessment.

Virtual reality (VR) technologies have emerged as promising tools for remote physical therapy showing promising results in improving balance, gait, and motor function in patients with neurological conditions [9,12,19]. Additionally, recent works take advantage of immersive analytics using VR [5,11] to enhance understanding of complex human motion data. However, it is unclear what visualization methods should represent hand motion and object interaction information. Our research aims to develop and evaluate a VR-based visualization method with increasing level-of-detail 3D visualization features to depict hand-object interactions. Our **contributions** include:

- The VRMoVi prototype, a VR visualization system incorporating multiple levels of 3D visualization layers to depict motion data.
- Empirical evidence of a two-stage study with 24 experts in visualizations showing that VRMoVi outperformed the traditional 2D display-based visualization, and adding fine-grained motion features, such as hand positions and rotations, improved their understanding of the data.

In the next section, we explore current visualization methods for human limb movement and highlight the importance of enhancing understanding of effective 3D visualizations for remote therapy.

2   Related Work

2.1   Remote Monitoring Strategies in Physical Therapy

Remote physical therapy (i.e., telerehabilitation) has gained attention, leveraging advancements in videoconferencing, wearable sensors, and VR/augmented reality (AR), offering promising results for clinical practice and research [9]. Telerehabilitation uses videoconferencing platforms to support real-time, high-definition video and audio patient and therapist interaction, enabling remote assessment and feedback [24]. However, they only have a 2D perspective, restricting therapists' ability to assess 3D spatial information of movements [13]. To overcome this limitation, depth cameras have been used to get a 3D interpretation of movements and postures [6], but they may still be affected by lighting, positioning, and occlusion [14]

Another approach to support remote physical therapy is using wearable devices incorporated into smartphones or wristbands. Those devices can be worn on different body parts to track movements [15] speed and joint angular motion, providing continuous and objective measures [18]. Once processed and interpreted, these insights shape personalized therapy plans [7].

Recently, VR/AR have emerged as promising tools for remote physical therapy allowing manipulation and control of the therapy environment [9] (e.g., adjust difficulty levels and provide real-time feedback [19]). VR/AR provides interactive and enjoyable experiences [12] and shows promising results in improving balance, gait, and motor function in patients with neurological conditions

[19]. This body of work has shown that these technologies can effectively support remote physical therapy; however, it is unclear what type of VR visualization can better represent hand-object interactions.

2.2 Motion Data Visualization Approaches

To facilitate evaluating patients' condition and progress remotely, it is essential to provide physical therapists with comprehensive and interpretable visualizations of patients' movements [17]. Traditional movement visualization includes 2D scatter plots dashboards [17,18,21] mainly of wearable sensors data, supplemented with information such as movement length, velocity, and joint angles. However, this visualization lacks 3D movement information, making it challenging to evaluate the patient's condition and progress [13].

In physical therapy, computer animations create a simplified animated human skeleton to mimic the patient's movements in 3D [13,19,20]. This can be useful for examining postures and exercises. For example, squat exercises have been represented with computer animations using spheres for joints and sticks for legs [20]. This approach allowed therapists to identify the correct positions of the knees. However, animations lack detailed information about movement trajectory and rotations. On the other hand, 3D trajectories have been used in motion analysis applications [11], mainly for player behavior analysis in sports [23] and computer games [10].

Immersive analysis using VR/AR has become an alternative to traditional human motion analysis, enabling 3D visualization of complex motion data beyond 2D displays [5,11,22]. For example, a VR-based visual analysis system using 3D trajectories and avatar animation has been used to represent large-scale movements (e.g., walking)[11]. However, it is unclear how VR visualizations could be applied to hand-object interactions (e.g., picking up, tossing). These interactions demand detailed insights into hand trajectory, rotation, and object interaction, crucial for understanding upper limb movement rehabilitation [17]. In this work, we fill this gap by proposing novel ways of 3D visualization with levels of detail to show the hand-object interactions.

3 VRMoVi

This work introduces VRMoVi (Virtual Reality Motion Visualization), a VR-based visualization system with increasing level-of-detail 3D visualization layers: 3D trajectories, avatar animation, and fine-grained visual symbols showing hand positions and rotations to depict hand movements and various hand-object interactions. VRMoVi consists of three modules to assist the user (e.g., physical therapists) in visualizing human motion data (Fig. 1):

Data visualizer. Currently, the data visualizer provides three 3D visualization layers with increasing levels of detail:

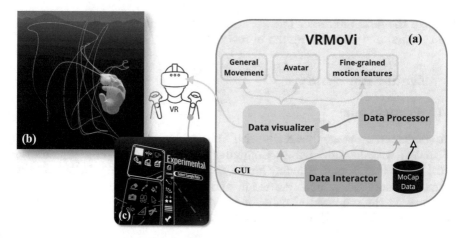

Fig. 1. (a) VRMoVi consists of three major modules: 1) a data visualizer, 2) a data processor, and 3) a data interactor. (b) illustrates hand motion and object interaction using general movement (GM) and hand-object avatar (A) layers. (c) Shows the GUI (data interactor) allowing the user to interact with the data visualizer and processor.

1. **General-Movement (GM)** layer for demonstrating the general movement trajectories of the hand/object (Fig. 1 (b));
2. **Hand-object Avatar (A)** layer for animating the interaction between the 3D models of hands and objects over time to provide context (Fig. 1 (b));
3. **Fine-grained motion (F)** layer for showing the detailed position and rotation of the hand at each time step to provide specific patterns on a smaller scale (Fig. 2 (d))

   Visualization layers are developed on top of an open-source VR painting program, OpenBrush [1], based on Unity [3]. We use OpenBrush's painting brushes to realize general movement and fine-grained motion layers. To render the hands' motion trajectories more clearly and smoothly (see Fig. 1(b)), we use translucent 3D curves. The hand position and rotation for each time step are rendered using visual symbols like dots and arrows. The animated hand-object avatar layer is realized by rigid body transformation of 3D models of two hands and objects. Both hands and objects are transformed based on the input hand and object position and rotation data.

Data Interaction. Users can interact with the VR environment and 3D visualizations through data interactors consisting of hand-held controllers and a graphical user interface (GUI; Fig. 1 (c)). The system allows users to adjust the density of input motion data or choose any visualization layer(s) to draw the data.

Data Processor. To allow the user to import motion-capture data into VR, we implement a data processor module based on Python data analysis libraries (e.g., Scikit-learn [16]), allowing users to process the data (e.g., down-sampling) through the data interactor (e.g., dragging a slider to update the resolution of data points) and visualize results in real-time.

4   Evaluation Study

4.1   Participants

We conducted a 2-stage user study with 24 individuals (ages 18–23, 9 female and 15 males) with backgrounds in Computer Science (100%), Data visualization (16%), VR (42%), and Human-Computer Interaction (100%). In this project stage, we recruited technology design and visualization experts to get initial feedback on the proposed visualization and to build a robust prototype before conducting sessions with clinicians and experts in physical therapy.

4.2   Study Design

The study aimed to explore the following research questions (RQ):

- **RQ1**. *Can participants differentiate hand and object trajectories?*
- **RQ2**. *Can participants interpret the hand-object interaction through visualizations?*
- **RQ3**. *What challenges and opportunities can VR visualization offer for hand-object interaction representations?*

We conducted a within-subject quantitative evaluation to respond to RQ1 and RQ2 (**Stage 1**); and to address RQ3, we conducted a qualitative study consisting of a brief semi-structured interview of a VR visualization containing various visual symbols spanning from a macro overview to micro details of hand movement and object interactions (**Stage 2**).

Visualization Methods. We used traditional and VRMoVi visualizations for hand-object interactions that gradually increased the level of detail being presented:

- **Traditional 3D (T3D).** Visualization of hand-object interaction using static 3D scatter plots displayed on a 2D screen. An interactive, Python-based data visualization library, Plotly [2], is used, which allows the user to interact with the 3D data using their mouse (zooming or rotating). The motion trajectories of hands and objects are shown in blue and red (Fig. 2(a)).
- **VRMoVi-GM (VR-GM).** Visualization of hand-object interaction using animated 3D curves (trajectories) displayed on VR, implemented using the animated general-movement (GM) layer of VRMoVi (Fig. 2(b)).
- **VRMoVi-GM+A (VR-GMA).** Visualization of hand movement and object interaction using animated 3D curves and hand and object models displayed simultaneously on VR. Animated general movement (GM) and hand-object avatar (A) layers of VRMoVi were utilized (Fig. 2(c)).

– **VRMoVi-GM+A+F (VR-Full).** Visualization of hand movement and object interaction using animated 3D curves, hand and object models, and detailed hand spatial information. We used the full visualization layers available in VRMoVi, including animated general movement (GM), hand-object avatar (A), and fine-grained motion (F) layers. In this method, GM and A layers are displayed simultaneously, followed by the F layer displayed at last to show the exact hand position and rotation at each time step (Fig. 2(d)).

Hand-Object Interaction Scenarios. We selected three hand-object interaction scenarios frequently used in upper limb rehabilitation [17]. These motion data were selected from an open-source VR dataset, OpenNEEDS [8], where human activities with object interactions were collected in VR. The hand-object interaction scenarios include two goal-oriented movements, 1) **Picking-up** an object and 2) **Tossing** an object, and one open-ended 3) **Drawing** using an object. Each scenario was represented by each visualization method.

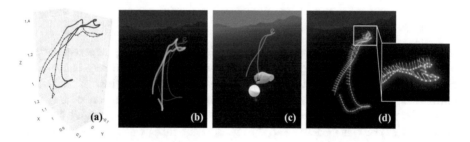

Fig. 2. Visualizations of hand movement and object interaction for the "Picking-up" scenario showing both hand and object motions using T3D (a), VR-GM (b), VR-GMA (c), and VR-Full (d) visualization methods.

4.3   Data Collection

Stage 1. For T3D, we built interactive static scatter data plots for each scenario (Fig. 2 (a)). For VR-GM and VR-GMA, we created short videos recorded within a VR environment to present a remote monitoring scenario where the patient recorded a video of their progress in VR and sent it to the physical therapist for feedback. In this manner, we can also compare these VR visualizations with T3D on the screen (Fig. 2 (b-c)). For each of the visualization methods, to quantify the accuracy of hand-object interaction interpretation across different scenarios, participants conducted the following tasks:

– **Task 1 Hand/object:** determine which line represents the hand (as opposed to the object) (RQ1)
– **Task 2 Action:** determine what action is being performed (e.g., writing, catching, holding) (RQ2)

After visualizing each condition (i.e., each visualization in each scenario), participants filled out a worksheet to respond to each prompt. Since we gradually added more details each time in the visualization, we did not counterbalance by randomizing the order of using those visualization methods to avoid learnability. Therefore, all participants experienced the visualization methods in the same order, with each condition lasting 10 min.

Stage 2. To address RQ3, we employed the VR-Full visualization across all scenarios. We created a 20-second video depicting the hierarchical structure of the three layers available in VRMoVi. These layers span from a macro overview of hand-object interaction to the micro details of hand orientation and position at each time step (Fig. 2 (b-d)). In the video, the animated general movement (GM) and hand-object avatar (A) layers were concurrently displayed, with the fine-grained motion (F) layer shown subsequently. Following this, in a brief semi-structured interview, participants were asked questions about their comprehension of each VRMoVi layer, difficulties deciphering the motion, and potential enhancements.

4.4 Data Analysis

In Stage 1, we assessed each participant's response as correct, incorrect, or missing. For Task 1, the participant's response was considered correct if they accurately distinguished hand trajectories from object trajectories. In Task 2, we marked a participant's answer as correct if their interpretation of the hand-object interaction aligned with the actual scenario (e.g.,"picking-up," "tossing," or "drawing") or was closely related to it. Subsequently, we carried out an analysis considering each task and visualization method individually. We first summarized the frequency of correct answers from all non-missing responses. Then, we performed inferential statistics. Given that our data were normally distributed (as per the Shapiro-Wilk test), we employed a repeated measure ANOVA, and a Tukey correction test. $p < 0.05$ was considered to be statistically significant.

During Stage 2, all participant responses were de-identified, documented, and subsequently transcribed. We then implemented methods derived from the thematic analysis [4]. Participants' responses were coded and grouped in an affinity diagram to uncover themes.

5 Results

Figure 3 shows the participants' overall accuracy for Task 1 (left) and Task 2 (right) across all hand-object interaction scenarios.

5.1 Task 1: Differentiating Hand and Object Trajectories (RQ1)

Participants better differentiated the hand from the object trajectories (independently of the action) in the VR+GMA (68%), than in the VR-GM (56%) and the T3D (51%); however, no significant difference was observed (p=0.108).

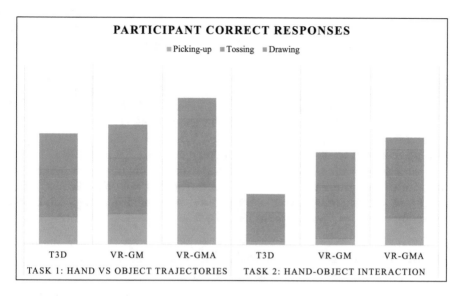

Fig. 3. The number of correct responses from the participants on Task 1 (left) and Task 2 (right) for each hand-object interaction scenario (picking-up, tossing, and drawing) across visualization methods: T3D, VR-GM, and VR-GMA.

Specifically, in the picking-up scenario, nearly twice the number of participants were capable of correctly discerning the object's trajectory from the hand when utilizing VR-GMA (79%) in contrast to using T3D (38%; p=0.009) or the VR-GM (42%; p=0.021). For the tossing scenario, 96% of participants differentiated the hand and object trajectories using VR-GMA, compared to 67% with T3D (p=0.029) and 79% with VR-GM visualization (p=0.298). Conversely, for the drawing scenario, there was no significant disparity among the visualization methods (p=0.310), with almost 50% of participants able to distinguish the hand and object moving trajectories using T3D, followed closely by VR-GM (46%), and then VR-GMA (29%).

Overall, the VR-GMA provides better feedback for differentiating hand/object trajectories, especially for task-oriented interaction scenarios, owning to the visual cues it provides (i.e., hand and object avatars), as illustrated in Fig. 3 (left).

5.2 Task 2: Interpreting Hand-Object Interactions (RQ2)

Regarding participants' performance in interpreting hand-object interactions, our findings indicated a significant difference across the visualization methods (p=0.022). Nearly twice as many participants accurately identified the hand-object interaction scenarios using VR-GM (44%) and VR-GMA (54%) in comparison to T3D (25%; p=0.031). This suggests that VR-based visualizations provide visual cues about interaction context better than T3D(Fig. 3(right)).

Upon analyzing the data for each scenario, significant differences were observed in the tossing (p=0.002), where all participants accurately identified this interaction using VR-GMA (100%), in contrast to VR-GM (78%; p=0.348) and T3D (43%, p=0.001). However, in the drawing and picking-up scenarios, less than half of the participants correctly recognized those interactions. For the drawing scenario, 46% correctly interpreted it using VR-GM, followed by T3D (26%) and VR-GMA (17%). Conversely, in the picking-up scenario, a higher proportion (43%) correctly identified it using VR-GMA, while only 2 and 1 participants managed to identify this interaction using VR-GM and T3D.

In summary, owning to the animated trajectory and hand-object interaction incorporated into the VR-based visualization methods with VR-GMA superior to VR-GM, participants were more successful in recognizing the hand-object interaction scenarios than those utilizing T3D.

5.3 Challenges and Opportunities for VR Visualization for Hand-Object Interaction (RQ3)

Our findings in Stage 2 suggest that VR visualizations of hand-object interaction effectively convey information about representations, particularly when the animation, visual cues, and level of detail are carefully crafted.

Translating Visualization into Meaningful Information. Participants identified certain motion characteristics they considered important for the visual representation of hand movement. Some of these features, such as the trajectory and angles of movements, can be effectively depicted in traditional 3D and VR visualizations, as they said: *"[Visualization can help] to understand the path hands take for various actions."*

As expected, participants highlighted features that VR animations could represent effectively, including velocity, direction, and depth: *"You could see how speed and positioning change the visuals."* Particularly for hand direction, participants discussed the potential for showing the orientation of a hand, object, or both synchronized: *"What direction the hand is facing,"* and augmented with VR capabilities *"Show hand movement and "ball" movement then both at the same time."*

Although VR-Full visualization uses arrows overlaying the trajectory path to indicate hand rotations during movement, participants associate them with features like force, energy, or velocity. Complementing the animation of arrows, the motion of the dots was linked to frequency or velocity: *"Dots that are slower represent slower movement..."*

Overall, features such as direction, velocity, and movement synchronization should leverage the unique capabilities of VR environments to enhance visualization. This approach could enable clinicians to offer remote advice to patients more effectively.

Improvements for Representation. Participants identified that the contrast in the existing visualization might not be accessible: *"Environment is really dark,* *hard to see stuff, could be in a lighter setting."* Therefore, suggestions included adopting a lighter setting or offering options to alter the colors. Additionally, to enhance accessibility further, participants recommended: *"Provide more context to what the action is by using different objects and colors. Add a timer so you can see how much time elapsed and how quickly the action is occurring."*

On the other hand, animation was perceived as a beneficial and simplified understanding of the movement: *"The animation made the motions easier to understand rather than a still diagram."* Participants concurred that real-time animation could better delineate the movement's start and end points, the intended action to be performed, and the movement's speed. However, they suggested these detail levels could be augmented by: *"Add additional views for motion, size, and angle that can be interpreted for what the user is achieving with or without the object."* Furthermore, explicit symbols could enhance trajectory representation, such as hand and object symbols: *"Add a recording of VR hand and real-life hand,"* as well as clearly defined start and end points, ensuring a one-to-one correspondence of movements.

Based on participant feedback, aspects such as accessibility, animation, visual cues, and detail levels should be refined to provide clinicians with an accurate representation of the movements.

6 Discussion

VR visualizations have shown great promise in offering insightful data about movements [5,11,22], holding the potential to enhance the process of remote monitoring in physical therapy by supplying clinicians with detailed information about patient movements, thereby facilitating remote feedback.

Our study demonstrated that integrating animation and 3D symbols significantly improved the understanding of movement actions and differentiated hand and object trajectories, particularly for goal-oriented movements like tossing or picking up an object. Therefore, delivering intelligible and precise visualizations of these movements is crucial. However, adding details might confuse open-ended movements, such as drawing, suggesting a simpler animation approach. An alternative strategy might entail incorporating multiple perspectives, thus offering a "zoom-in" function to display intricate motion details for open-ended or complex hand actions such as "drawing." This would allow the user to examine a specific portion of the data closely.

Our qualitative analysis showed that enhancing hand-object interaction trajectories with detailed visual symbols facilitates the translation of movement data into meaningful information. Therefore, using symbols [13,20] delivers visual cues about the position, direction, angle, speed, and depth, enabling physical therapists to make informed decisions when providing remote patient care. When designing these visualizations, it is crucial to consider factors such as the contrast between the background and 3D visualizations to ensure clear differentiation. Moreover, offering explicit labels and allowing user customization of displayed information can further improve motion understanding.

Despite achieving our objectives, we acknowledge some inherent limitations. Our findings are based on a 2-stage study conducted with participants from visualization-experienced backgrounds who were asked to conduct tasks related to specific movements. Therefore, to generalize the outcomes for the physical therapy field, in future work, we plan to engage physical therapists to gain feedback, refine our current visualization design, and assess the visualization's effectiveness in delivering meaningful information via qualitative and quantitative studies. Furthermore, we plan to increase hand-object motion scenarios and incorporate technical metrics such as accuracy and system performance.

Overall, considering our overarching goal of providing effective 3D visualizations for assessing object manipulation skills in remote physical therapy, we believe our research results signify an invaluable preliminary step in demonstrating the potential of VR visualization in supporting remote physical therapy.

7   Conclusion

This paper presents the development of VRMoVi, a VR-based visualization system that incorporates multiple levels of detailed 3D visualization features to depict human motion. Our results show that VR visualizations can potentially improve the understanding of hand-object interactions and enhance the interpretation and analysis of physical therapy remotely. The future work will evaluate the effectiveness of the proposed VRMoVi visualizations for assessing object manipulation skills in remote physical therapy, working with physical therapists.

Acknowledgements. This research was funded by the 2022 Research Seed Fund of Dr. Trudi Qi, the 2020–22 Research Startup Fund of Dr. Franceli Cibrian, and the Robert A. Day Undergraduate Research Grant of Meghna Raswan at Chapman University.

References

1. Openbrush. https://openbrush.app/. Accessed 21 Sep 2023
2. Plotly. https://plotly.com/. Accessed 21 Sep 2023
3. Python for unity. https://docs.unity3d.com. Accessed 21 Sep 2023
4. Braun, V., Clarke, V.: Thematic Analysis. In: APA Handbook of Research Methods in Psychology. American Psychological Association (2012)
5. Büschel, W., Lehmann, A., Dachselt, R.: MIRIA: a mixed reality toolkit for the in-situ visualization and analysis of spatio-temporal interaction data. In: Proceedings of the 2021 CHI Conference on Human Factors in Computing Systems (CHI) (2021)
6. Clark, R.A., et al.: Validity of the microsoft kinect for assessment of postural control (2012)
7. Dobkin, B.H., Dorsch, A.: The promise of mHealth: daily activity monitoring and outcome assessments by wearable sensors. Neurorehab. Neural Repair **25**(9), 788–798 (2011)
8. Emery, K.J., Zannoli, M., Warren, J., Xiao, L., Talathi, S.S.: OpenNEEDS: a dataset of gaze, head, hand, and scene signals during exploration in open-ended VR environments. In: ACM Symposium on Eye Tracking Research and Applications (2021)

9. Jun, H., Shaik, H., DeVeaux, C., Lewek, M., Fuchs, H., Bailenson, J.: An Evaluation Study of 2D and 3D Teleconferencing for Remote Physical Therapy. Virtual and Augmented Reality, PRESENCE (2023)

10. Kepplinger, D., Wallner, G., Kriglstein, S., Lankes, M.: See, feel, move: player behaviour analysis through combined visualization of gaze, emotions, and movement. In: CHI Conference on Human Factors in Computing Systems (CHI) (2020)

11. Kloiber, S., et al.: Immersive analysis of user motion in VR applications. Vis. Comput. **36**(10), 1937–1949 (2020). https://doi.org/10.1007/s00371-020-01942-1

12. Levin, M.F., Weiss, P.L., Keshner, E.A.: Emergence of virtual reality as a tool for upper limb rehabilitation: incorporation of motor control and motor learning principles. Phys. Ther. **95**(3), 415–425 (2015)

13. Moya, S., Grau, S., Tost, D., Campeny, R., Ruiz, M.: Animation of 3D avatars for rehabilitation of the upper limbs. In: 2011 Third International Conference on Games and Virtual Worlds for Serious Applications (2011)

14. Obdržálek, S., et al.: Accuracy and robustness of kinect pose estimation in the context of coaching of elderly population. Annu. Int. Conf. IEEE Eng. Med. Biol. Soc. **2012**, 1188–1193 (2012)

15. Patel, S., Park, H., Bonato, P., Chan, L., Rodgers, M.: A review of wearable sensors and systems with application in rehabilitation. J. Neuroengineering Rehabil. **9**, 21 (2012)

16. Pedregosa, F., et al.: Scikit-learn: machine learning in python. J. Mach. Learn. Res. **12**, 2825–2830 (2011)

17. Ploderer, B., et al.: How therapists use visualizations of upper limb movement information from stroke patients: a qualitative study with simulated information. JMIR Rehabil. Assist. Technol. **3**(2), e9 (2016)

18. Ploderer, B., et al.: ArmSleeve: a patient monitoring system to support occupational therapists in stroke rehabilitation. In: Proceedings of DIS 2016 (2016)

19. Postolache, O., Hemanth, D.J., Alexandre, R., Gupta, D., Geman, O., Khanna, A.: Remote monitoring of physical rehabilitation of stroke patients using IoT and virtual reality. IEEE J. Sel. Areas Commun. **39**(2), 562–573 (2021)

20. Rado, A.S.D., Plasek, J., Nuckley, D., Keefe, D.F.: A real-time physical therapy visualization strategy to improve unsupervised patient rehabilitation. In: Poster: IEEE Visualization (2009)

21. Rawashdeh, S.A., Reimann, E., Uhl, T.L.: Highly-individualized physical therapy instruction beyond the clinic using wearable inertial sensors. IEEE Access **10**, 14564–14574 (2022)

22. Reipschläger, P., Brudy, F., Dachselt, R., Matejka, J., Fitzmaurice, G., Anderson, F.: AvatAR: an immersive analysis environment for human motion data combining interactive 3D avatars and trajectories. In: Barbosa, S., et al. (eds.) Proceedings of the 2022 CHI Conference on Human Factors in Computing Systems (CHI) (2022)

23. Sacha, D., et al.: Dynamic visual abstraction of soccer movement. Comput. Graph. Forum **36**(3), 305–315 (2017)

24. Winters, J.M., Winters, J.M.: Videoconferencing and telehealth technologies can provide a reliable approach to remote assessment and teaching without compromising quality. J. Cardiovasc. Nurs. **22**(1), 51–7 (2007)

Creating Personalized Verbal Human-Robot Interactions Using LLM with the Robot Mini

Teresa Onorati[1]([✉])[iD], Álvaro Castro-González[2]([✉])[iD], Javier Cruz del Valle[1], Paloma Díaz[1][iD], and José Carlos Castillo[2][iD]

[1] Computer Science Department, Universidad Carlos III de Madrid, Av. de la Universidad 30, 28911 Leganés, Spain
{tonorati,pdp}@inf.uc3m.es, jacruzde@pa.uc3m.es
[2] Robotics Lab, Systems Engineering and Automation Department, Universidad Carlos III de Madrid, Av. de la Universidad 30, 28911 Leganés, Spain
{acgonzal,jocastil}@ing.uc3m.es

Abstract. Social robots are intended to establish natural interactions with humans. In most cases, human-robot communication is predefined and results in monotonous interactions in the long term that lead the user to cease the interaction. In this paper, we propose a robotic application to generate verbal interactions dynamically. However, if the users do not perceive these dialogues as interesting, they will not engage in the interaction with the robot. To mitigate this problem, we propose generating verbal dialogues considering the user's interests and preferences. To this end, we present a social robot application for conducting personalized conversations using data from social media accounts of interest for the user and large-language models to build the dialogue. After evaluating the proposed application, participants rated it very positively regarding its usability.

Keywords: Social Robotics · Social Media Mining · Human-Robot Interaction · Personalized Dialogues

1 Introduction

In the last decades, social robots have become synthetic companions with applications in several domains [4]: entertaining elderly people living alone or children in hospitals, helping out with the daily care of dependent people, or robots supporting children in educational tasks, to name a few. The success of such robots stands mainly in the possibility of establishing long-term relationships with the humans they share space with. On many occasions, social robots fail because their interactions are perceived as repetitive and non-natural, leading people to get bored and lose interest. For this reason, it is crucial to improve such interactions if we want a successful introduction of social robotics in different fields. The improvements can affect different aspects of the interaction, for example, the expressiveness, the gestures, the services, or the dialogue.

© The Author(s), under exclusive license to Springer Nature Switzerland AG 2023
J. Bravo and G. Urzáiz (Eds.): UCAmI 2023, LNNS 835, pp. 148–159, 2023.
https://doi.org/10.1007/978-3-031-48306-6_15

This paper will focus on human-robot dialogues and how they can benefit from personalizing the content. In literature, several examples exist of personalizing the experience with the robot to adapt its behaviors to the context and the users' needs and preferences, as described in Sect. 2. Our idea is to develop a so-called *personalized conversation skill* based on information of interest to the user, extracted from Twitter[1], and fed into Large-Language Models (LLM, from now on) to generate the dialogue. The social network Twitter guarantees access to updated news from official accounts. Personalization concerns selecting topics and source accounts depending on the user's personal data, like location, interests, and preferences. More details are included in Sect. 3.

The personalized conversation skill has been integrated into the social robot Mini (see Sect. 4) and used for a usability evaluation where 17 participants interacted freely with the robot (see Sect. 5).

2 State of the Art

Social robots are intended to establish natural human interactions and create long-lasting relations with them. Lately, they have been used in a wide range of applications in different real-life contexts, like rescue, education, and daily care. In these contexts, the robots should not be considered just sophisticated tools to carry out complex tasks but synthetic companions able to establish fruitful and trustable relationships with the users around them [4]. To establish these relationships, it is crucial for the users to accept the robots, considering both the offered functionalities and how the interaction works to design the robot behaviors to maximize the user experience [11].

One of the key aspects of designing natural interactions between humans and robots is the personalization to adapt the robot's behavior to the real needs and preferences of the user [11]. The advantages of personalizing the human-robot interaction are clear in different domains. For example, in manufacturing, the personalization of the interface allows to reduce the amount of information required based on the operators' experience and the complexity of the tasks, improving the operators' trust in the system and reducing the number of errors while interacting [1].

Another interesting example is Brillo, an autonomous bartender robot able to serve multiple customers at the same time and adapts its behavior depending on different aspects related to the users' profile, including their interests, moods, previous interactions, and preferences [13].

Churamani et al. have proposed a comparative study about how personalized conversations affect robot acceptance [8]. In particular, they design a learning scenario where the user teaches the robot several objects on a table between them. The study compares the same learning scenario with and without an initial personalized conversation between the user and the robot. In particular, the robot recognizes the face of the person interacting and adapts the dialogue to the user data stored in a database, like name and nationality. The results

---

[1] The work described in this paper was conducted before the July 2023 update of the Twitter Terms of Service and Privacy Policy.

show that even if the acceptance doesn't improve, the users tend to perceive the robot as more intelligent and likable with personalization [8].

When engaged in a conversation with the robot, the user can perceive the dialogue as more relevant and persuasive for her scope if it has been adapted considering the information included in her personal profile [16]. This personalization can be done mainly by applying two models [12]: static and dynamic. The static model consists of collecting information about the user just before starting the interaction [8]. The dynamic model allows the inclusion of multiple users with an initial mechanism to recognize which one of them is interacting with the robot [16], using different techniques like face recognition or asking for personal details.

In this paper, we will generate personalized dialogues based on a static model. In particular, we will process the information included in the user profile to adapt the topic of the dialogue. To this scope, the robot is going to base the conversation on updated news of interest for the user that is fed into LLM to generate the dialogue. In literature, there are already examples of dialogue systems built employing these language models to improve the response generation [24] or the usage of different input sources depending on the desired output [7].

The pre-trained language models have led to significant advances in the area of Natural Language Processing, such as machine translation, text classification, summary generation, and question answering, among others. As explained in detail in the next section, we are interested in models able to summarize texts and answer questions in the Spanish language.

3   Generating Personalized Dialogues

This paper proposes a method for allowing a social robot to generate different dialogues depending on the user's preferences. The personalization of the dialogues will rely on data obtained from social media sources, allowing the robot to keep the dialogues updated according to the most recent information posted in predefined public profiles.

The pipeline of this system is formed by three phases: first, we create a profile of the user that will interact with the robot; then, we fetch information from social media that matches the user's preferences; and, finally, we feed this information into LLM to generate the dialogue. Following, we detail each one of these three steps.

3.1   Creating User Profile

To adapt the conversation to the user, we rely on a profile that is based on the user's preferences. In an offline and face-to-face interview, we collect the user's interests and match them with official accounts on social media that are then stored locally. This data is used to search for the appropriate content to generate the dialogues.

In particular, we ask users about their interests in different categories of news: politics, sports, business, and current affairs. We associated several official Twitter accounts with each one of them. These accounts are selected considering the user's personal information, such as her location and preferences. For example, in the case of a female user interested in sports and living in Madrid, we can link the account *@realmadridfem*, the official profile of the Real Madrid C.F. women's team, to the topic of sports; in the case of a user supporting Barcelona football team, the topic sport can be associated to the accounts *@mundodeportivo* and *@sport*, two official Twitter accounts of sports journals oriented to the Barcelona fans. If an account matches multiple topics, we consider it in all categories.

3.2   Fetching Social Media Data

In this work, we propose to use Twitter as a social media channel to provide dynamic information. Once the topic for the conversation has been decided, we collect the Twitter accounts associated with the topic and the user and use them to search for the most recently posted *tweets* that match the following criteria:

- The tweet is written using the user's language. This information has been provided during the offline interview with the user.
- The tweet contains a link. Due to the limited length of the text in a *tweet*, its content lacks details and is usually very noisy (e.g., using slang, abbreviations, and emoticons). Frequently, official accounts post short tweets that include few words and a link to the full-body news with more detailed information.
- The tweet is from a verified account. To minimize the risk of getting fake news, we rely on verified accounts from sources the user trusts.
- We avoid replies and retweets.

As a result of the previous criteria, we obtain a collection of tweets that are processed to obtain the URLs of the included links. Then, the HTML code pointed by the URLs is parsed to obtain the main body of the news, filtering out irrelevant content. Finally, we randomly select one of the news that will be used for generating the dialogues.

3.3   Generating the Dialogue

Once we have the body of the news the user is interested in, the robot takes the initiative and starts the dialogue in two steps: first, the robot presents a summary of the news to attract the interest of the user; then, in case the user is interested, the robot invites the user to ask questions related to the news. We use specialized LLM for these tasks: a summarization model and a Long-Form Question-Answering (LFQA, from now on) model. Due to the limited computational resources available in the robot, we opted for using the models running on proprietary external servers (e.g., from Hugging Face or OpenAI). Also, considering that our potential users live in Spain, all considered models work in Spanish.

Table 1. Description of the prompt for summarization, translated from Spanish.

Prompt fragment	Description
Block1: Description of the task	
Your task is to summarize in Spanish the block of text that is placed between < >	General description of the task that we are looking for. Also, we detail how the text to be summarized is inputted
Block2: Additional instructions	
To respond, use your own words	We do not look for any particular style in the answer of the model
Use the number of words that you need to respond, but it cannot be longer than 100 words	We limit the response to 100 words maximum
Do not use introductory statements, such as 'The summary is' or similar	We want to avoid sentences that hinder the flow of the interaction
Block3: Input of the text	
The text to be summarized is: < [text] >	"[text]" should be replaced by the text that is being summarized

Summarization Model. Initially, we considered three summarization models that are well-known in the area: BERT (Bidirectional Encoder Representations from Transformers) [15], RoBERTa (Robustly Optimized BERT approach) [18], and Davinci (based on GPT-3.5)[21]. After analyzing their latency and response quality performance, we used the Davinci model. In particular, the model *text-davinci-003* can do language tasks following detailed instructions, such as summarization. The input to the Davinci model is a prompt where we detail the summarization task and contain the text to summarize. Table 1 details this prompt.

To avoid repetitive conversations, we adjusted other parameters used with this model. We assigned a temperature of 1.0 so the answers from the model are more dynamic and less deterministic. Also, the presence parameter has been set to 1.0 to minimize repetitive text in the model response.

Question-Answering. We evaluated different models oriented to generate a response giving a text and a question about it: two LFQA models, BART (Bidirectional and AutoRegressive Transformers) [17] and mT5 (Multilingual Text-to-Text Transfer Transformer) [23], and the same Davinci model we used for the summarization.

As it happened in the summarization process, after analyzing the responses provided by the models, we opted for the Davinci model (text-davinci-003). In this case, the prompt sent to the model requires the question and the text containing the response. Table 2 describes the prompt used for this task. In this case, the *top p* parameter of the model was set to 1.0 to provide a wider variability in the responses of the model.

Table 2. Description of the prompt for question-answering, translated from Spanish.

Prompt fragment	Description
Block1: Description of the task	
Your task is to respond in Spanish to the question that is included between ''' ''' using the information in the text that is placed between < >	General description of the task that we are looking for. It details how the question and the text are inputted
Block2: Additional instructions	
To respond, use your own words	We do not look for any particular style in the answer of the model
Use the number of words that you need to respond to the question, but it cannot be longer than 50 words	We limit the response to 50 words maximum
If you know the answer, respond without introductory statements like 'The response is' or similar	We want to avoid sentences that hinder the flow of the interaction
If you do not know the answer, respond with a sentence similar to 'I am sorry, but I cannot answer your question with the information I have'	We want to minimize the fake responses so the robot will inform you when it is not sure about an answer
Block3: Input of the text	
The question is: '''?[question]?'''	"[question]" should be replaced by the text of the question the user is asking
The text related to the question is: < [text] >	"[text]" should be replaced by the content where the model has to find the answer

4   Integration in the Social Robot Mini

The proposed system has been integrated into the social robot Mini [22], shown in Fig. 1. It is a social robot that assists elderly people with mild cognitive impairment. Mini has five degrees of freedom (one in each shoulder, two in the neck, and one more in the base), OLED screens that act as eyes, a LED heart that beats and can light up with different colors, speech synthesis and recognition modules, and a touch screen that can be used both to provide information to the user (such as showing videos or images) and to capture user responses through different menus.

The software architecture of the robot follows a modular structure. At the top level, a decision-making system orchestrates each robot's skill (see the top

Fig. 1. The robot Mini

of Fig. 2). These skills represent the robot's tasks or functionalities (such as playing games or performing cognitive stimulation exercises). Here is where our conversational skill is located: the *personalized conversation skill*. Skills use the Human-Robot Interaction (HRI) Engine (see the bottom of Fig. 2) to manage each interaction between the robot Mini and a user. These interactions are based on so-called Communicative Acts (CA, from now on), which represent atomic interaction units that the skills combine to achieve more complex human-robot interactions [10]. Inside the HRI Engine, the HRI Manager executes the requested CAs and generates the appropriate communicative actions based on the information gathered by the Perception Manager and provided by the sensors. These actions are a sequence of commands that the Expression Manager orchestrates to guarantee that the proper communicative message is conveyed by the robot through its actuators.

In our case, the *personalized conversation skill* uses CA that requests information from the user using Google Automatic Speech Recognition or the tablet menus. Also, a different CA allows Mini to convey the proper utterance. This skill implements the system proposed in Sect. 3: first, reads the user profile; then, selects news according to the user's profile; and, finally, uses the summarization and query-answer models to chat with the user. The main drawback of LLM is the delays introduced in the HRI flow. To deal with it, we have used conversational fillers to avoid a *frozen* robot while waiting for the results of the models. The conversational fillers include gestures of the robot thinking or sentences like *"mmm...Let me see"* or *"Let me analyze the results before giving you an answer"*.

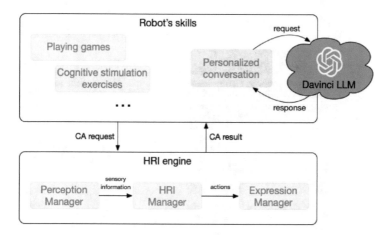

Fig. 2. The software architecture of the robot Mini.

5 Evaluation

In this paper, we are interested in assessing the usability of the proposed conversation skill to interact with the robot. Usability is an umbrella concept that includes aspects related to effectiveness, efficiency, and overall ease of use. In the literature, several questionnaires have been used for this scope [14]. In this paper, we opted for the System Usability Scale (SUS) [6], a list of 10 items evaluated by a 5-point Likert scale, from 1 corresponding to *strongly disagree* to 5 *strongly agree*, that have been extensively used in the area of social robotics [2,5,9,19,20]. The responses to the SUS questionnaire are used to compute a score from 0 to 100 that represents the overall usability performance.

5.1 Participants

We involved 18 participants, almost equally distributed among women (47,1%) and men (52,9%), most of them (64,7%) between 23 and 27 years old. All the users have a high experience with the technology, the majority work in the area of robotics and are willing to have a social robot at home. One participant suffered a malfunction of the robot, and its data have been removed from the data analysis. Data collection was conducted in compliance with the Data Protection Regulations of Universidad Carlos III de Madrid.

5.2 Procedure

After a brief introduction about the test's purpose, each user is asked to sign an informed consent form and fill out a demographic questionnaire to collect data about her age range, gender, education, experience with the technology, and, in particular, the area of robotics. At this point, the participant is ready to start interacting freely with Mini.

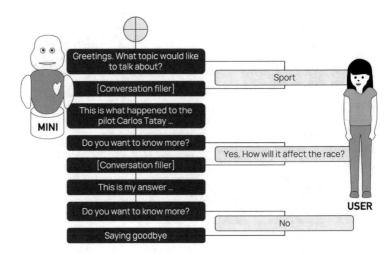

Fig. 3. The flow diagram for the use case.

As shown by the flow diagram in Fig. 3, Mini greets the user, proposes a list of topics, and waits for the user's choice. For example, if the user selects Sports, Mini looks for the latest news on Twitter and tells the user about a recent accident involving the motorcycle pilot Carlos Tatay during the Portimao race. After telling the news, the user asks about the consequences of this accident for the next races, and the robot answers that it is difficult to say, but it depends on how the injury will evolve. At this point, the user has no more questions to ask, Mini says goodbye, and the interaction ends.

After testing the personalized dialogue skill, the participants are asked to fill out a questionnaire with the 10 SUS items and two open-ended questions to indicate positive and negative aspects of their interaction with Mini. The tests were conducted in the presence of a researcher to support the participants in case of problems or doubts during the interaction and to take notes of any noteworthy situation, like errors or inconsistencies.

5.3 Results

To analyze the data collected from the SUS questionnaire, we computed the SUS score for each user, ranging from 1 to 100, which can be interpreted as the level of usability of the personalized conversation skill. To this scope, first of all, we calculated the average of the 17 SUS scores, obtaining $\mu = 80$, $\sigma = 14,40$. Although the value for the standard deviation denotes an important variability among the ratings from the participants, all of the scores are greater than 50, which is considered the lower limit for considering the usability acceptable [3].

In the literature, several scales can be used to interpret the SUS scores in terms of usability. In [3], the authors have proposed the adjective ratings that associate the SUS scores with a specific adjective for usability: *worst imaginable, poor, ok, good, excellent,* and *best imaginable.* In this evaluation, we can observe that the positive adjectives, from *ok* to *best imaginable,* are equally distributed among the collected SUS scores: $\mu = 64, 38$, $\sigma = 8, 00$ as *ok*; $\mu = 78, 5$, $\sigma = 3, 35$ as *good*; $\mu = 88, 75$, $\sigma = 1, 44$ as *excellent*; $\mu = 96, 25$, $\sigma = 3, 23$ as *best imaginable.*

Looking at the open-ended questions, the users have highlighted several positive and negative aspects of their experience with the robot. Among the positive aspects, we found the usage of breaking news to give updated information about a topic, the expressiveness of the robot, and the diversity of topics used for the conversation. Regarding the negative aspects, the main users' concerns are the delays due to the time needed to process the information and give answers.

6   Conclusions and Future Works

In this paper, we have presented a novel application that allows a social robot to maintain a personalized conversation about topics the user is interested in. Our application uses LLM to generate a natural, coherent conversation that makes the interaction more appealing. We considered different models for summarization and question-answering, and the Davinci model showed the best responses in both tasks. After evaluating the personalized conversation skill, participants rated its usability with high values.

Although the results are promising, we should consider several limitations. First, most of the participants have a similar profile. Also, since we consider a static user's profile, the topics and social media accounts available cannot be updated during the interaction.

In the near future, this skill will be applied to social robots supporting seniors living alone. We expect that thanks to interactions like the one provided by our personalized conversation skill, seniors perceive robots as companions and minimize their loneliness. To this end, further evaluation is needed to assess how users perceive the robot and how it affects the user's quality of life.

Acknowledgments. This work has been supported by the Madrid Government (Comunidad de Madrid-Spain) under the Multiannual Agreement with UC3M ("Fostering Young Doctors Research", SMM4HRI-CM-UC3M), and in the context of the V PRICIT (Research and Technological Innovation Regional Programme). This work has been partially supported by the project "Robots sociales para mitigar la soledad y el aislamiento en mayores (SOROLI)", funded by Agencia Estatal de Investigación (AEI), Spanish Ministerio de Ciencia e Innovación (PID2021-123941OA-I00), and the project sense2MakeSense, funded by the Spanish State Agency of Research (PID2019-109388GB-I00).

References

1. Andronas, D., Apostolopoulos, G., Fourtakas, N., Makris, S.: Multi-modal interfaces for natural Human-Robot Interaction. Procedia Manuf. **54**, 197–202 (2021)
2. Asl, A.M., et al.: The usability and feasibility validation of the social robot MINI in people with dementia and mild cognitive impairment; a study protocol. BMC Psychiatry **22**(1), 760 (2022)
3. Bangor, A., Kortum, P., Miller, J.: Determining what individual SUS scores mean: adding an adjective rating scale. J. Usability Stud. **4**(3), 114–123 (2009)
4. Breazeal, C.: Designing Sociable Robots. MIT Press (2004)
5. Broadbent, E., Montgomery Walsh, R., Martini, N., Loveys, K., Sutherland, C.: Evaluating the usability of new software for medication management on a social robot. In: Companion of the 2020 ACM/IEEE International Conference on Human-robot Interaction, pp. 151–153 (2020)
6. Brooke, J.: SUS: a quick and dirty'usability. Usability Eval. Ind. **189**(3), 189–194 (1996)
7. Cao, Y., Bi, W., Fang, M., Tao, D.: Pretrained language models for dialogue generation with multiple input sources. arXiv preprint arXiv:2010.07576 (2020)
8. Churamani, N., et al.: The impact of personalisation on human-robot interaction in learning scenarios. In: Proceedings of the 5th International Conference on Human Agent Interaction, pp. 171–180. HAI 2017, Association for Computing Machinery, New York, NY, USA (2017). https://doi.org/10.1145/3125739.3125756
9. Di Nuovo, A., et al.: Usability evaluation of a robotic system for cognitive testing. In: 2019 14th ACM/IEEE International Conference on Human-Robot Interaction (HRI), pp. 588–589 (2019). https://doi.org/10.1109/HRI.2019.8673187, ISSN: 2167-2148
10. Fernández-Rodicio, E., Castro-González, A., Alonso-Martín, F., Maroto-Gómez, M., Salichs, M.: Modelling multimodal dialogues for social robots using communicative acts. Sensors **20**(12) (2020). https://doi.org/10.3390/s20123440, https://www.mdpi.com/1424-8220/20/12/3440
11. Fronemann, N., Pollmann, K., Loh, W.: Should my robot know what's best for me? Human-robot interaction between user experience and ethical design. AI Soc. **37**(2), 517–533 (2022)
12. Hellou, M., Gasteiger, N., Lim, J.Y., Jang, M., Ahn, H.S.: Personalization and localization in human-robot interaction: a review of technical methods. Robotics **10**(4), 120 (2021)
13. John, N.E., Rossi, A., Rossi, S.: Personalized human-robot interaction with a robot bartender. In: Adjunct Proceedings of the 30th ACM Conference on User Modeling, Adaptation and Personalization, pp. 155–159 (2022)
14. Jung, M., Lazaro, M.J.S., Yun, M.H.: Evaluation of methodologies and measures on the usability of social robots: a systematic review. Appl. Sci. **11**(4), 1388 (2021)
15. Kenton, J.D.M.W.C., Toutanova, L.K.: BERT: pre-training of deep bidirectional transformers for language understanding. In: Proceedings of naacL-HLT, vol. 1, pp. 2 (2019)
16. Lee, M.K., Forlizzi, J., Kiesler, S., Rybski, P., Antanitis, J., Savetsila, S.: Personalization in HRI: a longitudinal field experiment. In: Proceedings of the Seventh Annual ACM/IEEE International Conference on Human-Robot Interaction, pp. 319–326 (2012)
17. Lewis, M., et al.: BART: denoising sequence-to-sequence pre-training for natural language generation, translation, and comprehension. arXiv preprint arXiv:1910.13461 (2019)

18. Liu, Y., et al.: RoBERTa: a robustly optimized BERT pretraining approach. arXiv preprint arXiv:1907.11692 (2019)
19. Louie, W.Y.G., Nejat, G.: A social robot learning to facilitate an assistive group-based activity from non-expert caregivers. Int. J. Soc. Robot. **12**(5), 1159–1176 (2020)
20. Keizer, O., et al.: Using socially assistive robots for monitoring and preventing frailty among older adults: a study on usability and user experience challenges. Health Technol. **9**(4), 595–605 (2019). https://doi.org/10.1007/s12553-019-00320-9
21. Ouyang, L., Wu, J., Jiang, X., Almeida, D., Wainwright, C., Mishkin, P., Zhang, C., Agarwal, S., Slama, K., Ray, A.: Training language models to follow instructions with human feedback. Adv. Neural. Inf. Process. Syst. **35**, 27730–27744 (2022)
22. Salichs, M.A., et al.: Mini: a new social robot for the elderly. Int. J. Soc. Robot. **12**(6), 1231–1249 (2020). https://doi.org/10.1007/s12369-020-00687-0
23. Xue, L., et al.: mT5: a massively multilingual pre-trained text-to-text transformer. arXiv preprint arXiv:2010.11934 (2020)
24. Zhao, X., Wu, W., Xu, C., Tao, C., Zhao, D., Yan, R.: Knowledge-grounded dialogue generation with pre-trained language models. arXiv preprint arXiv:2010.08824 (2020)

Automation of Error Recognition in Therapies Executions Based on ECogFun-VR

Francisco J. Celdrán[1]([✉])[iD], Antonio del Pino[2][iD], Sonia Pérez-Rodríguez[2][iD],
José J. González-García[1][iD], Dulce Romero-Ayuso[2,4,5][iD],
and Pascual González[1,3][iD]

[1] LoUISE Research Group, University of Castilla-La Mancha, 02071 Albacete, Spain
{FranciscoJ.Celdran,JoseJesus.Gonzalez,Pascual.Gonzalez}@uclm.es
[2] Department of Physical Therapy (Occupational Therapy Division), Faculty
of Health Sciences, University of Granada, Granada, Spain
{AntoniodelPino,Soniafis,DulceRomero}@ugr.es
[3] CIBERSAM-ISCIII (Biomedical Research Networking Center in Mental Health),
28016 Madrid, Spain
[4] CIMCYC, Mind, Brain and Behavior Research Center, University of Granada,
Granada, Spain
[5] Instituto de Investigación Biosanitaria ibs. Granada, 18012 Granada, Spain

Abstract. Attention deficit hyperactivity disorder (ADHD) is a neurodevelopmental disorder commonly diagnosed in children and adolescents that can impair academic performance and daily living skills. Therapies based on activities of daily living (ADLs) aim to promote patient independence in self-care, including communication and mobility. ECogFun-VR is a virtual reality tool that allows children with ADHD to perform therapies in a controlled environment. This application is based on the multiple errands test paradigm and consists of four instrumental cognitive activities. It includes five scenarios 1) Room: the starting and ending point of the tasks, connected to all other scenarios; 2) Bedroom: a room where the child must prepare the backpack and store the necessary material 3) Kitchen: a place where they must prepare a snack 4) Street: a scenario that recreates a city with traffic and pedestrians, through which the child must navigate to reach the bookstore 5) Bookstore: a place where they must choose and pay for books and school supplies.

Evaluations of the patient's performance are typically performed manually by a professional who is present during the session or by later reviewing a video recording of the session for errors. In this context, we propose a tool that allows professionals to define the tasks that patients must perform and to collect and analyze patient interactions during the session. The tool can compare patient interactions to a reference set of interactions to identify errors, using the same categorization currently used by professionals. This automation of the evaluation process can save time and improve the accuracy and consistency of evaluations.

Keywords: Virtual Reality · Activities of Daily Living · ADHD · occupational therapy

J. Bravo and G. Urzáiz (Eds.): UCAmI 2023, LNNS 835, pp. 160–171, 2023.
https://doi.org/10.1007/978-3-031-48306-6_16

1   Introduction

ADHD is a neurodevelopmental disorder that is usually diagnosed in childhood and adolescence. It can be considered an executive dysfunction which affects the ability to perform tasks of daily living, as well as academic performance [1].

Activities of Daily Living (ADLs) are those activities common to cultures and times that are necessary for survival and self-care, which includes communication, social skills and mobility [2]. So ADLs are important and closely linked with self-care and independence.

Some ADHD treatments are based or includes pharmacological treatment, but cognitive approaches are also needed to improve executive functions, which are linked to ADLs performance [7]. Some cognitive approaches focus on providing patients with planning skills. In this context, the ecological validity of treatments is important, referring to the treatment's ability to be translated into the real-life situations it is intended to impact [6]. The utility of using virtual reality (VR) devices for ADHD interventions has been proven, as VR can provide a more realistic and engaging environment for practicing and improving planning skills [9]. Thus, Virtual reality (VR) and augmented reality (AR) technologies are increasingly being used in healthcare interventions, including for ADHD treatment. These technologies allow us to create scenarios and dynamics that resemble those of real life more than with other technologies, offering several advantages: *Repeatability*: Experiments, sessions, and therapies can be repeated as needed, allowing for more systematic data collection and analysis; *Controlled environment*: VR and AR can provide a controlled environment for practicing and improving skills, reducing the risks associated with errors in real-life settings; *Transferability*: VR and AR can help patients transfer their skills to real life by providing them with a more immersive and engaging learning experience; *Adherence*: VR and AR can promote adherence to treatment by providing patients with more playful and motivating environments.

2   State of the Question

Virtual reality (VR) offers a promising approach for conducting neuropsychological tests, particularly those related to attention, memory, and executive function, with increased reliability. By combining VR with interactive tests that provide focused stimuli, distractions can be minimized, enabling patients, including children with ADHD, to maintain attention and concentration for extended periods. Research has shown the potential of VR in both assessing and training attention, as individuals with ADHD have exhibited improved performance on cognitive tasks when using VR compared to traditional assessment techniques [3].

There are already some studies which used this technology to help ADHD childrens. [8] developed a virtual classroom in which the child sits in a desk and performs some task designed to assess their attention while a series of typical classroom distractors are systematically controlled. Another example of VR usage in children with ADHD is "IAmHero" [10], which developed three types of tasks within their platform. The first task involves situating the child in a virtual space and providing instructions related to topological categories, such as retrieving a ball and placing it under a desk. The second task utilizes a motion sensing tool, requiring the child to navigate an infinite moving game while avoiding obstacles and collecting rewards. The third scenario presents puzzle-like tasks that guide the child in reconstructing sequences or objects. In addition, the platform includes a therapy session management system that allows therapists to monitor progress, evaluate goal achievement and customise therapy sessions.

Neither of the two examples uses ADLs in therapy, which could enhance the ecological validity of the tool. Additionally, the tasks are relatively simple, with either a single scenario or non-interconnected scenarios.

Regarding the process of retrieving interaction information of the execution, there are also some proposals about how to do it in Unity. For example, Toggle Toolkit [12] is a tool for designing and conducting experiments in VR using Unity. It provides the researcher with predefined scripts that can track users interactions, generating dynamic responses if needed and keeping a log of such interactions. Unity Experiment Framework [4] is also a proposal that tries to generate in Unity a series of pre-designed objects that make easy the realization and design of experiments in VR, dividing these experiments in trials, blocks and sessions, and allowing to define the dependent and independent variables for each of them. Unity Technologies also has some packages, all of them adequately documented and usually accompanied by a tutorial on their use, which can also be useful to facilitate both the management of interactions and their registration.

3   ECogFun-VR

ECogFun-VR is a VR tool designed to be used by children with ADHD therapies based on the multiple errands test paradigm. It consists of a total of five interconnected scenarios in which children with ADHD can move and interact with the environment to carry out the tasks indicated by the therapists. The existing scenarios are: (1) Hallway; (2) Bedroom; (3) Kitchen; (4) Street; (5) Bookstore.

The tasks to be carried out by the children make it necessary to go through all the rooms, as well as the need to organize correctly the order in which they are going to carry them out. The tasks will consist of packing a backpack for a school day, having to store the corresponding material in it, which includes buying the necessary material in the Bookstore (for this they have to previously recognize which material is missing and which is not), and also preparing a snack to eat during the recess.

The tool is being validated in children with ADHD through a manual video evaluation with the execution of the therapy. To facilitate this evaluation, we propose in this article the automation of the evaluation process.

3.1   Scenes

The scenarios have been designed to be visually pleasing and at the same time realistic, with the intention of facilitating immersion in the virtual environment, achieving greater ecological validity for the tool and generating motivation of use. With this same idea in mind, the interactions have been designed to mimic as closely as possible the way they would take place in real life. At the same time, alongside the objects needed to perform the different tasks, other objects have been included that act as *distractors* and therefore they should not be used by the children.

(1) Hallway. This scenario consists of a corridor with three doors, connecting each of them with a different scenario, the *Kitchen*, the *Bedroom* and the *Street*. To access each of these scenarios the user will have to open the door and pass through it. Movement is performed by moving in real life (to obtain maximum immersion).

Fig. 1. Image of Hallway scenario

In addition to connecting to the other scenarios, the corridor contains a small chest of drawers next to the door where the house keys are kept. The user can take the keys by approaching the controls and pressing the rear trigger. Once the user has acquired the keys, they can simulate putting them in their pocket by making a similar movement. The system has a collider located approximately where the user's legs would be. When the keys come into contact with this collider, they disappear and the system registers that they have been placed in the user's pocket.

(2) Bedroom. This scenario contains the following furniture: bed, desk (with two drawers which can be opened), bedside table (with three drawers that can be opened), wardrobe and a bookshelf.

In addition, the scenario contains a backpack, a computer with a keyboard and mouse, a soccer ball, a smartphone, a lamp, school supplies, and other decorative elements commonly found in a child's room, such as books and board games. The tasks in this scenario require the child to gather the materials they will need for the next school day and place them in their backpack. The child

Fig. 2. Image of Bedroom scenario

must also identify any missing materials so they can purchase them in the *Bookstore* scenario.

The object grabbing mechanics are the same as in the other scenarios and have already been explained for the hallway scenario. To place an object in the backpack, simply move the object so that it collides with the side of the backpack where the opening is located.

Some objects in the room are used as *distractors* and children should avoid interacting with them (e.g., the soccer ball). There will also be stimuli such as an alarm that will sound on the smartphone when the child has been in the room for a while. To turn it off, the child must press the lower trigger and then touch the phone with the finger.

(3) Kitchen. The following objects are available in this scenario: backpack, cooking fires, refrigerator, cutting board, cooking utensils (pot, pan, cutlery...), ingredients, crockery, drawers and distractors.

Fig. 3. Image of Kitchen scenario

The main task to be carried out in this scenario would be to prepare a sandwich with the indicated ingredients, wrap it in aluminum foil and put it in the backpack. To do this, the ingredients have to be placed in the proper order on the interaction space (the cutting board), and then the child need to touch the area with the aluminum foil. This will generate a new object, a wrapped sandwich that can be stored in the backpack in the same way as the objects have been stored in the room. In addition, the user can also cook using the hobs and other utensils. The hob top have buttons like a real ceramic hob and the interaction with them is the same as explained in the case of the smartphone.

The refrigerator has an alarm in case it is left open for too long. Inside it we have fruit, vegetables, drinks and sausages, which are in packages that can be opened by pressing the secondary trigger when we are in contact with them. When we open it, a slice of the corresponding sausage will appear in 3D that we can take and use for the recipe.

As in a real scenario, in order to obtain slices of bread it is necessary to use a knife and make contact with the bread, after which we can grab the slice and use it as an ingredient.

(4) Street. The following elements are present in this scenario: pedestrians, parked cars, cars running, motorcycles, dogs, buildings, sidewalks, roads, playground, traffic lights, crosswalks, bookstore and home.

Fig. 4. Image of Street scenario

This scenario, based on the proposal presented in [5], places users in a city where they must travel from an initial point (their house) to the *Bookstore* store. To move they will have to press the main trigger of the left controller while pointing in the direction in which they want to move. In the city there will be traffic and pedestrians moving around independently, and they will have to navigate the space respecting the rules of the road and without colliding with other people or animals and avoiding distractors such as signs of a circus or a playground. Once they arrive at the *Bookstore* store, they can enter the store just by approaching the door.

The user can also return to the house, as once they have purchased the necessary materials, they will need to travel from the *Bookstore* to the house. In that case, to access the *Hallway* of the house the user will have to take the keys out of the pocket by pressing the main trigger of the right controller, and then approach the door with them in hand. If the children have not taken the keys before leaving the house, they will not be able to perform this action.

(5) Bookstore. In this scenario the user has the following elements: shelves, counter, sales clerk, cash register, textbook (all subjects), notebooks (all subjects), books and tales, pencils, pens (red and blue), erasers and colored pencils.

Fig. 5. Image of Bookstore scenario

In this scenario the child must gather the required materials and place them on the counter. The cash register will show the price of the selected items. After selecting all products, the user can take out a coin purse by pressing the main trigger with his left hand. When you bring your right hand close to the purse, a series of coins and bills will be shown floating in the air, and you will have to place them on the counter until you reach the exact amount required. After that, the objects on the table will transform into a closed bag representing all the objects purchased, which they can pick up before leaving the room and returning to the *Street*.

3.2   Errors

As we have seen, the application contains a variety of scenarios that can be combined to design complex tasks closely related to activities that a child may perform on a daily basis. This complexity is transferred to the control of the errors made by the child during the execution of the programmed task. Until now, errors made during the sessions were collected manually, which required a therapist to supervise the session and record errors as they occurred, or the session video was saved for later review and annotation.

During the performance of the assigned tasks, the child is exposed to committing errors of different types that are derived from different cognitive deficiencies that this type of children with ADHD have. Errors are labeled in the following categories [11]: (1) omission; (2) commission; (3) sequence; (4) estimation; (5) question/comment; (6) breaking the rules

Errors of *omission* refer to tasks that should have been performed but were not. Errors of *commission*, on the other hand, refer to tasks that have been performed and should not have been performed. *Sequence* errors refer to tasks that are performed in the wrong order. *Estimation* errors are those in which the amount of ingredients needed for a recipe or the space needed to lift or set down an object is misestimated. Errors by *question/comment* refer to any type of interaction children may have with the therapists present, as they are instructed to act as if they were alone. Errors due to *breaking the rules* refer to those errors that happen due to not following previously given instructions (not respecting the traffic rules, colliding with other pedestrians, trying to get out of the bounds of the scenario...).

Of these errors, errors of *omission, commission, sequence* and *estimation* are most closely related to the tasks that children must perform with the tool. These tasks can be divided by scenario as follows:

- Room: Gather materials and store them in the backpack.
- Kitchen: Take ingredients to the interaction area and store food in the backpack.
- Hallway: Take the key before leaving the house.
- Bookstore: Buy materials.

Therefore, all these tasks are performed in the same way. Carrying the different objects (throughout the simulation) and using them in the specific interaction zones.

Regarding the errors by *question/comment*, any comment, even if it is not a direct question to the therapists, is currently considered an error of this type. In this case, the microphones in some VR glasses can be used for detection purposes.

The errors labeled as *breaking the rules* have been defined individually by the therapists based on the errors that they have been able to observe so far during the execution of the sessions, being: leaving the limits of the scene (all scenes); not crossing at the crosswalk (Street); crossing with the traffic light red (Street); colliding with animals or people (Street); not ending up in the corridor, which is the starting point (General); dropping to the ground (all scenes); not checking the already available material in the bedroom before going shopping (General).

3.3 Automation

To assess errors, a simple graphical tool with a straightforward user interface (see Fig. 6) was designed, allowing therapists to define the tasks for children to complete in each session. This facilitates the creation of complex tasks, such as those involved in following a recipe. Once the task is defined, the tool generates a JSON file in which a list of actions to be performed is established. This will allow us to recognize errors of *omission, commission, sequence* and *estimation.* Since we will be able to compare the interactions performed with those that actually define the tasks to be performed.

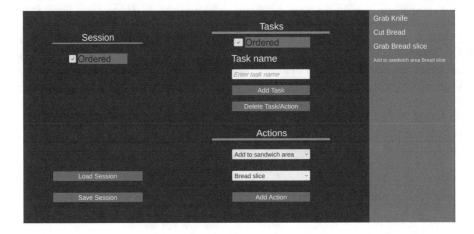

Fig. 6. Session generator interface

In order to understand the actions performed by users, it is important to recognize different types of interactions. These detected interactions will be recorded in a log and exported in a JSON file. This file will then be compared with the one previously generated by the therapist when defining the task, resulting in the automatic detection of the errors described in the previous section.

Based on the classification of interactions proposed by [12]), in our case, we have determined that the actions and therefore the triggers to control are: (1) Collider trigger; (2) Collider-Key trigger; (3) Distance trigger.

Fig. 7. Example of Triggers: A) Collider; B) Key-Collider; C) Distance

(1) Collider triggers are those that are activated when there is a collision between two objects. Unity allows us to check which objects collide, which makes

it possible for us to use them to register which objects are being added to the interaction zones of the different rooms (the backpack and the counter). It also allows us to record some of the errors for *breaking the rules*, such as crossing on red, as we can create a collider for such areas and collect when the user has collided with a part of the road that does not have a zebra crossing. In the same way, it allows us to record when entering a crosswalk to check the color of the traffic light. It also allows us to check if the user collides with other pedestrians, cars, animals and other objects in the environment (such as the walls that form a boundary to the stage).

As for the *(2) Collider-Key Trigger*, they are activated when, in addition to a collision, a key or button is pressed. This is the method we use to grab objects (collide with it and press the trigger). Therefore, it is also the method we will use to record such interactions. In the same way, there are also some interactions (pressing buttons such as those on the cooktop, turning off the smartphone alarm, or pressing a switch, see Fig. 7) that require touching those objects while pressing another trigger, simulating the action of pressing them with a finger. In this case, it is also necessary to detect the collision but with the only caveat that the button pressed is different.

We also use *(3) Distance Triggers*, as can be seen in Fig. 7) these are used to record whether a drawer is open or closed (depending on its distance from its initial position), to know whether the refrigerator and other furniture has been closed or opened, as well as to record the error for breaking the rules of lying on the floor during execution.

Error Recognition. Once we have the data with the expected actions (those created by the therapist) and the actual actions performed, we can compare them using a rule-based system. This rule-based system has as input each of the actions performed and the instant of time at which they occurred. The actions are associated with the type of action and the object on which the action is performed.

With these information, we can list the executed actions ordered by time, and check if that is the expected action. For this, it is not enough just to compare actions one by one, but we must take into account the complete trace of actions, both executed and expected, in order to check what type of error it is. Thus, *estimation* errors require checking whether the action has been performed correctly but has been repeated more than necessary. On the other hand, *sequence* errors require knowing whether the action would be correct in the future, or in the past. In regard to *omission* errors, the system must check whether the child has already started a future task without having finished the current one. In this case, when detecting an error of omission the system must be able to register it and go ahead checking the rest of the tasks, avoiding getting stuck comparing the rest of the execution against an action that has not been performed (but maybe the rest of the session has been executed correctly).

Once this process has been conducted, we generate a table containing the errors committed (see Fig. 8), i.e. the character string that defines them, the

action	time	error_type
Grab Juice milk	-	Omission
Store in backpack Juice milk	-	Omission
Grab donut	9.5605	Comission
Cross with red light	20.6300	Break the rules

Fig. 8. Error Table of an execution example

time in which they were committed (with the exception of omission errors, which have no associated time because they have not been carried out) and the type of error involved. We will also generate a table with performance metrics (see Fig. 9) such as: total number of errors (by type and in general).

```
[1] "Omission Errors: 2"
[1] "Comision Errors: 1"
[1] "Sequence Errors: 0"
[1] "Estimation Errors: 0"
[1] "Question/comment errors: 0"
[1] "Break the rules errors: 1"
[1] "Total Errors: 4"
```

Fig. 9. Error Metrics of an execution example

4   Conclusions and Future Work

In this paper, we present ECogFun-VR, a virtual reality application for treating children with ADHD by exposing them to activities of daily living in an ecologically valid manner. A tool for automatic analysis of common types of errors made by these ADHD children has been added to this application. To do this, the system incorporates a task manager and a tool for controlling and recording user interactions. This interaction control is based on recognizing different triggers that the user activates and that the system can register.

As we have seen, the tool is able to recognize the errors made by children. However, it would be crucial to test the accuracy of the system in locating and categorizing the errors made. Therefore, we are designing an experiment to analyze the precision of automatically detected errors. This experiment will compare the automatic results to those manually recorded by multiple experts who will be present during the activities and subsequently analyze the video recording. We also propose improving the way errors are displayed by associating them with specific actions recorded in the video recording of the execution (from the user's perspective). This will provide experts with information not only about the error committed but also about the context in which it occurred. It would also be interesting to add the possibility of modifying scenarios before the execution to adapt them to the needs of each therapy and give it greater variability

Acknowledgements. This paper is part of the R+D+i projects PID2019-108915RB-I00 and PID2022-140907OB-I00 funded by MCIN/AEI/ 10.13039/ 501100011033, it has also been funded by the University of Castilla-La Mancha (2022-GRIN-34436) and by 'ERDF A way to make Europe'. Finally, this work was also partially supported by CIBERSAM of the Instituto de Salud Carlos III.

References

1. Arnold, L.E., et al.: Long-term outcomes of ADHD: academic achievement and performance. J. Atten. Disord. **24**(1), 73–85 (2020)
2. Romero Ayuso, D.M.: Actividades de la vida diaria. Ann. Psychol. **23**(2), 264–271 (2007)
3. Bashiri, A., Ghazisaeedi, M., Shahmoradi, L.: The opportunities of virtual reality in the rehabilitation of children with attention deficit hyperactivity disorder: a literature review. Korean J. Pediatr. **60**(11), 337–343 (2017)
4. Brookes, J., et al.: Studying human behavior with virtual reality: the unity experiment framework. Behav Res **52**, 455–463 (2020)
5. Juan-González, J., et al.: UrbanRehab: a virtual urban scenario design tool for rehabilitating instrumental activities of daily living. J. Ambient Intell. Hum. Comput. **14**(3), 1339–1358 (2023)
6. Kim J, Gabriel U, Gygax P.: Testing the effectiveness of the Internet-based instrument PsyToolkit: a comparison between web-based (PsyToolkit) and lab-based (E-Prime 3.0) measurements of response choice and response time in a complex psycholinguistic task. PLoS ONE **14**(9) (2019)
7. Ramsay, J. Russell.: Nonmedication Treatments for Adult ADHD: Evaluating Impact on Daily Functioning and Well-Being. American Psychological Association, 2010
8. Rizzo, A.A., et al.: The virtual classroom: a virtual reality environment for the assessment and rehabilitation of attention deficits. CyberPsychol. Behav. **3**(3), 483–499 (2000)
9. Romero-Ayuso, D., et al.: Effectiveness of virtual reality-based interventions for children and adolescents with ADHD: a systematic review and meta-analysis. Children (Basel). **8**(2), 70 (2021)
10. Schena, A., et al.: IAmHero: preliminary findings of an experimental study to evaluate the statistical significance of an intervention for ADHD conducted through the use of serious games in virtual reality. Int. J. Environ. Res. Public Health **20**, 3414 (2023)
11. Schmitter-Edgecombe, M., et al.: The night out task and scoring application: an ill-structured, open-ended clinic-based test representing cognitive capacities used in everyday situations. Arch. Clin. Neuropsychol. **36**(4), 537–553 (2021)
12. Ugwitz, P., et al.: Toggle toolkit: a tool for conducting experiments in unity virtual environments. Behav. Res. Methods **53**, 1581–1591 (2021)

Walkability for Active Aging: Factors and Features for an Adapted Trip Planning

Sahar Tahir[1,2(✉)], Bessam Abdulrazak[1,2], Dany Baillargeon[2,3], Catherine Girard[2,4], and Véronique Provencher[2,4]

[1] AMbient Intelligence Lab, Département Informatique, Faculty of Science, University of Sherbrooke, Sherbrooke, QC, Canada
{sahar.tahir,bessam.abdulrazak}@usherbrooke.ca
[2] Research Center on Aging, Centre Intégré Universitaire de Santé et de Services Sociaux de l'Estrie, Centre Hospitalier Universitaire de Sherbrooke, Sherbrooke, QC, Canada
[3] Department of Communication, University of Sherbrooke, Sherbrooke, QC, Canada
[4] School of Rehabilitation, University of Sherbrooke, Sherbrooke, QC, Canada

Abstract. Mobility, especially walking, is crucial for older adults' independence, social engagement, and overall well-being. Walkability encompasses safe, accessible, and pleasant environments that enhance mobility and foster social interactions. While various mobility factors assess urban pedestrian-friendliness, they are rarely integrated into trip planning tools for older adults. Therefore, we study the significance of these factors within the context of older adults living in a medium-sized Canadian city. We combine a comprehensive literature review with workshops and interviews involving older adults to study the mobility factors and develop a web-based trip planning tool. We operationalize and measure walkability factors to understand older adults' specific requirements and preferences. This study enabled us to identify crucial walkability factors and demonstrates how these factors can be integrated into a multicriteria optimization algorithm to create a trip planning tool for older adults.

Keywords: Active aging · walkability · older adults · mobility · trip planner tool · multicriteria optimization

1 Introduction

Aging brings natural declines in physical and sensory capabilities, affecting mobility [1]. Reduced muscle strength, flexibility, and balance, along with potential sensory impairments, hinder older adults' movement [2]. This decline in mobility profoundly impacts independence, limiting essential tasks, service access, and social engagement [3]. Despite these challenges, mobility remains crucial for older adults, with walking as a primary mode for shorter community trips. Walking facilitates engagement with the environment, access to local services, and social connections [3]. While promoting physical activity, it also fosters mental stimulation. However, it's important to note that aging-related frailty can lead to slower walking speed and balance issues, thereby

© The Author(s), under exclusive license to Springer Nature Switzerland AG 2023
J. Bravo and G. Urzáiz (Eds.): UCAmI 2023, LNNS 835, pp. 172–177, 2023.
https://doi.org/10.1007/978-3-031-48306-6_17

increasing the risk of falls [3]. Our study aims to examine the relevance of walkability factors for older adults and translate them into features for a multicriteria optimization algorithm. We explore key walkability factors and their impact on mobility and well-being of older adults (Sect. 2) by reviewing relevant literature and analyzing insights from workshops and interviews with older adults (Sect. 3). Our approach involves operationalizing and incorporating these factors into a trip planning tool (Sect. 4), with the goal of promoting active aging and enhancing the overall quality of life for older adults in urban environments.

2 Walkability Factors and Their Impact on Older Adults

Walkability, a pivotal concept in urban planning and public health, encompasses the extent to which the built environment supports pedestrian-friendly transportation [4]. This section delves into seven pivotal factors contributing to walkability and their specific implications for older adults. **1) Adequate street connectivity**, characterized by well-connected streets, sidewalks, and pedestrian crossings, significantly influences older adults, encouraging them to engage in more active trips from home, offering safer and more accessible walking options [5, 6]. **2) Diversity of land use types** within a neighborhood significantly influences older adults' walking behavior, encouraging them to walk longer distances and opt for active transportation [5–7]. **3) Adequate infrastructure**, as well-maintained sidewalks, crosswalks, and benches, is essential, creating safe and convenient walking environments, promoting walking as a preferred mode of transportation among older adults [8]. **4) Ease of reaching key amenities** like grocery stores, parks, and healthcare centers within a neighborhood significantly impacts older adults' walking behavior, making walking an attractive and viable option for their daily activities [6, 9]. **5) Safety and security** are critical for older adults' perception of walkability, instilling confidence and encouraging them to walk, even during nighttime, through well-maintained and safe streets [8, 9]. **6) Visual appeal** of the environment significantly influences older adults' choice to walk, making walking more enjoyable with visually pleasing surroundings, including greenery, water features, and well-designed streetscapes [10, 11]. **7) Elevation changes** within a neighborhood or along walking routes can impact the comfort and feasibility of walking for older adults, influencing their walking habits, emphasizing the importance of considering slopes [12]. These walkability factors collectively exert a significant influence on the mobility and physical activity of older adults and provide valuable insights for urban planning and public health strategies aimed at promoting active and healthy aging.

3 MOBILAINES Study on Walkability/Walking Experiences

The research presented in this paper is part of the MOBILAINES[1] project that adopts a Living Lab approach, which facilitates user-centric and iterative development through collaboration with stakeholders in real-world settings [13]. MOBILAINES project is

---

[1] Mobilainés is a French contraction of "mobilité" (mobility) and "ainés" (older adults).

divided into three phases, each contributing to the creation and evaluation of a one-stop platform for older adults' transportation planning: Phase 1: Identification of needs and preferences of older adults through reviews of scientific and gray literature and workshops in collaboration with older adults. Phase 2: Development of a mobility aid prototype tailored to the needs and preferences of older adults. Phase 3: Evaluation and testing of the prototype with older adults presenting various profiles in terms of needs and preferences. This paper focuses on results from the exploration phase (phase1) where the research team aims to gain comprehensive insights into the realities, needs, and expectations of older adults and transportation service providers. To achieve this, we undertake two primary activities. 1st, the team conducts extensive reviews of scientific and gray literature to gather up the most recent knowledge and evidence related to older adults' mobility, facilitators, and barriers. This literature review informs the subsequent stages of the project, ensuring that the platform aligns with the latest research findings and best practices. 2nd, workshops were designed based on the results and insights from the literature review. These workshops were organized in collaboration with older adults to actively engage them in the platform's conceptualization process. Workshops create an interactive environment where participants can share their experiences with trip planning and mobility, as well as express their needs and preferences. The insights gained from these workshops played a crucial role in defining the criteria for the ideal MOBILAINES platform. More specifically for the subject of this paper, these workshops helped identify the main facilitators and barriers of mobility.

4   Facilitators and Barriers: Insights from Our Workshops

To gain deeper insights into the mobility experiences and preferences of older adults, the MOBILAINES team conducted workshops and phone interviews with six participants in the region where the project takes place. These workshops aimed to achieve three primary objectives: 1) document and describe participants' mobility experiences, 2) identify and understand the barriers and facilitators they perceive in mobility and mobility planning, and 3) capture their needs and preferences to inform the development of an ideal solution. The analysis of these workshops and interviews focused in part on exploring the participants' experiences related to walking as a mode of transport, revealing key insights into the significance of walkability for older adults. Significant barriers and facilitators emerged during the discussions. Participants emphasized the importance of *incline avoidance*, expressing concerns about the *physical challenges* posed by uneven and hilly terrain. They also highlighted their preference *for pleasant paths* and *routes passing by parks*, enjoying scenic routes along the lake enriched with amenities such as *benches*, *toilets*, and *shops*, which played a crucial role in enhancing the overall walking experience. Additionally, participants noted that they could comfortably *walk distances up to 500 m*, with longer distances posing challenges. *Weather conditions* significantly impacted their willingness to walk, especially in winter, due to slippery and icy sidewalks, and extreme heat discouraged daytime walking during summer. Furthermore, *traffic congestion* and the *lack of sidewalks* in certain areas affected their walking choices, making them *feel unsafe in heavy traffic locations*. Finally, older adults acknowledged that their walking pace had slowed with age and emphasized the need to consider this reduced pace when planning walking trips.

5 Translating Walkability Factors to Features in the Multicriteria Optimization Algorithm

Having gained valuable insights into the mobility experiences and preferences of older adults through our workshops, we were able to focus on the translation of the barriers and facilitators into features that should ideally be integrated in trip planning tools. Thus, we compiled our results and cross validated them via a series of in-person workshops with transportation service providers, community-based and public health stakeholders (n = 8), and older citizens (n = 8). These collaborative efforts led us to identify ten essential functionalities that trip planning tools should offer, with seven directly addressing walkability and walking preferences for older adults:

a) **Walkability**: Considering participants' feedback, MOBILAINES trip planning algorithm incorporates two functionalities to cater to older adults' walking preferences: *1) Walking Speed Adaptation*: The algorithm adapts walking speed to align with older adults' comfort and capabilities, Considering the findings from the literature [14, 15], MOBILAINES multicriteria optimization algorithm adopted a walking speed of 1 m/s for older adults. This speed is chosen to provide a balance between ensuring a comfortable pace for walking and promoting beneficial physical activity. *2) Optimizing Distance*: To align with literature [16] and accommodate the preferences and physical capabilities of older adults, we have set in MOBILAINES project a limit of approximately 400 m for the walking distances in our algorithm.

b) **Crowd / traffic Avoidance:** This functionality addresses concerns about heavy traffic by recommending less congested routes, utilizing open-source city data, a trip planning tool can identify and recommend routes that steer clear of highways and roads with higher speed limits, prioritizing safer and less congested pathways. Implementation is planned for a future version of MOBILAINES.

c) **Incline Avoidance:** To provide an enjoyable walking experience, the algorithm incorporates Tobler's Hiking Function [17] to adjust walking distances based on terrain incline. In line with our MOBILAINES study results, we specifically avoid recommending routes with slopes exceeding 5%, as this threshold is deemed suitable for all-inclusive older adults [18].

d) **Weather Consideration:** While not impacting recommended paths directly, weather information is offered through OpenWeather data [19], helping users plan their trips around weather conditions.

e) **Dark Avoidance**: Considering the lack of accessible street lighting data, to enhance safety and comfort, MOBILAINES trip planning tool considers the time of day, by providing information about nightfall time helping users plan their trips accordingly.

f) **Winter Obstacle Avoidance**: Slippery and icy sidewalks during winter pose significant barriers to walking. Our algorithm aims to recommend ice-free routes, although real-time snow removal data is unavailable. We plan to use Sherbrooke city's snow removal priority data [20] to help older adults avoid lower-priority routes.

g) **Amenities Inclusion:** Essential amenities along walking routes, such as toilets, benches, and bus shelters, are incorporated into the planning tool algorithm, promoting convenience and accessibility for older adults.

6 Discussion and Conclusions

We aimed in this study to explore the significance of walkability factors for active aging among older adults. Through a comprehensive literature review and engaging workshops and interviews, we gained valuable insights into their mobility preferences and challenges related to walking. Our results emphasize walkability's crucial role in older adults' physical and mental well-being. They prioritize avoiding slopes, enjoying scenic paths with parks, and shorter walking distances, aligning with walkability measures by Gell et al., 2015 [21]. Our MOBILAINES algorithm adapts walking speeds, optimizes distances, and avoids inclines to cater to their needs. By integrating these features, our algorithm offers tailored, walkable routes, promoting independence, well-being, and social interaction. Weather conditions, especially in winter, and traffic congestion emerged as significant barriers (consistent with Moran et al., 2017 [22]). We prioritize user-centric design to meet older adults' unique needs. However, our study's limitations include a small workshop sample and reliance on available data sources for algorithm recommendations. Future research should include larger, more diverse samples for validation and refinement. In conclusion, this study underscores the importance of walkability for active aging and offers user-centric design solutions to enhance older adults' mobility and well-being.

References

1. Vieillissement et santé. https://www.who.int/fr/news-room/fact-sheets/detail/ageing-and-health. Accessed 22 Nov 2021
2. Tedros Adhanom Ghebreyesus, Liu Zhenmin, Michelle Bachelet, Natalia Kanem, Global report on ageism. https://www.who.int/publications-detail-redirect/9789240016866. Accessed 11 Nov 2021
3. Robnett, R.H., Chop, W.C.: Gerontology for the Health Care Professional. Jones & Bartlett Publishers (2013)
4. Frank, L.D., Engelke, P.O.: The built environment and human activity patterns: exploring the impacts of urban form on public health. J. Plan. Lit. **16**(2), 202–218 (2001). https://doi.org/10.1177/08854120122093339
5. Fan, P., et al.: Walkability in urban landscapes: a comparative study of four large cities in China. Landsc. Ecol. **33**(2), 323–340 (2018). https://doi.org/10.1007/s10980-017-0602-z
6. Deng, C., et al.: A data-driven framework for walkability measurement with open data: a case study of Triple Cities, New York. ISPRS Int. J. Geo-Inf. **9**(1), 36 (2020). https://doi.org/10.3390/ijgi9010036
7. Taleai, M., Yameqani, A.S.: Integration of GIS, remote sensing and multi-criteria evaluation tools in the search for healthy walking paths. KSCE J. Civil Eng. **22**(1), 279–291 (2018). https://doi.org/10.1007/s12205-017-2538-x
8. Lee, S., Lee, C., Nam, J.W., Abbey-Lambertz, M., Mendoza, J.A.: School walkability index: application of environmental audit tool and GIS. J. Transp. Health **18**, 100880 (2020). https://doi.org/10.1016/j.jth.2020.100880
9. Golan, Y., Wilkinson, N., Henderson, J.M., Weverka, A.: Gendered walkability: Building a daytime walkability index for women. J. Transp. Land Use **12**(1) (2019). https://doi.org/10.5198/jtlu.2019.1472

10. Kaczynski, T., Henderson, K.A.: Environmental correlates of physical activity: a review of evidence about parks and recreation. Leis. Sci. **29**(4), 315–354 (2007). https://doi.org/10. 1080/01490400701394865

11. Sugiyama, T., Francis, J., Middleton, N.J., Owen, N., Giles-Corti, B.: Associations between recreational walking and attractiveness, size, and proximity of neighborhood open spaces. Am. J. Public Health **100**(9), 1752–1757 (2010). https://doi.org/10.2105/AJPH.2009.182006

12. Keskinen, K.E., Rantakokko, M., Suomi, K., Rantanen, T., Portegijs, E.: Hilliness and the development of walking difficulties among community-dwelling older people. J. Aging Health **32**(5–6), 278–284 (2020). https://doi.org/10.1177/0898264318820448

13. Provencher, V., et al.: Developing a one-stop platform transportation planning service to help older adults move around in their community where, when, and how they wish: protocol for a living lab study. JMIR Res. Protoc. **11**(6), e33894 (2022). https://doi.org/10.2196/33894

14. Abe, T., et al.: Association between hilliness and walking speed in community-dwelling older Japanese adults: a cross-sectional study. Arch. Gerontol. Geriatr. **97**, 104510 (2021). https:// doi.org/10.1016/j.archger.2021.104510

15. La vitesse de marche à pied, volet 1 : à quelle vitesse devrais-je traverser la rue en toute sécurité ? les faits saillants sur la vitesse de marche. https://www.mcmastervieillissementop timal.org. Accessed 19 Jul 2023

16. Hess, D.B.: Walking to the bus: perceived versus actual walking distance to bus stops for older adults. Transportation **39**(2), 247–266 (2012). https://doi.org/10.1007/s11116-011-9341-1

17. Higgins, CD: Hiking with tobler: tracking movement and calibrating a cost function for personalized 3D accessibility. Findings 28107 (2021)https://doi.org/10.32866/001c.28107

18. Alves, F., Cruz, S., Ribeiro, A., Bastos Silva, A., Martins, J., Cunha, I.: Walkability index for elderly health: a proposal. Sustainability **12**(18), 7360 (2020). https://doi.org/10.3390/su1 2187360

19. Current weather and forecast - OpenWeatherMap. https://openweathermap.org/. Accessed 19 Jul 2023

20. Données ouvertes - Ville de Sherbrooke. https://donneesouvertes-sherbrooke.opendata.arc gis.com/. Accessed 19 Jul 2023

21. Gell, N.M., Rosenberg, D.E., Carlson, J., Kerr, J., Belza, B.: Built environment attributes related to GPS measured active trips in mid-life and older adults with mobility disabilities. Disabil. Health J. **8**(2), 290–295 (2015). https://doi.org/10.1016/j.dhjo.2014.12.002

22. Moran, M.R., et al.: Exploring the objective and perceived environmental attributes of older adults' neighborhood walking routes: a mixed methods analysis. J. Aging Phys. Act. **25**(3), 420–431 (2017). https://doi.org/10.1123/japa.2016-0165

Predicting User Adoption and Attrition of Digital Health

Garbhan Harrison[✉] and Ian Cleland

School of Computing, Ulster University, Belfast, UK
{harrison-g3,i.cleland}@ulster.ac.uk

Abstract. Digital technologies, such as smartphones and wearable technologies, are transforming the way we deliver healthcare. Many studies have demonstrated the potential benefits of digital health, including wider access to care, and improved clinical outcomes. User participation, however, is crucial for these benefits to be seen. The ability to predict participant attrition before it occurs would provide an opportunity to intervene, potentially keeping the participant engaged for a longer period. Additionally, modelling of adoption provides insight into the reasons why a user may or may not engage with a digital solution. This paper utilises data from six digital health evaluation studies to create two classification models: one predicting if users stay in past day 16 and the next predicting if those that do stay past 16 will stay in past day 84. These models returned accuracies of 79.1% and 85.2% and F1- Measures of 0.795 and 0.736 respectively. Six different classifier algorithms were evaluated and the best results for both models came from a Neural Network classifier, with an AdaBoost classifier being the next best. Survival analysis was undertaken on the dataset and indicated that those who engaged in the study from day 16 onward were more likely to stay in the study long term. Survival analysis was undertaken on the dataset and indicated that the majority of participants survive until at least day 16.

Keywords: Digital Health · Adoption Modelling · Attrition · Survival Analysis

1 Introduction

In recent years, the widespread availability of smartphones and advancements in digital health technologies have revolutionised the healthcare industry. Digital health apps have emerged as a popular tool for individuals to track their fitness, monitor chronic conditions, and access personalized healthcare resources. However, despite the increasing number of available apps, the challenge lies in understanding and predicting user behavior, specifically adoption and attrition rates, to ensure the success and sustainability of these applications [1].

It is well known that studies involving digital technology can have challenges with user retention [2], which can negatively affect a trial's results. The potential impact of predicting user adoption and attrition of digital health apps is therefore substantial. Understanding why some users adopt and continue to engage with the app while others

© The Author(s), under exclusive license to Springer Nature Switzerland AG 2023
J. Bravo and G. Urzáiz (Eds.): UCAmI 2023, LNNS 835, pp. 178–189, 2023.
https://doi.org/10.1007/978-3-031-48306-6_18

discontinue their usage is crucial for developers, healthcare providers, and policymakers alike. By identifying factors that influence user behaviour, we can enhance user experience, optimize app functionalities, and tailor interventions to mitigate attrition rates.

This project aims to address this challenge by leveraging machine learning techniques to predict user adoption and attrition of digital health apps based on engagement data, and utilise usage data from 6 different digital health evaluation studies to answer the following questions:

1. If a participant's initial engagement reflects long term use, i.e., if they will continue to engage with the technology, and
2. If it is possible to predict when a participant will stop engaging with the study.

This paper is formatted as follows: Sect. 2 describes existing related work. Section 3 presents the methodology to be used including description of the dataset and the machine learning techniques to be used. Section 4 contains the results from this paper and a discussion on them. Section 5 has the conclusion, discussing the takeaways from this paper, and future work/improvements to be made.

2   Related Work

There is a large body of work related to technology adoption. The main inspiration for this paper comes from [3], which seeks to understand participant attrition in their respective studies; however, it aims to find a more in depth explanation as to why participants are not engaging with a study. It achieves its goals through the use of clustering, placing participants into five different clusters of engagement ranging from "dedicated users" to "sporadic users". These clusters were analysed, and it was found that participants older than 60 were more likely to be in a high engagement cluster, and people from minority groups were more likely to be in clusters with less engagement.

Survival analysis is also used to investigate the probability of when participants will leave a study. By calculating the median value, it can be found at which day in the study the majority (50%) of participants are likely to have dropped out. This day ranges between 2–26, with the Brighten study having the highest day at 26, and both Start and SleepHealth having the lowest median days at 2. The median across all studies is 5.5 days. The study notes the reason the Brighten study has such a large median day is the monetary incentive for the participants to remain in the study. It is also found that studies including the "clinicalReferral" feature, meaning that the participant was referred by a clinician, retained only half of the participants who were not referred by day 4. This is opposed to half of those who had been referred dropping out by day 44.

A takeaway from the study was that more than half of the participants dropped out within the first week, and that age, disease status, clinical referral and monetary incentive were the main factors behind those that stayed in longer. Another key takeaway from this study was every study's inability to recruit an ethnically diverse sample of participants representative of the US, raising questions about the reliability of the data. It also finds that the data underrepresented those from the southern United States.

The paper, however, notes that despite issues with the data, it is still one of the largest when it comes to participant retention with digital health and offers solutions to these issues. The paper suggests the use of a run-in period [4] could help to alleviate issues with the bias in the data. This would work by introducing a period before the study formally started to analyse the level of participant engagement and then either intervene or exclude those who were not as engaged, therefore leaving the study with a smaller group more likely to engage long term. This, however, would not solve representation bias issues in the data. It also suggests that a monetary incentive attached to future studies would enhance participant engagement.

The authors have previously investigated the use of machine learning to predict adoption of reminding solutions by people living with dementia [5]. Using 7 different classification models, a Neural Network, Decision tree, SVM, Naïve-Bayes, AdaBoost, KNN and CART, it was determined that the best model for the problem was the KNN model with 99.41% accuracy. The Neural Network model was next best at 94.08% accuracy. Methods such as univariate and multivariate analysis were used to determine relevant features, and SMOTE was used on the existing dataset to rebalance the minority adopter class. This paper had a comparatively small sample size, with 21 adopters and 152 non adopters for a total sample size of 173. Metrics such as accuracy, F-measure and ROC are used to evaluate the classifiers.

Several conclusions can be drawn from the literature. Firstly, regarding ethics of this project and how any solutions created had to have a participant's best interests in mind. This would take the form of making it a priority that the system developed had the highest possible accuracy metrics, ensuring that if the system was implemented, participants would not miss out on intervention. This is especially necessary with more vulnerable participants, such as those suffering from Alzheimer's. The next takeaway was how the solution to be developed should be approached in terms of methodology. Having analysed the problem tackled in the paper by Chaurasia et al. [5] it is similar to the one identified in this paper. It would therefore make sense to consider the approaches that were taken within that paper. This involves decisions such as balancing the data when the minority class is low, and the classifiers used to be appropriate for the proposed solution within this paper. There were multiple takeaways from the paper by Pratap et al. [4]. The first being that from the datasets evaluated in that paper, there is a large range in the 'median' day of user drop out, and that these needed to be identified in the datasets looked at within this paper. Another takeaway is the bias identified within this paper, regarding ethnicity and location. These were important to consider when developing the solution and processing the results within this paper.

A gap has been identified from the study by Pratap et al. [4] and Laursen et al. [3]. In the paper by A. Pratap et al. participants have been classified into clusters based on engagement activity, however there was no attempt to make predictions on whether participants were likely to stay active in the studies. In the paper by D. Laursen et al. they identified how run-in periods were not used by more than 5% of the randomly selected 470 studies explored, and those that did use run-in trials tended to incompletely report on their run-in periods, invalidating the assessment of how useful they are. The solution proposed in this paper could potentially make the process for this run-in period much easier as personnel will be able to identify more readily who is at risk of not engaging in the study.

3 Methods

This section describes the work performed on the dataset to make it suitable for classification modelling, as well as detailing the models developed. The datasets were sourced from Synapse [6], titled "Participant Retention in Digital Health Studies". Ethical considerations had to be made before any work was completed. To access the data, an intended data use statement was written, outlining how the data was to be utilised in this paper, and this statement then had to be approved by Synapse. The dataset used contains data that was given consensually by participants and has been anonymised. Additionally, each of the studies has been reviewed by their respective institution's ethics board. There is however a potential for bias in the results. As identified in [4], the dataset does not proportionally represent minority groups or those from the southern states in America. Although, as will be discussed later in the paper, both race and state will be removed from the dataset before classification occurs, racial and geographic bias could still be present in the final results, and it is important to be aware of that.

Two classification models were built. The first used features extracted from the first two weeks to predict which participants would stay past day 16. The second model would then predict from those that did stay past day 16 who would stay past day 84, using features extracted between day 16 and week 12. The following sections present the various study datasets in detail and describe how the data was processed and the modelling approaches used.

3.1 Description of Datasets

The whole dataset used for modelling is made up of data from six different studies. Each of these studies tracks participants completing a task within an app. Each study had two separate data sources attached; 1) a metadata and 2) an engagement data. The metadata consisted of personal information such as the participant's age group, gender and whether they self-reported to have a disease. The engagement data contained information such as the day a user was active, the time of day engagement occurred, and how they logged information into the app. The metadata and engagement data for each study were merged to create two datasets, a metadata set and an engagement set. The compiled metadata dataset had 76,730 entries, indicating the number of participants across all studies. The compiled engagement dataset had 3,100,724 entries, meaning that ~ 3.1M engagements were made between the 76,730 participants. Table 1 shows the population count of each study and the count after the data has been 'cleaned', detailed in the next section of this paper.

Table 1. The Population Count of the 6 Studies

Study	Total No. of Participants (Before cleaning)	Total No. of Participants (After cleaning)
Brighten	883	577
ElevateMS	605	302
mPower	12,236	5,170
Phendo	7,532	4,229
SleepHealth	12,770	6,792
Start	42,704	22,310
All studies except Start	34,026	17,070
All Studies	76,730	39,380

3.2   Cleaning the Dataset

Before feature engineering and model creation could be performed, the dataset must first be cleaned. The metadata was explored to ensure that each study that was incorporated into the metadata set had the same features. It was found that both the mPower and ElevateMS studies had a feature called "clinicalReferral" that was a Boolean value showing whether or not that participant had been referred to the study by a clinician. As only two studies had this feature the decision was made to remove it from the dataset.

The feature "race_ethnicity" is not present in the Start study, and as it comprised a large portion of the dataset, this feature was removed. Both Start and SleepHealth did not indicate which state in America the participant was from, so the feature "state" was cut also. Each participant's ID was also removed, leaving the following demographic features in the metadata set: "study", "age_group", "gender" and "diseaseStatus". At this point, any participant that has no value for a feature (a NaN value) is removed from the dataset. Next, each of the demographic features underwent ordinal encoding.

Analysis was then made as to when participants stopped engaging with the study. It was found that the majority of participants do not stay in the study past the first week – the number of dropouts by day 7 being 49,083. Of that number, 36,249 do not make it past the first day. It could be argued that many of these are not true participants, possibly just investigating the app with no intention of participating long term. A decision was made to remove all participants that did not make it past the first day. This left the dataset with a total number of 40,821 participants – and with NaN values removed this number drops to 39,380. The dataset after those who do not stay past the first day are removed can be seen in Fig. 1.

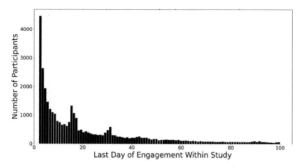

Fig. 1. Histogram of participants that stay past the first day, zoomed in to day 100.

3.3 Survival Analysis

With the dataset cleaned, survival analysis using the Kaplan-Maier method was performed to visualise participant retention, showing along a curve when a percentage of the participants were still active. The 'maxDayInStudy' feature was created to identify each participant's last day of activity. A slightly zoomed in version of the curve generated can be seen in Fig. 2. The curve shows that by day 5, 25% of the participants have dropped out, by day 16, 50% have dropped out, and 75% by day 45. The day of most note is day 16. If it can be predicted who will still be in by day 16, it can then be known that those people have a 50% chance of remaining in the study. With this knowledge, a conclusion can be drawn that activity before this day will considered initial engagement, as during this time period the majority of participants were engaging in their study.

One challenge in identifying drop out was the ambiguity around what characterises long term continued engagement. Based on when a participant drops out or how often a user engages during that period is a factor. For example, if a user engages once on the first day, then once again on day 80, is this long term use? The paper which originally compiled the dataset [414] only looks at data within the first 12 weeks as this was the scope of the Brighten and ElevateMS studies. Although not incentivised, both of these studies include a reasonable number of participants that stayed in beyond this date – nearly half the participants for each, which means 12 weeks/84 days appears to be a reasonable cut off. The survival analysis shows that by day 84 only 14% of participants are still active in their respective study, a reasonable qualifier for "long term" use.

3.4 Creating Classes

With a clean dataset and initial survival analysis undertaken, the next step was to create the classes for classification. Based on the survival analysis, it was decided that the classification model would be used to predict if a participant would stay in the study for 16 and 84 days. To do this two features are created, "stayInStudy16" and "stayInStudy84". These features are Boolean, whether the participant has an active day past that point or not. The day 16 class has 20,638 participants, 52.4% of the total dataset. The day 84 class has 5,639 participants, making up 14.3% of the 39,380 population. Each study's

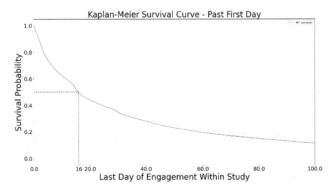

Fig. 2. Kaplan-Meier Survival Curve. It can be seen with the dotted line that 50% drop out by day 16.

number of adopters, and that number's proportion to the total population of the study after the data has been cleaned can be seen in Table 2.

Table 2. Each study's day 16 and day 84 adopter count

Study	Adopters past day 16	Adopters past day 84
Brighten	504 – 87.3%	284 – 49.2%
ElevateMS	224 – 74.1%	143 – 47.3%
mPower	2,402 - 46.4%	1,160 - 22.4%
Phendo	2,306 - 54.5%	796 - 18.8%
SleepHealth	2,361 - 34.7%	788 - 11.6%
Start	12,841 - 57.5%	2,468 - 11.0%
All studies except Start	7,797 – 45.6%	3,171 - 18.5%
All Studies	20,638 - 52.4%	5639 - 14.3%

3.5   Feature Engineering

The goal for feature engineering is to use the engagement dataset to create appropriate information for each participant. As discussed in the survival analysis section it was decided to look at how initial engagement impacts adoption, therefore a decision was made to look at user data from the first two weeks of participation before day 16 for the day 16 adopter model.

Features were made over the first week and the first two weeks. The first features made were the number of engagements made in these time periods. This was achieved through totalling the number of occurrences of a user's ID within the engagement set. The next features created were the number of active days a user had over the two periods. This was achieved by grouping the unique active days a user had in each time period and taking the total count of them. The next features created were the number of days each participant was inactive over the two time periods.

However, when looking at the correlation matrix created for these features, it was seen that these two inactive days' features produced a negative correlation with the classes being predicted for – days 16 and 84. For this reason these features were left out of modelling. The next features created were the maximum period of inactivity for a participant in the first week and first two weeks. The reasoning behind this feature was that it would be intuitive to believe that a participant with a larger period of inactivity may not be as committed to the study and may drop out sooner. However, when looking at the correlation table, the first week of inactivity negatively correlated with the day 16 class. The final features implemented for the first model are the count of each participant's "taskType", i.e. how they engaged with the study within the first week and first two weeks. A list and description of the features derived from the demographic/metadata datasets used in this day 16 adopter model can be seen in Table 3, as well as features derived from engagement data. Both sets of demographic and engagement features combined make up the data that will be used for classification. A correlation matrix with reduced features so as to be visible in this paper, can be seen in Fig. 3. Each feature is created from data generated by each participant.

Features were derived for the second model that would predict those that stayed or 'adopted' past day 84. These would be similar to the day 16 adopter model, but instead looking at engagement between day 16 and 84. These features are: "numEng16to84", "numActiveDays16to84", "maxPeriodOfInactivity16to84", "avgEngagementsPerActiveDay16to84", "activeSensorCount16to84", "passiveSensorCount16to84, surveyCount16to84". All features listed are created identically to their day 16 model counterpart.

3.6 Classification Models

Six traditional classification models where utilised, KNN, SVM, Gaussian Naïve-Bayes, Decision Tree, AdaBoost and an MLP Neural Network. This was decided after considering the work from Chaurasia et al. [5],

The classifications were performed on the final dataset, comprising 39,380 participants for the day 16 model and 20,638 for the day 84 model. A test train split was performed on the dataset at a split of 60/40. This leaves the training set for the day 16 model at a size of 23,628 and the size of the day 84 model at 12,382. For the day 84 model, SMOTE [7] was used to increase the size of the dataset used for classification by synthetically adding new participants that are part of the minority class. A sampling strategy of 0.8 was used bringing the size of the training set for this model up to 16,272. The day 16 model was also run to test if it could predict for day 84, though it was not expected to succeed. SMOTE was applied to the dataset at a sampling strategy of 0.8, bringing the size of the training set from 23,628 to 36,486. Additionally, before the classification models were run, the dataset was processed through a 'MinMax' scaler to normalise the data.

The models were evaluated with three different metrics, accuracy, F1-measure and Area Under Curve (AUC). Ideally the model should have as high an accuracy as possible. An ideal accuracy would be 90% or above [8] and would be in line with models from

Table 3. Name and description of features extracted from metadata set for day 16 model.

Feature	Description
Study	The study the participant was a part of
Age_group	Age group present: 17–29, 30–39, 40–49, 50–59, 60–120
Gender	Possible genders: Male, Female and Prefer not to answer
diseaseStatus	Whether self reports to have a disease. A Boolean of True or False
numEngFirstWeek	Total engagements a participant has made in the first week
numEngTwoWeeks	Total engagements by a participant in the first 2 weeks
numActiveDaysFirstWeek	Total days a participant has engaged in the first week
numActiveDaysTwoWeeks	Total days a participant has engaged in the first 2 weeks
firstWeekRatioActiveToInactiveDays	Ratio of active to inactive days in the first week
twoWeeksRatioActiveToInactiveDays	Ratio of active to inactive days in the first 2 weeks
maxPeriodOfInactivityTwoWeeks	Longest number of days between making an engagement in the first 2 weeks
activeSensorCountOneWeek	Total engagements with an active sensor in the first week
passiveSensorCountOneWeek	Total engagements with a passive sensor in the first week
surveyCountOneWeek	Total engagements with a survey in the first week
activeSensorCountTwoWeeks	Total engagements with an active sensor in the first 2 weeks
passiveSensorCountTwoWeeks	Total engagements with a passive sensor in the first 2 weeks
surveyCountTwoWeeks	Total engagements with a survey in the first 2 weeks

papers that have made similar predictions [9]. However, an accuracy of 0.8 or over would still be considered good and 0.7 and over would be considered acceptable, with anything under that unacceptable. The accuracy score is complemented by the F1-measure and AUC metrics; both must have an acceptably high score alongside the accuracy for the model to be a success. The F1-Measure should be as high as possible, with a value of 0.7 or higher being considered acceptable. Similarly looking at "Applied Logistic Regression" [10] it can be safely said that an AUC score of 0.7 or above would be an

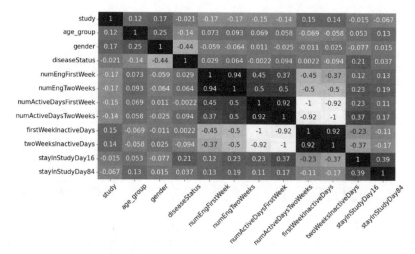

Fig. 3. Reduced Features Correlation Table

acceptable score and 0.8 or higher would be excellent. If all three metrics are above an acceptable level for a classifier, then that model can be deemed a success.

4 Results

This section will detail the results of both the day 16 model and the day 84 model. The classification models were run, and the results recorded. The results for the day 16 model, predicting when participants will drop out after day 16 can be seen in Table 4. The classifier that produced the best scores across the board was the Neural Network. The worst performing classifier was the Naïve Bayes.

The day 16 model was run again to test if it could predict the participants who would make it past day 84. Those results can be seen in Table 4. From this test, the best accuracy was returned by the Naïve Bayes classifier, the best F1-Measure produced by the AdaBoost classifier, and the best AUC score was shared by the AdaBoost and Neural Network. The worst accuracy came from the Neural Network classifier, and the worst F1-Measure and AUC score came from the Naïve Bayes classifier.

The day 84 model was run to test if the participants that remained after 16 days would still be staying in past day 84. The results for this can be seen in Table 4. The best results for all three metrics came from the Neural Network classifier. The Decision Tree classifier had the worst accuracy and the Naïve Bayes classifier had the worst F1-measure and AUC score.

5 Discussion

Positive results can be seen from both models. With the day 16 model, the results show that when predicting for those who will stay past day 16 all classifiers except for Naïve Bayes are able to produce acceptable accuracies, combined with reasonably high F1-measure and AUC scores. The Neural Network produced the best results, followed by

Table 4. Results for the model for predicting 16 and 84 days.

Classifier	Day 16 model			Day 16 on day 84			Day 84 model		
	ACC	F1	AUC	ACC	F1	AUC	ACC	F1	AUC
KNN	0.772	0.778	0.772	0.769	0.346	0.624	0.821	0.689	0.787
SVM	0.708	0.686	0.713	0.771	0.325	0.608	0.838	0.714	0.804
NB	0.591	0.442	0.604	0.791	0.275	0.576	0.824	0.619	0.728
DT	0.723	0.728	0.724	0.782	0.300	0.591	0.810	0.658	0.763
AdaB	0.784	0.790	0.785	0.756	0.424	0.700	0.847	0.724	0.808
MLP	0.791	0.795	0.791	0.745	0.420	0.700	0.852	0.736	0.818

the AdaBoost, then KNN classifier. These results are not as high when compared to the work of similar papers, such as that by Chaurasia et al. [5] with accuracies over 90%, and F1-measures and AUC scores over 0.9. However, the models used within this paper work with a much larger dataset than many others, so it is understandable that producing such high scores might be difficult to achieve, especially when using a relatively small window of time.

The day 16 model, however, does not perform as well when predicting which participants will stay in their study by day 84. The accuracies from all classifiers are acceptably high, never dropping below 74%, however, each classifier is paired with unacceptably low F1-measures and AUC scores. Whilst the AdaBoost and Neutral Network classifiers do have an AUC score of 0.7, which is considered acceptable by the criteria laid out previously, neither classifier has acceptable F1-measures. As all three metrics are unacceptable, the classifiers cannot be considered a success.

The day 84 model can be seen to give good results when predicting who will stay in beyond day 84. The Neural Network provides the best results across the board, with the best accuracy, F1-measure and AUC score. Its F1-measure is acceptable at 0.736, its AUC can be considered excellent under the criteria outlined at 0.818, and although the accuracy is not ideal, it is still good at 85.2%. This accuracy, whilst not quite as high as that seen in Chaurasia et al. [5], it is better than that seen in S. O'Neill et al. [1] at 82.5% accuracy. The KNN, Naïve Bayes and Decision Tree classifiers returned unacceptable F1-measures, each below 0.7. This leaves the SVM, AdaBoost and the Neural Network classifiers as the best for this model. The day 84 model performs better again than the day 16 model for Accuracy and AUC. It could be derived from this that looking at engagement and activity over a longer period of time is more beneficial to predicting which participants will stay in long term.

6   Conclusion and Future Work

Digital health studies are very important and with the negative consequences that occur due to those who drop out, it is crucial to predict who is likely to drop out and when. As such, the models developed in this paper have demonstrated that they are capable of utilising features based on participants' demographics and engagement, and predict with good metrics who will and won't drop out of the studies at days 16 and 84. These models could be implemented into a study or used during a run-in period before a study formally starts.

Whilst the results for the models were good, there is room for better results such as the model accuracies hitting the ideal target of 90% whilst still having acceptable F1 and AUC scores. It is possible that removing those who dropped out after the first day was not enough, and that removing those who dropped out within the first few days may have led to better results. It is hard to know if more features could have been extracted from the engagement data, however, a possible route for future work is a feature based around the time of day that a participant engages with their study.

There is still a lot of room for work to be done with this dataset and these models. With some demographic data needing to be cut due to it not being present in all studies, there is a potential to create models that are specific to studies sharing the same feature set. One major consideration would be to cluster participants similarly to that done in the work of A. Pratap et al. [4] and to make a model for each engagement cluster that predicts when participants from each cluster are likely to drop out.

References

1. O'Neill, S., et al.: Development of a technology adoption and usage prediction tool for assistive technology for people with dementia. Interact. Comput. **26**(2), 169–176 (2013)
2. Ramos, L., Blankers, M., van Wingen, G., de Bruijn, T., Pauws, S., Goudriaan, A.: Predicting success of a digital self-help intervention for alcohol and substance use with machine learning. Front. Psychol. **12**, 734633 (2021)
3. Laursen, D., Paludan-Müller, A., Hróbjartsson, A.: Randomized clinical trials with run-in periods: frequency, characteristics and reporting. Clin. Epidemiol. **11**, 169–184 (2019). https://doi.org/10.2147/clep.s188752
4. Pratap, A., et al.: Indicators of retention in remote digital health studies: a cross-study evaluation of 100,000 participants. npj Digit. Med. **3**(1), 21 (2020). https://doi.org/10.1038/s41746-020-0224-8
5. Chaurasia, P., et al.: Modelling mobile-based technology adoption among people with dementia. Pers. Ubiquitous Comput. **26**(2), 365–384 (2021). https://doi.org/10.1007/s00779-021-01572-x
6. i. Sage Bionetworks, "Synapse | Sage Bionetworks", Synapse.org (2020). https://www.synapse.org/#!Synapse:syn20715364/wiki/596097. Accessed 24 Aug 2022
7. Chawla, N., Bowyer, K., Hall, L., Kegelmeyer, W.: SMOTE: synthetic minority over-sampling technique. J. Artif. Intell. Res. **16**, 321–357 (2002). https://doi.org/10.1613/jair.953
8. Brownlee, J.: https://machinelearningmastery.com/failure-of-accuracy-for-imbalanced-class-distributions/(2020). Accessed 25 Aug 2022
9. Robillard, J., Cleland, I., Hoey, J., Nugent, C.: Ethical adoption: a new imperative in the development of technology for dementia. Alzheimer's Dement. **14**(9), 1104–1113 (2018). https://doi.org/10.1016/j.jalz.2018.04.012
10. Lemeshow, S., Sturdivant, R.X., Hosmer, D.W.: Applied logistic regression. John Wiley & Sons, Incorporated, New York, p. 177 (2013). ProQuest Ebook Central

A Distributed Cognition Approach to Understanding Compensatory Calendaring Cognitive Systems of Older Adults with Mild Cognitive Impairment and Their Care Partners

Tamara Zubatiy[1(✉)], Kayci L. Vickers[2], Jessica L. Saurman[2],
Felicia Goldstein[2], Amy Rodriguez[2], Niharika Mathur[1],
and Elizabeth Mynatt[1,2]

[1] Georgia Institute of Technology, Emory University, Atlanta, Georgia
tzubatiy3@gatech.edu
[2] Northeastern University, Boston, USA

Abstract. While consumer digital calendars are widely used for appointment reminders, they do not fulfill all of the compensatory functions that are supported by calendars designed for cognitive rehabilitation therapies (CRTs). To inform the development of digital compensatory solutions, we employed a Distributed Cognition framework to elucidate how older adults with mild cognitive impairment (MCI) and their care partners manage calendaring details when supported by a traditional rehabilitation calendar. Participants mapped out their calendaring cognitive systems, composed of people and artifacts, completed a chart detailing how they track specific types of information, and shared calendaring strategies with each other. We used a Distributed Cognition framing to articulate information flows and breakdowns in participants' calendaring systems, and we identified groups of participants with similar breakdowns in their calendaring systems. We close by suggesting design recommendations for digital calendaring approaches to support dyads of older adults with MCI and their care partners.

Keywords: Older Adults · Compensatory Technology · Aging in Place · Mild Cognitive Impairment · Compensatory Calendaring

1 Introduction

While consumer digital calendars are widely used for appointment reminders, they do not fulfill all of the compensatory functions in calendars designed for cognitive rehabilitation therapies (CRTs). These compensatory functions include journaling thoughts and feelings and tracking un-timed to-do's, in addition to traditional appointment management. However, as we later describe, even compensatory CRTs such as the Memory Support System (MSS), which includes all the requisite compensatory features, is a month-by-month paper booklet that

Supported by NSF Graduate Research Fellowship.

only a small percentage of people with MCI will come to use. In this work, we explore opportunities to design and develop digital tools to support compensatory calendaring. To inform the development of digital compensatory solutions, we employed a Distributed Cognition (DCOG) framework to elucidate how older adults with mild cognitive impairment (MCI) and their care partners manage calendaring details when supported by a traditional rehabilitation calendar. Participants mapped out their calendaring cognitive systems, composed of people and artifacts, completed a chart detailing how they track specific types of information, and shared calendaring strategies with each other. We analyzed the maps, charts and discussion data to cluster participants who reported similar systems, and identified the potential lapses and failure points of these system types. Based on this analysis, we conclude by recommending design features for future digital compensatory calendaring approaches.

2 Background

We present prior work about the challenges of cognitive decline for older adults with MCI as well as digital calendaring and information sharing solutions for families and people with traumatic brain injury.

2.1 Mild Cognitive Impairment (MCI)

Mild cognitive impairment, MCI, is a neurobehavioral syndrome [12] characterized by subtle cognitive decline in the context of intact functional skills [16]. Although MCI can be caused by a number of underlying causes, it is most frequently diagnosed when there is suspected or confirmed underlying neurodegenerative disease, and it is regarded as a stage between normal aging and dementia. Approximately 14.9% of individuals with MCI will progress to dementia over 2 years (AD Sheets) and 38% will progress to dementia in 5 years [5].

Individuals with MCI experience declines in one or more cognitive functions, including executive functioning (e.g., problem solving, decision making, mental flexibility), changes in attention and stimulus perception, and difficulties with memory. Current estimates indicate MCI affects 16.6% of adults over the age of 65 [12]. To offset these changes, or in some cases allay their progression, it is recommended individuals manage modifiable neuroprotective factors (e.g., exercise, diet, mood) and engage in cognitively stimulating therapies or activities. Cognitive training broadly focuses on two main methods of improving cognition: (1) cognitive remediation, or strengthening weak skills [4], and (2) compensation, or finding new strategies to support weaker functions [14]. There are various types of compensation that is possible [9] including compensatory calendaring solutions [3], the subject of this paper. Given the high rates of progression in individuals with MCI, Huckans et al. recommend compensatory cognitive rehabilitation therapies (CRTs) to support memory function rather than remediation-focused therapies. Past research to understand the effectiveness of such behavioral interventions [7] generally concludes that CRTs are most effective when combined with other behavioral interventions [2], such as regular exercise, social engagement, and mood management.

2.2   Digital Calendaring for Older Adults with MCI

While there are an abundance of digital calendaring services available (ie: Google Calendar, Outlook, Apple Calendar, etc.), people with MCI and more progressive cognitive decline may struggle with these traditional tools due to small screen size and complexity of interactions [1]. McDonald et al. explore the potential of digital calendars for people with acute brain injury, but they do not specifically address specific needs of people with MCI and their care partners, who have to prepare for cognitive decline [8]. Furthermore, a traditional digital calendar is designed for appointment tracking and/or reminding of upcoming appointments. However, as Greenaway et al. point out, older adults with MCI require support beyond just appointments scheduling [3]. They consider calendaring to include all the components in the MSS: time-dependent appointments, to-do lists, and memorable thoughts and moments [3]. Shin et al. has explored the design requirements of digital calendaring systems for older adults with MCI [13]. They present challenges associated with using traditional digital calendars including information input, and describe a potential solution in the form of an augmented pen. They underscore that one of the difficulties is switching between paper and digital, and making sure that both sources reflect the same information. However, Shin et al. also limit their investigation to appointment scheduling, while we focus on all three components of the MSS in this current study. Furthermore, we consider the entire compensatory calendaring system, including the care partner. Neustaedter et al. present their version of a collaborative family digital calendar, targeted at families with children, and not including older adults [11]. They also present the following design requirements: support simple and flexible interaction, be publicly visible and always-on, provide at-a-glance awareness of content, and allow access outside the home [11]. The PenCoder is a great example of supporting simple and flexible interaction [13]. However, it does not facilitate an always-on display and doesn't directly address external access to the calendar. Mennicken et al. build on Neustaedter's ideas of an always-on, whole-family display, and add in the dimension of the smart home, bringing in controls for household systems directly into the digital calendar [9]. Mennicken et al. also focus on younger families with children, and do not work with older adults, or with people who do not have normal cognition [9]. In this work, we attempt to bring together these insights from technologies with the key theories behind the effectiveness of the paper MSS [3] in order to support collaborative digital calendaring for older adults with mild cognitive impairment and their care partners.

2.3   Digital Communication Support Tools for Older Adults and Care Partners

Because compensatory calendaring goes beyond event tracking, we can also draw on broader literature addressing information sharing and communication within families, and specifically between older adults and their care partners. Tee et al.

explored how extended families manage information sharing with the immediate family by encouraging family members to draw out family trees and identify the communications preferences and systems for each member [15]. Tee et al. considered information sharing as broadly as well as photo sharing within the family, while in this study we focus on how specifically scheduling or calendar-related communication occurs [15]. Mynatt et al. present a system for communicating between older adults and their children or other family support systems who do not live with them [10]. They present a set of always-on displays that represent different aspects of the status of the older adult, and communicate that status to the family members in a different home via visual encodings in the display. This research gives us insights into understanding the unique requirements of creating technologies for older adults to use and we build on these in our research. Furthermore, Mynatt et al. address many aspects of family communication including health, environment, relationships, activity and also events [10]. In the current study, we apply these insights to understanding how older adults with MCI and their care partners communicate to manage their calendars, and we use DCOG to elucidate how the information is exchanged between partners and mediated by the artifacts in their calendaring cognitive systems.

2.4 Distributed Cognition (DCOG)

Distributed cognition (DCOG), a framework that comes from Cognitive Science, uses cognitive ethnography in order to understand how information is represented and transformed in a cognitive system [6]. DCOG takes as its unit of analysis entire cognitive systems; these are collaborative systems of people and artifacts that exchange information via internalization and externalization. Internalization involves representing information from the outside world in the brain or internally in an object, while externalization involves taking information out of the brain and representing it externally by writing or speaking, for example [6]. In this paper, we use DCOG to analyze the calendaring cognitive systems of dyads composed of older adults with MCI and their care partners.

3   Methods

In this study, we applied the framework of distributed cognition (DCOG) to identify design considerations for sustainable digital calendaring solutions to support older adults with MCI and their care partners. This work was conducted in the context of a larger empowerment program for older adults with MCI which includes lifestyle interventions, cognitive rehabilitation therapies (CRTs) including the MSS, and research opportunities across domains. We refer to our participants with MCI as program members. We had 18 total participants with MCI (n=18, mean age=75) and 12 of their care partner(n=12, mean age=68). Participants with MCI 13 through 18 were program members who attended the sessions without their care partners, so we do not have inventory or mind map data from these care partners.

Researchers used DCOG to describe how to procedurally use the MSS, how information is internalized into the MSS, and how information is externalized. This review helped us capture how the MSS is effective for people with MCI, with the goal to bring those advantages into digital forms as desired by the MSS users. We also organized bi-weekly, optional "Group" meetings for participants who wanted to start learning more about the digital calendaring techniques, beginning at the "adaptation" phase of training.

3.1 Mind Mapping Exercise

We asked participants to draw how they used paper and digital calendars to remember activities in their daily life, ranging from appointments to journaling past events. We gave them blank pieces of paper, cut out icons for cell phones, paper calendars, the MSS, digital assistants, and wall calendars, as well as a pen and tape. We told all participants to start by drawing themselves and their care partners, then to add their tools, and then to draw arrows to indicate the direction that information flows in those interactions. After the sessions, one researcher analyzed the mind maps using a DCOG framing, and presented the analysis to the research and participants in subsequent meetings to refine the analysis and begin to cluster similar calendaring approaches.

3.2 Compensatory Calendaring Inventory Worksheet

We facilitated a worksheet with participants that was structured to elicit the artifacts and tools participants use to externalize specific types of information including: appointments, to-do lists, daily tasks, partner schedule, daily thoughts, notes, medical information and other. These are the types of information that people with MCI require compensatory systems to support. In addition, appointments, to-do lists, daily tasks, daily thoughts and notes can be considered as components of the MSS [3]. The tools and artifacts represented include: personal day planner, wall calendar, phone, post it notes, notebook or journal, whiteboard, somebody, something else and nothing. One of the program staff developed this worksheet to isolate how people with MCI and their care partners are compensating for keeping track of different information streams, and to identify if there is a primary source of truth for each dyad.

We conducted these inventories after the mind-mapping exercise so as not to bias the participants into thinking about their mind-maps in such a structured way. We then used the chart data to bolster our DCOG analysis, support our discussions, and assist us in our analysis of common lapses in these dyadic calendaring systems.

3.3 Group Discussion Session

Finally, after everybody had completed their individual inventories, we led focus group discussions with our participants, with groups of about 20 people / 10 dyads. We discussed the DCOG analysis with participants, walked them through their own cognitive systems, and identified what breakdowns they experienced in their calendaring systems. First, we encouraged participants to present summaries of their calendaring systems. Then the researcher would ask guiding questions that the group would discuss such as: "Please describe your strategy for remembering what you need to get done throughout the day", which refers to one of the information types the chart is tracking. We also asked participants with MCI to reflect on the division of labor in their household associated with planning for and executing calendaring. Finally, we ideated together with the group about potential solutions that they would consider helpful.

4   Results

4.1   Analysis of Compensatory Calendaring Inventory Worksheets

We gathered 30 inventory chart data from 18 people with MCI (PwMCI) and 12 of their respective care partners.

Participants with MCI preferred to track their appointments in phone calendars or wall calendars, closely followed by personal paper calendars, in this case the MSS. Notably, participants with MCI self-reported that they did not rely on others to track this information, while care partners indicated they worked together with their partners with MCI. 15 of 18 participants with MCI reported using a single tool for recording four or more of the 7 information types, and 8 out of 18 participants with MCI tracked more than half of the information types using just two different tools. One person with MCI reported using 5 tools to simultaneously track 4 different information types.

In contrast, care partners strongly preferred the digital phone calendar for appointments, and four of the other information types: daily tasks, partner schedule, medical info, and notes. Seven of the 12 care partners relied on their phones more than any of the other tools. Among care partners, medical information was least often recorded using these tools.

4.2   DCOG Analysis of Dyads' Calendaring Systems

We collected 21 mind maps from program members and 10 from care partners. We analyzed the cognitive-system mind maps that participants created using several approaches. Because of the diversity of the compositions of these mind maps, certain types of direct comparisons were not possible. Here we present three canonical examples of calendaring cognitive systems in which participants report on having single and multiple ground truth references, and we also share examples of mind maps in which the person with MCI and her/his care partner are not telling the same story.

The Power and Pitfalls of Simplicity. One calendaring cognitive system example stood out among the rest as the system with the fewest components. In this case, M15, a participant with MCI, externalizes everything to her MSS. As a result, the MSS can externalize everything back to her. However, this system is not without its failure points. For example, what would happen if this participant forgot her MSS at home and did not have it, or if she did not transfer notes from month to month. Furthermore, the participant with MCI is the only person represented here, and this system does not take her care partner, or any larger care network, into account. This may make planning and coordination challenging. [3].

"Double Check." "I Didn't Realize This Was so Complicated". On the opposite side of the simplicity spectrum, we have some complicated calendaring cognitive systems in the mix. Some participants, including M18, M5 and M6 reported needing to double-check certain information that they had externalized across multiple tools. Double-checking can turn a five-step system into a ten-step system.

M18's complicated compensatory calendaring system composed of two humans, a computer, a wall calendar, two cell phones and an MSS. M18 exchanges calendaring information by sending invitations via digital calendars over email. However, M18 shares that there is ambiguity in the system, and that he has to double check between what is on the cell phone, what is on the computer, and what is on the wall. The mind-map indicates that his care partner is responsible for entering all the information onto the wall calendar and cell phone calendar, and it looks like the computer is externalizing information to the MSS, which is not possible.

In this next example, M6, a participant with MCI and his care partner, CP6, present their web of tools including post-it notes, a conversational agent, a wall calendar, a whiteboard, phones and M6's MSS. Notably, the pair's diagrams are very similar, indicating that they both know their systems well.

Finally, consider the case study of M5, a participant with MCI, and CP5, her care partner. CP5 is a busy working professional. The pair's calendaring cognitive systems include three humans: each other and CP5's assistant, in addition to 2 phones, a wall calendar, a personal calendar and an iPad. In this case, the phone is the ground truth, not an MSS, but the assistant must be diligent to maintain and upkeep the system to ensure that it is always up to date with the latest changes from both partners. Separately from the phone, M5 also maintains a wall calendar at home with general family dates and activities. The assistant cannot access this home calendar, so there is a risk that there are conflicts between the information represented in the wall calendar and the information on the phone, potentially causing confusion.

One System, Two Different Stories. Looking at these calendaring systems through the lens of distributed cognition allows us to articulate the differences between the perceived information flows of participants with MCI and their care partners. We encountered at least three sets of mind maps M11/cP11, M3/CP3, M1/CP1 with conflicting narratives about the information flow in their homes. Across the three, the mind maps drawn from the point of view of participants with MCI tended to be more egocentric, and focused on the person with MCI, largely out of context of the bigger system. In the cases when the member presented a bigger system, they presented themselves as the dominant actor in the system, which was not consistent with their care partners' mind maps. Furthermore, they under-represent the amount of work their care partners must be doing to compensate, keep the systems in sync, and propagate information through the system. This finding echoes the discovery from the calendar inventory analysis, which found that program members perceived themselves as being more independent than they really are. This observation reveals that either members are unaware of this additional compensation labor their care partners put in, or they don't recognize it as much as they perhaps should.

5   Discussion

Compensatory calendaring strategies make it possible for people with MCI and their care partners to reliably track the key details needed to manage their schedules. However, as we have seen, it is tough to set up a complete system that can be 100% dependable as an external memory source. As a result, many dyads have a patchwork of solutions, creating still more opportunities for error. Furthermore, as the partner with MCI's cognitive abilities decline, the system will increasingly need to compensate for this decline, and the care partner will often step in to provide support. We discuss common lapses in compensatory calendaring systems across participants, explore the tensions resulting from care partners double-compensating for failing systems, and suggest design criteria for digital compensatory calendaring solutions.

5.1   Key Lapses Identified

We identified several common failure points in participants' compensatory calendaring systems that point to opportunities for future digital and hybrid approaches. The first key failure point was the challenge of having too many possible sources for there to be a ground truth. One sub-case of this is the disconnect between paper and digital tracking tools reported by pairs 18, 11, 5, 6, 3 and 1. A second critical theme challenge was lack care partner centralization. Whether acknowledged by the pwMCI or not, based on the data it was care partners responsible for most of the externalization and internalization of information in the home. Because decline is often gradual, sometimes care partners start internalizing one small thing at a time, and then suddenly find themselves responsible for managing both her own and her partner's schedules. Such behavior is considered ill-advised by clinicians, because it has been shown to negatively affect care partners' mental health and increase caregiver burden [3].

A third lapse was that the neither pwMCI nor care partners tended to track thoughts and feelings that were not immediate actions or to-dos, despite this being a key part of the MSS that many participants claimed to use. Finally, the systems that we considered were all static and did not take into consideration the progressive nature of cognitive decline. This potentially makes participants vulnerable to breakdowns in their information management systems that they are not prepared for. The whole idea of setting up compensatory systems is to externalize all the information one wants to remember so the system can provide reliable support in the face of continued cognitive decline. However, the complex systems we have discovered may not actually hold up during decline as well a as a more simplified system.

5.2   Design Criteria for Digital Compensatory Interventions

Based on the lapses and population-specific characteristics identified, the first design criteria for digital compensatory calendaring solutions is that they must address both people with MCI and their care partners/care network. The system must be accessible to both participants, and may have variable configurations in order to ensure it is usable by all parties. The second criteria is that such solutions should address the digital - paper divide directly, and design a way to keep the two in sync, minimizing manual labor. In terms of DCOG, the tools must be able to internalize and externalize information amongst themselves and to always keep a synchronized externalized display available for the user with MCI. The system should not only support user queries, but also be proactive, engaging the user, and reducing the cognitive load required to reminisce and plan for the future. Finally, the system should be aware of and responsive to potential errors including: mis-entry of event details, changes in events, and cancellations.

Address both People with MCI and Care Partners. While there has been research about supporting families calendaring and managing a shared calendaring system, literature about compensatory calendaring focuses primarily on people with MCI, and often not on their whole system. Care partners are warned that they may need to start compensating for failing systems, but care partners' schedules and existing systems are largely not a part of the conversation. We propose that future design approaches in this space focus on the intersection of the patient's calendaring needs and her/his care network's needs. To this end, the system must be accessible to both people with MCI and their care partners/care network. In order to accomplish this integration, the system must have variable access controls that can be adjusted alongside cognitive decline. These access controls will, for example, allow the care partner to turn off the person with MCI's ability to edit events if they have been making mistakes, or conversely give them access if they have been more independent. Over time, the needs of a person with MCI may change dramatically, so it is important to have a system that can be responsive to the needs of declining cognition.

Address the Digital-Paper Divide. As we saw in the inventory data, there is a disconnect between the paper MSS systems that people with MCI are trained to use in order to compensate for their declining cognition, and the more traditional digital calendaring tools their care partners use such as Google or Outlook Calendar. Beyond having been designed for different functions: digital calendars focused on scheduling while MSS books include journal sections and to-dos, the paper and digital media are difficult to keep in sync and can require considerable manual upkeep. This maintenance is also a potential source of human input error that can confuse people with MCI who may not be able to recognize the error.

Shin et al. designed PenCoder to both address the paper to digital gap by automatically digitizing all things written using a custom hardware pen, and to tackle the query gap by having all of the written information available to search digitally [13]. The conceptual system can analyze when information is written about a calendar appointment and add it to the digital calendar automatically, for example. Such a system could intelligently digitize all three sections of the MSS, and could directly integrate those sections into common digital applications such as calendars and notes.

We also draw on literature about how blind people reminisce, without the ability to look back at photos or videos. Yoo et al. present an audio system to support blind people in capturing memories with their loved ones, and then being able to query and search for memories to replay again and again [17]. However, both of these studies assume that a participant with MCI will always remember how to query and be aware of what to query. As a minimum we suggest that any digital capture of compensatory calendaring information come with an always on display visible to all parties [9]. However, an ideal system would go beyond this physical presence and also aim to be proactive in capturing memories for later reflection.

A Proactive Assistant. Managing the calendars and schedules of even just two people can require considerable work. As we saw with M5 and CP5, some even utilize an assistant, whose job it is to keep up with the partners' calendars and ensure each of them is up to date. When getting a human assistant is impractical or impossible, care partners often step in maintain the system. However, any human assistant, even care partners and professionals, are prone to human error. Ideally, a digital assistant could offset this compensatory burden from people with MCI and their care partners. Such a digital assistant would function like M5 and CP5's human assistant: internalize each party's schedule, reconcile them, and then ensure all parties can access the updated and current schedule. This system should be robust in terms of error-handling, and should be able to learn the routines of its users in order to predict when events have been entered correctly. It should also notify its users when it makes any changes, to ensure that all parties have received the updates. Finally, such a digital system should be able to faithfully represent the same information across multiple form factors including mobile phone, tablet or desktop. Beyond simply visualizing the "day" or week, in terms of appointments, to-dos and thoughts/notes, and ideal system would also analyze the information encoded to put together "memories". The system would present these memories to people with MCI when they are in the mood to reminisce. Currently, therapists recommend that people with MCI review their notes in the MSS, however so few participants reported tracking their daily thoughts and memories. Such a system would do the work of aggregating memorable moments and present them to users so they can access them again and again.

6   Conclusion

While people with MCI and their care partners have unique calendaring needs, future digital compensatory calendaring systems may decrease burdensome care partner calendar maintenance and may enhance quality of life for people with MCI. Unfortunately, traditional digital calendaring tools do not consider all of the necessary compensatory components. In this study we asked people with MCI and their care partners to map out and describe the compensatory calendaring systems they have built together to compensate for declining cognition. Using a DCOG framework, we described these compensatory calendaring systems as well as the functionality of a common cognitive rehabilitation therapy (CRT), the Memory Support System (MSS). We identified key lapses present across the participants, as well as key ways that current digital calendars do not easily fulfill all of the requisite compensatory functions. Some lapses we discovered include: a disconnect in the system between program members' self-compensating using the MSS and their care partners double compensating using their phones; care partners acting as a bottleneck for new information entering the system; having too many ground truth sources; and not tracking thoughts and feelings. We triangulated the insights from our mapping and inventory activities, along with feedback from collective discussions with people with MCI and their care partners to understand the design requirements for future hybrid digital/physical approaches. We then presented our design criteria which call for future digital compensatory calendaring solutions to address both people with MCI and their care networks, to address the digital paper divide, and to be proactive in initiating interactions with users.

Acknowledgements. This material is based upon work supported by the National Science Foundation Graduate Research Fellowship under Grant No. 2021280804.

References

1. Baldauf, M., Bösch, R., Frei, C., Hautle, F., Jenny, M.: Exploring requirements and opportunities of conversational user interfaces for the cognitively impaired. In: Proceedings of the 20th International Conference on Human-computer Interaction with Mobile Devices and Services Adjunct, pp. 119–126 (2018)
2. Chandler, M.J., et al.: Comparative effectiveness of behavioral interventions on quality of life for older adults with mild cognitive impairment: a randomized clinical trial. JAMA Netw. Open **2**(5), e193016–e193016 (2019)

3. Greenaway, M., Duncan, N., Smith, G.: The memory support system for mild cognitive impairment: randomized trial of a cognitive rehabilitation intervention. Int. J. Geriatr. Psychiatry **28**(4), 402–409 (2013)
4. Hampstead, B.M., Gillis, M.M., Stringer, A.Y.: Cognitive rehabilitation of memory for mild cognitive impairment: a methodological review and model for future research. J. Int. Neuropsychol. Soc. **20**(2), 135–151 (2014)
5. Heister, D., Brewer, J.B., Magda, S., Blennow, K., McEvoy, L.K., Initiative, A.D.N., et al.: Predicting MCI outcome with clinically available MRI and CSF biomarkers. Neurology **77**(17), 1619–1628 (2011)
6. Hollan, J., Hutchins, E., Kirsh, D.: Distributed cognition: toward a new foundation for human-computer interaction research. ACM Trans. Comput.-Human Interact. (TOCHI) **7**(2), 174–196 (2000)
7. Huckans, M., Hutson, L., Twamley, E., Jak, A., Kaye, J., Storzbach, D.: Efficacy of cognitive rehabilitation therapies for mild cognitive impairment (mci) in older adults: working toward a theoretical model and evidence-based interventions. Neuropsychol. Rev. **23**(1), 63–80 (2013)
8. McDonald, A., Haslam, C., Yates, P., Gurr, B., Leeder, G., Sayers, A.: Google calendar: a new memory aid to compensate for prospective memory deficits following acquired brain injury. Neuropsychol. Rehabil. **21**(6), 784–807 (2011)
9. Mennicken, S., Kim, D., Huang, E.M.: Integrating the smart home into the digital calendar. In: Proceedings of the 2016 CHI Conference on Human Factors in Computing Systems, pp. 5958–5969 (2016)
10. Mynatt, E.D., Rowan, J., Craighill, S., Jacobs, A.: Digital family portraits: supporting peace of mind for extended family members. In: Proceedings of the SIGCHI Conference on Human Factors in Computing Systems, pp. 333–340 (2001)
11. Neustaedter, C., Brush, A.B., Greenberg, S.: A digital family calendar in the home: lessons from field trials of LINC. In: Proceedings of Graphics interface 2007, pp. 199–20 (2007)
12. Petersen, R.C., et al.: Practice guideline update summary: mild cognitive impairment: report of the guideline development, dissemination, and implementation subcommittee of the american academy of neurology. Neurology **90**(3), 126–135 (2018)
13. Shin, Y., Barankevich, R., Lee, J., Kalantari, S.: Pencoder: design for prospective memory and older adults. In: Extended Abstracts of the 2021 CHI Conference on Human Factors in Computing Systems, pp. 1–7 (2021)
14. Sohlberg, M.M., Mateer, C.A.: Training use of compensatory memory books: a three stage behavioral approach. J. Clin. Exp. Neuropsychol. **11**(6), 871–891 (1989)
15. Tee, K., Brush, A.B., Inkpen, K.M.: Exploring communication and sharing between extended families. Int. J. Hum Comput Stud. **67**(2), 128–138 (2009)
16. Wiley, J., et al.: 2021 Alzheimer's disease facts and figures. Alzheimer's Dementia **17**(3), 327–406 (2021)
17. Yoo, M., Odom, W., Berger, A.: Understanding everyday experiences of reminiscence for people with blindness: Practices, tensions and probing new design possibilities. In: Proceedings of the 2021 CHI Conference on Human Factors in Computing Systems, pp. 1–15 (2021)

Using Generative AI to Assist with Technology Adoption Assessment

Chris Nugent[1]([⊠]) [iD], Ian Cleland[1] [iD], Luke Nugent[1], Macarena Espinilla Estevez[2] [iD],
Alicia Montoro Lendinez[2] [iD], David Craig[1,3], Francesco Agnoloni[4],
and Elena Tamburini[4]

[1] School of Computing, Ulster University, Belfast, Northern Ireland
cd.nugent@ulster.ac.uk
[2] Department of Computer Science, University of Jaén, 23071 Jaén, Spain
[3] Southern Health and Social Care Trust, Craigavon, Northern Ireland
[4] Medea S.r.l., Florence, Italy

Abstract. Adoption of innovative solutions in healthcare remains a challenge and is a contributing factor to the barriers of their large scale uptake in both private and public healthcare settings. Traditionally, the study of technology adoption has been limited to considering the patient's perspective, however, there is now an increasing appreciation that this should be expanded to consider adoption implications from a carer's perspective in addition to healthcare professionals and indeed on a larger scale, from a healthcare service provider's perspective. In this work we attempt to establish a proof of concept framework whereby technology adoption of innovative healthcare solutions can be built using generative AI. By considering established and validated clinical questionnaires for the purposes of assessing technology adoption for patients we have created a new suite of questionnaires that can be used for care givers. The approach was evaluated with a set of 28 patient focussed questions. All of the questions produced by the generative AI were deemed to be correct with an average Rouge-1 F1 score of 0.71.

Keywords: Technology Adoption · Generative AI · Healthcare Innovations

1 Introduction

It is well known that the introduction of new and emerging technologies within the healthcare sector are faced with a number of barriers. One such barrier is related to the issues linked with end user technology adoption. Several models and frameworks have been generated to consider the factors influencing the adoption of new technologies, and accordingly have been shown to assist in forecasting of whether such new technologies will be readily accepted amongst the intended cohort of end users [1]. Research endeavours that focus upon technology adoption commonly seek to recognise, predict

This research has been partially funded by the ARC (Advanced Research and Engineering Centre) project, funded by PwC and Invest Northern Ireland and the AGAPE Project funded through the AAL-Active and Assisted Living Programme.

J. Bravo and G. Urzáiz (Eds.): UCAmI 2023, LNNS 835, pp. 202–207, 2023.
https://doi.org/10.1007/978-3-031-48306-6_20

and define new variables that may have an impact on adoption behaviours. Gaining an understanding of these variables can assist in reshaping the design of the technology solution itself in addition to defining approaches for changing the behaviour and/or views of the user subsequently promoting the introduction, adoption and sustained use of innovative technologies [2].

From a healthcare perspective the focus of studying technology adoption has largely been focussed on the development of adoption models for patients. Nevertheless, when taking a holistic view of the challenges associated with the adoption of innovative healthcare solutions and services, barriers exist across a spectrum of stakeholders ranging from those governing budgets and resources at a municipality level, to healthcare professionals working within primary and secondary care facilities, to formal and informal caregivers and finally to the patients themselves.

To increase the likelihood of adoption it is therefore necessary to consider the full spectrum of stakeholders which subsequently requires the development of multi-stakeholder adoption models. This research aims to answer the following research questions:

1. Can a suite of questionnaires for care givers involved in the adoption of new services and solutions be created based on pre-existing clinically validated adoption models (questionnaire) for patients?
2. Is there an opportunity for generative Artificial Intelligence (AI) in the form of Large Language Models (LLMs) to be used to generate a new suite of technology adoption questionnaires?

2   Background

An understanding of the factors which influence technology adoption can be gained through the use of statistical and machine learning approaches. The aim of these approaches is to generate a prediction measure relating to the likelihood of a user adopting a new technology based on a diverse set of features describing the individual and their surrounding environment. Over the past 2 decades, there have been a relatively small number of studies undertaken which have developed technology adoption models.

The Technology Acceptance Model (TAM) is recognised as one of the most extensively used models within the technology acceptance domain. TAM explains technology adoption with 3 key factors, specifically the perceived usefulness of the technology, its perceived ease of use, and user attitudes towards use [1]. Proposals have been made to re-evaluate the extent of some factors related with TAM as it has been suggested that limited probabilities exist amongst the descriptions and predictions [2].

An extended version of the TAM model, namely the Psychosocial Impact of Assistive Device Scale (PIADS), was generated to incorporate personal factors influencing technology adoption. PIADS is a self-reporting survey with scores distributed amongst three subscales, specifically competency, adaptability and self-esteem to assess the extent of how an assistive technology affects quality of life [3].

The unified theory of acceptance and use of technology (UTAUT), was generated to overcome the limitations of both the TAMS and PIADS models. Specifically, their perceived constraints to consider explanatory behaviour and experimental evaluation [4].

The UTAUT model consists of four key components: effort expectancy, performance expectancy, facilitating conditions and social influence [5].

An extended version of the UTAUT model, UTAUT2, was generated to include a further 3 components: habit, hedonic motivation, and price value [5]. The Senior Technology Acceptance Model (STAM) was generated to assess technology acceptance specifically amongst the older generation due to the emergence of additional complexities, such as anxiety and self-efficacy concerns [6].

Whilst it is evident that the nature of technology adoption models have evolved and now consider a much broader range of features ranging from use contexts to user demographics, they largely only consider a single stakeholder in the complex process of introducing a new technology solution. There is therefore an opportunity to take a multi stakeholder perspective on technology adoption.

3   Experimental Design

The aim of the current work has been to create a framework to produce a suite of adoption questionnaires that are specific to the perspectives of each of the stakeholders in the end-to-end adoption process (Fig. 1). The initial phase has been a proof of concept for the creation of caregiver questionnaires based on validated patient questionnaires. The basis of our work stems from the UTAUT adoption modelling process [7]. The UTAUT model has been frequently applied to user adoption behaviours in a number of health-related studies. Although standardised and well-tested items exist for the TAM and the UTAUT, the way in which these variables are operationalised varies from study to study. Researchers typically choose to change the wording of one or more items or to add new items. As UTAUT was not originally formulated for use in the healthcare context these additions and modifications are generally seen as very justifiable [8].

A questionnaire for patient adoption was produced across 10 categories with a further suite of questions to collect demographic data and details of previous usage of technology. In total 43 questions were defined. The set of questions were subsequently reformulated through a manual process in an effort to translate the patient adoption questionnaire into questions presented from the perspective of a caregiver. This was used as the benchmark against which any automated process would be compared.

3.1   Use of Generative AI to Reformulate Questionnaires

To translate the questions to complement those re-formulated manually we investigated the use of the Large Language Model (LLM) ChatGPT-3.5. This was chosen as it was recognised as having a user friendly interface providing easy access to the LLM whilst also having a good reputation for providing quality responses. The prompt presented in Fig. 1 was used as the basis of reformulating the questions for the caregiver's perspective and was followed by a list of patient focused questions to be converted. The prompt was designed to contain the context for the type of data required. Details were then provided describing the context for the original statement followed by who was now going to be asked the question and who the new subject of the statement would be. The LLM was then told what quality of statement was expected and finally an example was provided.

I am going to provide some statements from a medical adoption questionnaire where the user has been asked to select a number from 1 (Strongly disagree) to 7 (Strongly agree). The statements I am going to provide are directed to a patient for them to answer about themselves. Rewrite these statements but with the perspective of the patient's caregiver answering them on behalf of themselves. Ensure that the generated statements are as concise as the source statements with the same level of recall and precision. Here is an example, original: "Overall, Digital Health is helpful to my life", generated: "Overall, using Digital Health technology is useful to me in managing my patient's healthcare". Convert the following:

Fig. 1. Details of the prompt presented to the LLM to convert the patient questions to those relevant to a caregiver.

3.1.1 Evaluation Metrics

The following evaluation metrics were used to compare the difference between the gold standard (manually re-formulated questions) and the results produced by the LLM:

Character Length Difference: Number of characters in Caregiver$_{Manual}$ minus the number of characters in Caregiver$_{LLM}$.

Word Length Difference: Total number of words in Caregiver$_{Manual}$ minus the total number of words in Caregiver$_{LLM}$.

Number of Same Words: The number of times the same word appears in both the Caregiver$_{Manual}$ and Caregiver$_{LLM}$ responses.

Number of Different Words: The number of words that appear in Caregiver$_{Manual}$ that do not appear Caregiver$_{LLM}$.

In addition we used Rouge-1 Recall, Rouge-1 Precision and Rouge-1 F1 Score.

4   Experimental Results and Discussion

In the first instance, the output from the LLM was compared with the gold standard response for the reformulation of each of the patient questions which was produced and validated by two members of the research team, for a sample of 28 questions. From the sample considered, 100% of the results produced by the LLM were deemed to be correct and had maintained the integrity of the original statement whilst changing who was being asked the question, in addition to the subject of the question.

Half of the questions produced by the LLM (14 from 28) were observed to contain the phrase *"as a caregiver"* somewhere in the reformulated response. This usually occurred in the statements that may be considered as being stakeholder agnostic (Table 1). The net effect of this happening was a less concise response than the manually written responses, however, was deemed as to not having made the statement incorrect. The number of words generated in the LLM response was therefore generally greater than the number of words written in the gold standard with an average of 2.75 more words than the manual response. Approximately 7% of the LLM responses contained fewer words than the manually rewritten response.

Table 1. Comparison of Caregiver$_{Manual}$ and Caregiver$_{LLM}$ in the instance when the original question is considered as being stakeholder agnostic.

Patient	Caregiver$_{Manual}$	Caregiver$_{LLM}$
I can easily learn to use a Digital Health technology	I can easily learn to use a Digital Health technology	I can easily learn to use Digital Health technology *as a caregiver*
I can imagine trying out a Digital Health technology	I can imagine trying out a Digital Health technology	I can imagine trying out a Digital Health technology *as a caregiver*

Results of the comparisons between Caregiver$_{Manual}$ and Caregiver$_{LLM}$ are presented in Table 2. It was found that the LLM created a response with an average of 18.2 more characters than the gold standard. Approximately 10% of the LLM results were shorter than the gold standard. A large majority of the LLM and manual responses contained similar words with the average being 8.57 of the same words within each statement. (The minimum number of same words was 2 and the maximum number of same words was 13.) Within the responses there were similar words being used between both the LLM and manual responses meaning that there were very few instances of different words being used with the average number of different words between the 2 statements being 7.4. The average recall of the responses was 0.64 with the highest recall being 0.86 and the lowest being 0.08. The average precision score was 0.81 with the highest score being 1 and the lowest being 0.14. A large number of the responses had a very high precision score meaning that the data generated by the LLM was extremely accurate and relevant to the manually produced response. The responses had a high F1-Score with the average being 0.71, the highest being 0.92 and the lowest being 0.11.

Table 2. Comparison between Caregiver_LLM and Caregiver_Manual for a set of 28 questions.

	Character Length Difference	Word Length Difference	Number of Same Words	Rouge-1 Recall	Rouge-1 Precision	Rouge-1 F1-Score
Average	−18.21	−2.75	8.57	7.39	0.81	0.71
Max value	6	1	13	24	1	0.92
Min Value	−45	−6	2	2	0.14	0.11
Standard Deviation	12.33	1.89	2.88	5.81	0.25	0.22

5   Conclusions

Within this research we undertaken a proof of concept study to define a suite of adoption questionnaires for caregivers involved in the innovation cycle based on a clinically validated questionnaire used to assess patient adoption. We have extended this concept by using generative AI in the form of an LLM to re-write the patient questions into the perspective of a caregiver.

Our results demonstrated the ability to both manually reformulate the questions and also leverage the power of an LLM. Results from the comparison between manually rewritten and those produced by the LMM showed a high degree of correlation with a Rouge-1 F1 average score of 0.71. This can be interpreted in two different manners. Firstly, it has demonstrated the ability of an LLM to produce new adoption questions. Secondly, the high degree of correlation between the manually rewritten and the LLM approach can, to a certain extent, be considered to have validated both approaches.

Whilst our methodology and results are sound the next important step in our work is to validate the suite of questions developed through in situ assessment and extend to other stakeholders involved in the technology adoption process.

References

1. Taherdoost, H.: A review of technology acceptance and adoption models and theories. Procedia Manuf. **22**, 960–967 (2018)
2. Dube, T., Van Eck, R., Zuva, T.: Review of technology adoption models and theories to measure readiness and acceptable use of technology in a business organization. J. Inf. Technol. Digit. World **02**(04), 207–212 (2020)
3. Martins, A., Pinheiro, J., Farias, B., Jutai, J.: Psychosocial impact of assistive technologies for mobility and their implications for active ageing. Technologies **4**(3), 28 (2016)
4. Chaurasia, P., et al.: Modelling mobile-based technology adoption among people with dementia. Pers. Ubiquitous Comput. **26**(2), 365–384 (2022)
5. Kamal, M., Subriadi, A.P.: UTAUT model of mobile application: literature review. In: Proceedings - IEIT 2021 1st International Conference on Electrical and Information Technology, pp. 120–125 (2021)
6. Chen, K., Lou, V.W.Q.: Measuring senior technology acceptance: development of a Brief, 14-Item Scale. Innov. Aging **4**(3), 1–12 (2020)
7. Liu, Y., Lu, X., Zhao, G., Li, C., Shi, J.: Adoption of mobile health services using the unified theory of acceptance and use of technology model: self-efficacy and privacy concerns. Front. Psychol. **13**, 944976 (2022). https://doi.org/10.3389/fpsyg.2022.944976. PMID: 36033004; PMCID: PMC9403893
8. Rouidi, M., Elouadi, A.E., Hamdoune, A., Choujtani, K., Chati, A.: TAM-UTAUT and the acceptance of remote healthcare technologies by healthcare professionals: a systematic review. Inform. Med. Unlocked **32**, 101008 (2022)

"Why Did You Say That?": Understanding Explainability in Conversational AI Systems for Older Adults with Mild Cognitive Impairment (MCI)

Niharika Mathur[1]($^{(\boxtimes)}$), Tamara Zubatiy[1], Agata Rozga[1], and Elizabeth Mynatt[2]

[1] Georgia Tech, Atlanta, USA
nmathur35@gatech.edu
[2] Northeastern University, Boston, USA

Abstract. As Conversational AI systems evolve, their user base widens to encompass individuals with varying cognitive abilities, including older adults facing cognitive challenges like Mild Cognitive Impairment (MCI). Current systems, like smart speakers, struggle to provide effective explanations for their decisions or responses. This paper argues that the expectations and requirements for AI explanations for older adults with MCI differ significantly from conventional Explainable AI (XAI) research goals. Drawing from our ongoing research involving older adults with MCI and their interactions with the Google Home Hub, we highlight breakdowns in conversational flow when older adults seek explanations. Based on our experience, we conclude with recommendations for HCI researchers to adopt a more human-centered approach as we move towards developing the next generation of AI systems.

Keywords: Conversational AI · Explainable AI · Mild Cognitive Impairment · Older Adults

1 Introduction

In this research, we discuss Conversational AI systems and their explanations in the context of supporting older adults with Mild Cognitive Impairment (MCI), an early stage of cognitive decline. These AI systems use Natural Language Processing and speech recognition to engage in real-time interactions. Notably, commercially available smart speakers, such as Google Home[1] offer natural language support to older adults by providing auditory stimuli, thus minimizing physical device engagement [6]. Older Adults with MCI have also reported feeling "empowered" through their use [8]. However, our focus in this paper is on

---

[1] Retrieved September 24, 2023, from https://home.google.com/welcome/.

J. Bravo and G. Urzáiz (Eds.): UCAmI 2023, LNNS 835, pp. 208–214, 2023.
https://doi.org/10.1007/978-3-031-48306-6_21

their ability to handle more complex, personalized and interpersonal conversations that go beyond basic tasks like information retrieval and entertainment. Complex tasks, such as calendaring and information retrieval, often lead to conversational breakdowns as AI systems struggle to remember and present information effectively. Problems arise when older adults dealing with the onset of changes in their cognitive abilities desire more information or explanations from the AI. This increased need for cognitive support as a result of MCI presents a challenge for the current capabilities of conversational AI systems.

We draw insights from ongoing work with this demographic, analyzing interaction logs and qualitative interviews. We argue that Explainable AI (XAI), which focuses on explaining AI decisions, holds promise in addressing these breakdowns. Our paper also critiques current approaches to designing and evaluating explanations in XAI and recommends a more inclusive approach to XAI development.

2 Background

2.1 Conversational Breakdowns in AI Systems

While Conversational AI systems show potential in assisting in aging in place, there has also been concurrent research that highlights the breakdowns in the interactions that users have with AI systems, thus bringing into question the truly conversational nature of such technologies. In [1], authors Clark et al. describe the current perception of conversational agents as merely task-focused entities, while highlighting the desire to incorporate the "dynamics of bond and trust" in order to make truly conversational agents that have the potential of extending their role from a functional to a social one. As we have observed in our work with older adults so far, current Conversational AI systems fall short of this expectation as interactions with them often break down or are abandoned when prompted for information besides the one that they have been programmed to store and retrieve.

2.2 Role of Explainable AI

Explainability in the context of AI systems is the ability of an AI system to provide reasoning and explanations for its decisions. A common example is when one receives recommendations with the option to know why those recommendations were made. This helps the user in understanding the context behind the recommendations and has shown to have an impact on their overall perception of the system [7]. However, most approaches to explainable AI today tend to be algorithm-centered and focused on models generated by the AI systems. The existence, type and nature of explanations provided to all users is objectively decided by the AI, with little to no understanding of an individual users' cognitive and social requirement for explanations. This highlights the need for a human-centered approach to explainability. In [2], Ehsan et al. advocate for the need for social transparency in explanations by acknowledging the socially situated nature of both the AI and the people interacting with it. Through our work, we also advocate for further extending this social perspective of the user to include an examination of the cognitive perspective of users and its impact on decision making. Two users can be socially situated in the same context, however, their cognitive contexts could be vastly different. The envisioned contribution of our work comes from the need to understand explanations in AI systems in more depth in the context of older adults with MCI and their carepartners who may have different perceptions of the system and have different explainability requirements given the difference in their individual cognitive abilities.

3   Previous Work: Analyzing the Use of Commercial Conversational AI Systems

The context of our argument in this paper is derived from our work described in detail in [4]. To briefly summarize, our aim is to study the long-term usage of a commercial conversational AI system, the Google Home Hub (GHH), for medication management. Our research team designed a medication assistant through exploratory user research with "dyads" (i.e., pairs) of older adults with MCI and their caregivers, who are often their spouses, but also adult children in some cases. Our team recruited the dyads for the study through a cognitive empowerment program at a local healthcare facility. The purpose of the medication assistant is to check-in with the participant at a pre-defined time for their medications and follow a conversational trajectory on the basis of its current technical abilities. As detailed in [4], the team deployed the medication assistant in the GHH of 7 dyads with an average age of 74.5 for patients with MCI and 68.5 for their carepartners for a period of 20 weeks. Throughout this 20 week study period, we collected interaction logs, and conducted qualitative interviews with the dyads at two different points during the study duration. The interaction logs list and categorize the verbal interactions between a user and the GHH in a spreadsheet (each cell is a text interaction). A quantitative analysis of the interaction logs collected over 20 weeks revealed an engagement rate of 67% for all participating dyads. We define the engagement rate as the ratio of the interactions (the "check-in") that the AI initiated at the pre-defined medication time versus the interactions that the user actually responded to (or engaged in a conversation with the AI). We also calculated a weekly engagement rate for 20 weeks for all the participating dyads which showed a steady increase in engagement with the AI as the study progressed, hinting at an overall acceptance of the system. The findings from the qualitative interviews also point towards the fact that in its present form, the system provided feelings of confidence and support to the participants, specially the caregivers, who called it an *"alternate way of monitoring their partner's medication"* (Caregiver quote).

As a result of the work described in [4], we have been positively encouraged by the results hinting at an overall acceptance of the system. Our published analysis so far focuses primarily on interaction design, system usage and engagement. However, inspired from this work and continuing through recent further analysis of interaction logs and additional interviews, not previously published, we have also begun to observe an emerging trend that hints at the expectations and requirements that older adults have from the AI system that currently go unfulfilled and lead to conversational breakdowns. We discuss some of these emerging conversational breakdowns in the next section.

4   Understanding Explainability Expectations for Conversational AI

Some of the observed conversational breakdowns that hint at specific explainability expectations from Conversational AI systems are:

1. Current Conversational AI design is constrained in what it can and cannot do and works primarily on templated interactions. In the context of our work, when prompted by a user with questions such as *"Are you sure I have not taken the medication? I think I have"* (patient quote), the system has no way of presenting an explanation for its decision to clarify user skepticism, instead answering with the same template response with no added information about an individual patient's medication status. This highlights the conversational limitation for the AI in its explanations to a user.
2. In most cases, the caregivers or the patients want to know more about their medication history for personal tracking or sharing medication records with clinicians. One caregiver asked the system *"Can you tell me if he took the medication this morning or no?"*, but was met with no response as the system has no internal knowledge about previous interactions, hinting at its lack of explainability and information storing capabilities.
3. The requirement for more information and explanation from the system varies across users. Some never asked a follow-up question and assumed the AI must be right at all times, while others wanted to know more about their medication data and for the AI to provide more explanation for its responses. For the latter case, a patient describes, *"Can it tell me for sure that I have not taken the medication because I think I have and there is no pill in the pillbox too"*. The system's non-response to user questions like this highlights the expectation for the AI to calibrate responses to a user's cognitive model in order to effectively build trust in it.

4.1   Conversational Breakdowns Through the Lens of Explainable AI

Through analysis of our results so far, we argue that the emergence of conversational breakdowns and user frustration is closely tied to a lack of explainability in Conversational AI systems. Our argument in this paper is rooted in the ways in which current conversational AI systems are inept at working with complex interactions and to contribute to the understanding of how they can be better designed in the future. There is currently a gap in the way that explanations are understood in the context of non-traditional users interacting with conversational AI systems. We further argue that not every explanation offered is a good one, the central concern is whether it serves the purpose of explainability for a specific user in the context of their social and cognitive abilities. We highlight that explainability requirements are different for older adults and we need to take a significantly different, human-centered approach to understanding explainability than what most XAI studies presently do.

Designing explanations for AI systems requires understanding aspects that relate to define what an effective explanation is, when it is delivered and how it can impact a user's mental model [3]. Currently, explanations in XAI studies are largely constructed on machine-generated inputs. The generated explanation in most XAI studies has not been trained on the system's internal knowledge of a user interacting with it. The nature of explanations provided to all users is objectively decided by the AI, with little to no understanding of an individual user's cognitive model. This mismatch in a user's expectation from an explanation and the actual explanation can influence trust in the system [7]. User frustrations resulting from ambiguity in explanations can lead to reduced feelings of autonomy and independence. These are two central concerns for older adults as they age with cognitive impairments [5]. We also highlight that in XAI studies, the generated explanations are largely tested with low-risk tasks that are not representative of real-world scenarios. The role of risk associated with a task has also not been explored so far. Additionally, these evaluations are often performed by standalone non-experts who test the system in isolation and not in collaborative contexts.

4.2   Recommendations for Explainable AI Studies

Concluding our argument so far, we offer three broad recommendations for HCI researchers working towards understanding explanations in the next generation of AI systems.

R1: Explanations in XAI studies need to be generated keeping in mind a user's individual cognitive and social context. Conducting exploratory user research with actual users of the system can be helpful in shifting the algorithmic-centeredness of explanations to the mental model of the user. Generation of different levels of explanations that vary by the level of information contained in them or by the type of information (visual, non-visual) is also important to calibrate explanations with real-world scenarios.

R2: Explanations in XAI studies need to be evaluated in the context of use-inspired scenarios that are representative of real-world situations. A diversity in tasks that includes understanding differing perceptions of risk involved in them is crucial in grounding explainability in more authentic, collaborative and networked contexts rather than in isolation. Here, we highlight that explanations for high-risk tasks such as medication reminders may require a different evaluatory plan than a low-risk such as movie recommendation.

R3: Finally, we offer the recommendation that XAI studies should engage more representative users for performing experiments rather than users that can be conveniently recruited. In this context, evaluations with experts in their life experience of living with MCI has the potential to lead to the generation of more actionable insights for the design of XAI systems in the future.

5   Conclusion and Future Work

In this paper, we have presented an overview of our ongoing work with older adults with MCI and their use of Conversational AI systems. As older adults continue to age, there is a pressing need to develop systems that can engage more robustly in order to truly provide support for aging in place. We argue that this robustness is closely tied to building effective explanations into AI systems that can enhance conversational interactions. We further argue that explainability for older adults requires a human-centered approach that keeps them at the center of the explanation. We conclude this work in progress by offering recommendations to generate cognitively situated explanations, to test them with diverse, risk-aware tasks and with users that truly represent the use context representative of the scenario in question. Our future work focuses on evaluating different levels and types of explanations with older adults and representative users. Through the use user-centered qualitative research, we plan to conduct an in depth analysis of what AI explainability means for diverse users.

References

1. Clark, L., et al.: What makes a good conversation? Challenges in designing truly conversational agents. In: Proceedings of the 2019 CHI Conference on Human Factors in Computing Systems, pp. 1–12 (2019)
2. Ehsan, U., Liao, Q.V., Muller, M., Riedl, M.O., Weisz, J.D.: Expanding explainability: towards social transparency in AI systems. In: Proceedings of the 2021 CHI Conference on Human Factors in Computing Systems, pp. 1–19 (2021)
3. Liao, Q.V., Gruen, D., Miller, S.: Questioning the AI: informing design practices for explainable AI user experiences. In: Proceedings of the 2020 CHI Conference on Human Factors in Computing Systems, pp. 1–15 (2020)
4. Mathur, N., Dhodapkar, K., Zubatiy, T., Li, J., Jones, B., Mynatt, E.: A collaborative approach to support medication management in older adults with mild cognitive impairment using conversational assistants (CAs). In: Proceedings of the 24th International ACM SIGACCESS Conference on Computers and Accessibility, pp. 1–14 (2022)
5. Rogers, W.A., Mitzner, T.L.: Envisioning the future for older adults: autonomy, health, well-being, and social connectedness with technology support. Futures **87**, 133–139 (2017)
6. Sengupta, K., et al.: Challenges and opportunities of leveraging intelligent conversational assistant to improve the well-being of older adults. In: Extended Abstracts of the 2020 CHI Conference on Human Factors in Computing Systems, pp. 1–4 (2020)

7. Shin, D.: The effects of explainability and causability on perception, trust, and acceptance: implications for explainable AI. Int. J. Hum Comput Stud. **146**, 102551 (2021)
8. Zubatiy, T., Vickers, K.L., Mathur, N., Mynatt, E.D.: Empowering dyads of older adults with mild cognitive impairment and their care partners using conversational agents. In: Proceedings of the 2021 CHI Conference on Human Factors in Computing Systems, pp. 1–15 (2021)

Proposal of a Device for Obstacle Detection Applied to Visually Impaired People

Marco Rodríguez[1,2]([✉]) [ID], Lilia Muñoz[1,2] [ID], Vladimir Villarreal[1,2] [ID], and Dimas H. Concepción[1,2] [ID]

[1] Universidad Tecnologíca de Panamá, Chiriquí Panama City, Panamá
{marco.rodriguez1,lilia.munoz,vladimir.villarreal,
dimas.concepcion}@utp.ac.pa
[2] Grupo de Investigación en Tecnologías Computacionales Emergentes (GITCE), Panama City, Panama

Abstract. Visual impairment poses significant and varied challenges in the daily lives of those affected. It is estimated that around 2.2 billion people worldwide have some form of visual impairment, with causes ranging from eye accidents to congenital conditions and acquired diseases. Assistive technologies (AT) therefore play a vital role in improving their quality of life, enabling them to lead active lives and participate fully in society. However, the adoption of these technologies is often hampered by factors such as the lack of public policy and the high costs associated with the development of such technologies. To address these challenges, the following article proposes the development of a low-cost device based on the use of printed circuit boards (PCBs) to assist people with visual impairment (PwVI). The device will incorporate an Ultrasonic Distance Sensor (UDS) that will trigger an audible alert when it detects an obstacle, allowing the PwVI to perceive and avoid potential risks.

The proposal is based on previous studies related to AT and aims to improve the independence and social inclusion of PwVI, especially in contexts where economic resources are limited. The implementation of this device is expected to overcome existing barriers and provide an affordable and effective solution to improve the quality of life of PwVI.

Keywords: Assistive technologies · Electronics · Obstacle detection · Ultrasonic distance sensor · Visual impairment

1 Introduction

Visual impairment, resulting from partial or total loss of vision, is a significant limitation in a person's life. This condition can be the result of a variety of factors, including eye accidents, congenital conditions and acquired diseases, underlining the complexity and diversity of its possible causes [1]. Thus, according to the World Health Organization (WHO), there are at least 2.2 billion people with some form of visual impairment [2]. In the same context, the Pan American Health Organization (PAHO) states that visual impairment has a high prevalence among eye diseases. Among these, uncorrected refractive errors and cataracts are the most prevalent and are most commonly found in people over the age of 50 [3]. On the other hand, in Panama, two studies have been carried out to know the situation of people with disabilities. Starting with the First National Survey on Disability (PENDIS), conducted in 2006, which presents the national results regarding the rate of people with disabilities, reaching a prevalence of 11.3% of the total population, that is, 370,053 Panamanians diagnosed with a disability. This means that according to the distribution of people categorised by age groups, it is determined that 11.1% have some type of limitation, where PwVI occupies the second position with a prevalence of 2.6% between the ages of 40 and 44, due to causes related to the organic ageing cycle [4]. Subsequently, during the 2010 census, a study was carried out on the situation of people with disabilities in Panama, with the aim of identifying the situation of this sector of the population in the Republic, which represents 2.9% of the total population, of which blindness is the second most prevalent (22.0%) [5]. In a different order of ideas, we can mention that AT are implemented with the aim of improving the abilities of people with disabilities, enabling them to lead proactive lives, increasing their participation in education and the labour market, making our society a more equal and diverse place where everyone, regardless of their physical or mental abilities, has the opportunity to reach their full potential and contribute to the common good of society [6, 7]. Therefore, the effective development of AT is limited by factors such as the absence of public policies aimed at this sector, as well as the high costs involved in developing this type of technology, the individual and collective benefits of improving access to AT are essential, since their integration contributes to fostering a more participatory population, allowing them to fully contribute to society [8]. Considering the above, the following article will present a proposal for a AT targeting PwVI, organised as follows: Sect. 1 introduction, Sect. 2 related works, Sect. 3 objective of the study, Sect. 4 study proposal, Sect. 5 conclusions and future work are presented.

2   Related Works

The increase in the number of visually impaired individuals necessitates a significant increase in the provision of AT to provide more accessible options for this population. While there have been recent advances in AT, there is still a need for affordable, effective solutions that improve the mobility of the visually impaired [9]. In this sense, some studies related to the topic addressed for the proposal will be presented below. The first one covers the development of glasses as an AT tool, through a Raspberry Pi 4, to address the condition of visual impairment. It establishes the development of an intelligent system that allows the identification of people and objects through a Pi camera, with audio feedback to notify the user of any obstacle [10]. CAPture, on the other hand, is the development of a visual assistance cap, based on detecting everyday objects, family members and social helpers (police, doctors, among others), based on the implementation of the Internet of Things (IoT), deep learning and text-to-speech and image processing [11]. Similarly, in the study developed by Yassine Bouterra [12], a portable assistive device is implemented that integrates a fuzzy decision support system based on the use of sensors and processing boards running in real time. This system provides decision support and a user interface, using the generated data as input to provide safety guidance. Decision information is communicated to the user through an interface that combines voice and haptic feedback. Meanwhile, in the Smart Electronic Assistive Device for Visually Impaired Individual through Image Processing, the development of a smart electronic device for the PwVI is proposed. This device operates through glasses that handle text detection and recognition. It includes a smart cane that performs obstacle detection through combined sensors and image processing [13]. Finally, we can mention the proposal for the design of electronic systems and communication for the mobility of PwVI inside buildings, where applications are developed in Android and iOS operating systems, with a design of radio frequency modules that have the purpose of providing support for mobility in indoor environments [14]. In this way, it becomes clear how current advances in the field of AT help to overcome existing barriers and provide significant support to optimize the mobility of PwVI.

Consequently, the studies previously addressed will be used as key references to inform, guide and establish a starting point in the development of the proposal that will be discussed in the following section.

3   Objective of the Study

The use of AT aims to implement technological solutions that enhance the independence and promote the inclusion of PwVI. Although the effective adoption and integration of these technologies can present challenges, they also offer significant opportunities to overcome barriers, promote inclusion, and ultimately improve the quality of life of PwVI [15].

Additional challenges arise when visual impairment is combined with socioeconomic deprivation. These circumstances directly affect the ability to acquire such assistive devices. Limiting communication, access to information, and active participation in society for PwVI [16]. It is important to be aware of these inequalities and to take this factor as a reference in the development of ATs, considering the specific sector they are aimed at.

In this context, we are trying to develop a device with a printed circuit board that will facilitate the connections between the electronic components. This device will incorporate a UDS that, upon detecting an obstacle, will emit an audible alarm to inform the PwVI of the relative distance of the obstacle.

Thus, the objective of this proposal is to create a low-cost device that can efficiently identify objects, whether fixed or mobile. This will make it possible to address specific challenges more effectively and provide affordable solutions that serve multiple purposes and bring significant benefits to PwVI, especially in contexts where economic resources are limited.

4   Study Proposal

According to the report of the WHO and the United Nations Children's Fund (UNICEF) it is considered that 1 billion people with disabilities and older adults are denied access to AT, this is a growing problem specifically in middle and low income countries, where the economic factor is an important cause for timely access to these technologies [17]. With respect to this, it is proposed to develop a PCB device, in which all the electronic components will be assembled in a unified way, contributing to the reduction of space, allowing to establish a controlled connection and provide stability in addition to contemplate the physical protection, which by means of a housing will allow to protect the internal electronic components of the device, besides contemplating slots to allow the access to ports and buttons. The device will be segmented according to the operability of each of the PCBs, dividing them as follows:

1. Power PCB: It will oversee providing the current and voltage necessary for the electronic components to perform their functions.
2. Processing PCB: It will oversee operating and manipulating the specific instructions assigned to the electronic components.
3. Control PCB: Operates the system to turn the device on and off.

After establishing the operational functions, the use of an UDS integrated into the control PCB is incorporated. This sensor has superior operational features in terms of efficiency compared to conventional ones, allowing for accurate and consistent determination of programmed distances, and establishing a wider range of precision capture, efficiently determining objects with symmetric characteristics by examining size and position. This will allow the identification of two types of objects, symmetric and asymmetric. The implemented UDS model can better identify irregular objects, obtaining more suitable measurements within the operating range, operating under the concept of spatial location, where the relative position of an object is determined, describing the physical location in relation to other elements or reference points. This makes it essential for understanding and analyzing the physical environment surrounding the PwVI.

As shown in Fig. 1, we will observe the concept of the device implemented in the upper transverse plane of the human body.

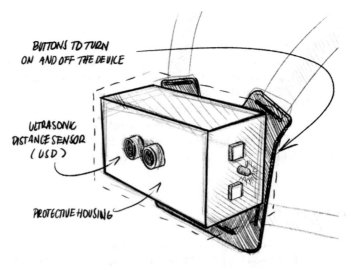

Fig. 1. PwVI device concept.

5 Conclusions and Future Work

Considering that individuals with visual impairment face various problems regarding timely access to information and mobility aspects, it is appropriate to promote equal opportunities to ensure their full integration. For this reason, as we observed in the section on related studies, AT have proven to be a significant resource in enhancing the adaptive capabilities of PwVI, allowing them to overcome their limitations and facilitating proactive participation in society. As a result, this proposal will develop a low-cost device, built on a PCB, to facilitate the integration of the electronic components that make it up, including the use of a distinctive and efficient UDS, for obstacle detection, providing greater autonomy to PwVI.

Acknowledgments. This project is funded by the National Secretariat of Science, Technology and Innovation (SENACYT), under merit-based contract No. 130–2022, supported by the Public Call for New Researchers 2021. Lilia Muñoz and Vladimir Villarreal are members of the National Research System (SNI). Dimas H. Concepción is supported by a grant from the SENACYT National Postgraduate Strengthening Program.

References

1. Dale, N., Sakkalou, E., O'Reilly, M., Springall, C., De Haan, M. Salt, A.: Functional vision and cognition in infants with congenital disorders of the peripheral visual system, Dev. Med. Child Neurol. **59**(7), 725–731 (2017). https://doi.org/10.1111/dmcn.13429
2. Vision impairment and blindness, Blindness and vision impairment, octubre de 2022. Accedido: 28 de junio de 2023. [En línea]. Disponible en: https://bit.ly/44DHq5l
3. Visual Health - PAHO/WHO | Pan American Health Organization. Accedido: 28 de junio de 2023. [En línea]. Disponible en: https://bit.ly/3D7nbBC
4. Grupo para la Educación y Manejo Ambiental Sostenible, Ministerio de Economía y Finanzas Secretaría Ejecutiva del Fondo de Preinversión, Ministerio de la Presidencia de la República de Panamá, y Secretaría Nacional para la Integración Social de las Personas con Discapacidad, Estudio sobre la prevalencia y caracterización de la discapacidad en la República de Panamá. octubre de 2006. Accedido: 19 de junio de 2023. [En línea]. Disponible en: https://bit.ly/3CC6CNQ
5. Ministerio de Economía y Finanzas, Sistema Integrado de Indicadores para el Desarrollo, y J. A. Rodríguez, Guerra, Atlas Social de Panamá, Situación de las personas con Discapaci-dad en Panamá. 2010. Accedido: 19 de junio de 2023. [En línea]. Disponible en: https://bit.ly/3NAZIP9
6. World Health Organization. Regional Office for the Eastern Mediterranean, Strategic action framework to improve access to assistive technology in the Eastern Mediterranean Region. World Health Organization. Regional Office for the Eastern Mediterranean, 2022. Accedido: 27 de junio de 2023. [En línea]. Disponible en: https://bit.ly/3pGgvXX
7. Andrich, R.: Towards a Global Information Network on Assistive Technology. In: 2020 International Conference on Assistive and Rehabilitation Technologies (iCareTech), ago, pp. 1–. (2020). https://doi.org/10.1109/iCareTech49914.2020.00009
8. Policy brief: Access to assistive technology. Accedido 27 de junio de 2023. [En línea]. Disponible en: https://bit.ly/3D60KN3
9. Manjari, K., Verma, M., Singal, G.: A survey on assistive technology for visually impaired. Internet Things **11**, 100188 (2020). https://doi.org/10.1016/j.iot.2020.100188

10. Mustafa, A., Omer, A., Mohammed, O.: Intelligent Glasses for Visually Impaired People. In: 2022 14th International Conference on Computational Intelligence and Communication Networks (CICN), dic, pp. 29–33 (2022). https://doi.org/10.1109/CICN56167.2022.100 08291

11. Bhati, N., Samsani, V.C., Khareta, R., Vashisth, T., Sharma, S., Sugumaran, V.: CAPture: A Vision Assistive Cap for People with Visual Impairment. In: 2021 8th International Conference on Signal Processing and Integrated Networks (SPIN), ago, pp. 692–697 (2021). https://doi.org/10.1109/SPIN52536.2021.9565940

12. Bouteraa, Y.: Design and development of a wearable assistive device integrating a fuzzy decision support system for blind and visually impaired people, Micromachines, 12(9), 1082 (2021) https://doi.org/10.3390/mi12091082

13. Flores, I., Lacdang, G.C., Undangan, C., Adtoon, J., Linsangan, N.B.: Smart Electronic Assistive Device for Visually Impaired Individual through Image Processing. In: 2021 IEEE 13th International Conference on Humanoid, Nanotechnology, Information Technol-ogy, Communication and Control, Environment, and Management (HNICEM), nov, pp. 1–6 (2021). https://doi.org/10.1109/HNICEM54116.2021.9731961

14. Montes, H., et al.: Conceptual Design of Technological Systems for the Mobility of Visual Impairment People in Indoor Buildings, oct, pp. 647–652 (2019). https://doi.org/10.1109/IES TEC46403.2019.00121

15. Global Report on Assistive Technology. Accedido: 28 de junio de 2023. [En línea]. Disponible en: https://bit.ly/3pAa1tM

16. de Witte, L.: Assistive Technology Provision in Low Resource Settings: Challenges and Opportunities. In: 2020 International Conference on Assistive and Rehabilitation Technologies (iCareTech), ago, pp. 16–18 (2020). https://doi.org/10.1109/iCareTech49914.2020.00038

17. Almost one billion children and adults with disabilities and older persons in need of assis-tive technology denied access, according to new report. Accedido: 4 de julio de 2023. [En línea]. Disponible en: https://bit.ly/3XJHGxw

Transforming Elderly Care Through Ethical and Social Evaluation of Intelligent Activity Recognition Systems in Nursing Homes

Alicia Montoro Lendínez[1]([✉]) [ID], Carmen Linares[1], Ana Perandres[3],
Alfonso Cruz[3], José Luis López Ruiz[1] [ID], Chris Nugent[2] [ID],
and Macarena Espinilla[1] [ID]

[1] Department of Computer Science, University of Jaén, Jaén, Spain
amlendin@ujaen.es
[2] Department of Computer Science, Ulster University, Belfast, UK
[3] Ageing Lab Foundation, Málaga, Spain

Abstract. With the continuous increase in life expectancy, the demand for nursing home professionals to care for the older adults is growing. To address this challenge, smart systems, including intelligent activity recognition systems, have been developed to monitor individuals and assist both the older adults and their caregivers. However, the ethical evaluation of these systems using frameworks such as the dignified and positive aging (DPA) model is often overlooked. This article presents a social evaluation of the ACTIVA system. The ACTIVA system is an intelligent multiple users activity and location recognition system, all in real time. This system makes use of fuzzy rules for activity monitoring and has been tested in a nursing home with activities such as going to the toilet, showering, sleeping, opening or closing wardrobes or being out of a room. This paper presents the social evaluation through rigorous evaluation of the ACTIVA system using the DPA model to monitor in real time the location and activities of multiple users. This research aims to improve older adults care in nursing homes through the responsible integration of intelligent activity recognition systems, emphasising the importance of considering ethics when deploying such technologies in healthcare. The study's findings have significant implications for the future development and implementation of these systems, promoting dignity, positivity, and improved well-being for the older adults population.

Keywords: Intelligent activity recognition system · Real-time monitoring · Social · Ethical evaluation · Healthcare technology

Grant PID2021-127275OB-I00 funded by MCIN / AEI / 10.13039/501100011033 and by "ERDF A way of making Europe".

1   Introduction

It is clear that life expectancy is increasing all the time. A consequence of this is that the number of older adults is also increasing. Therefore, nursing homes will be home to many older adults and the demand will be increasing [1].

However, it is important to look for solutions that help both older adults and care professionals with the aim of a dignified and positive ageing such as the ACTIVA system. The ACTIVA system is an intelligent multiple users location and activity recognition system [8] based on the use of fuzzy rules [2,3]. For this purpose it is necessary to use of different sensors and the ACTIVA app. The main objective of the ACTIVA system is to improve the quality of life of older adults who are in a situation of dependency, increasing their sense of security in their families, in the user and in the caregiver. It also aims to reduce the stress level of carers [6].

It is important that smart systems such as ACTIVA are evaluated from different approaches, usually from a technical point of view. In particular, the ACTIVA system has already been evaluated by Sustainable Development Goals (SDG) [9]. The ACTIVA system has been tested for one month with older adults and their carers in a nursing home in Alcaudete, Jaén for this social evaluation.

For this reason, this work presents a social and health care evaluation using the model of Dignified and Positive Ageing (DPA). The DPA model [4] is an evaluation with ethical overtones where the rational method, the uncertainty evaluation process, the results of the evaluation and aggregation process and the software tool are evaluated [5]. Therefore, this contribution is focused on glimpsing and evaluating the benefits and shortcomings that social services and home help service professionals find in the functionality of the ACTIVA intelligent system. The valuable feedback obtained from this evaluation will be utilized to inform and enhance the development of future activity recognition systems, ensuring they better cater to the needs and preferences of both older adults and their caregivers.

The article is structured as follows. In the Sect. 2 it is shown the ACTIVA system in detail. The Sect. 3 is shown using the DPA model. Finally, the conclusions extracted are presented in the Sect. 4.

2   ACTIVA: A Intelligent System for Activity Recognition

The ACTIVA system is an intelligent system for activity recognition. The system monitors in real time the location and activity of the user within a room and reports possible anomalies in the activity carried out by the user. Specifically, it performs location and time-based activity recognition of multiple users using fuzzy rules and time windows [2,3].

All the data collect from the sensors in the intelligent core is processed in real time, allowing each patient to be visualised from the ACTIVA app. In addition, the location, the activity being performed and the degree of danger of the activity and the battery percentage of the bracelet are displayed to alert the caregiver to charge it. It also has a history where you can consult the activities that have been carried out. The activities set to be monitored are the following: showering, using the toilet, opening or closing the wardrobe in the room, sleeping or being out of the room. To carry out the monitoring of the activities, motion, vibration and opening and closing sensors have been installed in the different locations of the room such as the shower, the toilet, the wardrobe or the bed. In addition, a wearable device, such as a wristwatch, has been used to locate the user [7].

3   Evaluation with DPA Model

This section shows the results of the evaluations carried out using the DPA model. The DPA model [4] allows the evaluation of initiatives and practices of companies, organisations and professionals, which aim to promote a dignified and positive ageing process in the population.

The principles and criteria that make up the DPA model are guidelines for the practices and attitudes of professionals and organisations that promote the goal of creating dignified and positive ageing. There are 5 principles and each principle has certain criteria associated with it as shown in Fig. 1.

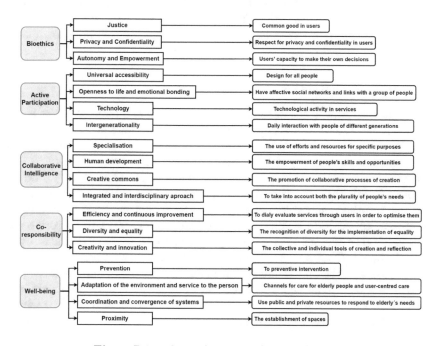

Fig. 1. Principles and criteria of the DPA model

A total of 5 professionals have evaluated the incorporation of this ACTIVA system in the nursing home using the DPA model. The quantitative evaluation, where 1 is indicated if the criterion is totally fulfilled, 0 if it is not fulfilled and 0.5 if it is partially fulfilled and the qualitative evaluation, where feedback is given by the professionals (social workers, caregivers and nursing home directors) on the intervention of the ACTIVA system in relation to the principles of the DPA model.

3.1   Bioethics

The criteria of *justice and privacy* and *confidentiality* have been assessed at a quantitative level with index 1 indicating full compliance with the intervention. The criterion *autonomy and empowerment* has been evaluated with 0.5 which indicates that it is partially fulfilled and could be encouraged .

Regarding the qualitative level, the opinions of the people interviewed in relation to the criterion *justice* indicate that the ACTIVA system complies with it because it "respects the framework of the Organic Law on Personal Data Protection", guaranteeing the right of all people to have their rights respected. In the criterion of *privacy and confidentiality*, they also consider it to be met, highlighting "individualised attention". In the criterion of *autonomy and empowerment*, respondents consider the "technology invasive and limits the autonomy of users" due to the suggestions they make about changing habits so that the sensors can correctly collect information on the activity carried out by the user. Suggestions include, for example, raising or lowering the toilet seat or opening or closing cupboard doors.

3.2   Active Participation

On a quantitative level, the criterion *openness to life and emotional ties* is fully met with index 1. The criterion *intergenerationality* is not met at all with index 0. And the criteria *universal accessibility* and *technology* are almost fully met with respect to the criteria with indicators of 0.9 and 0.8 respectively.

In relation to the qualitative level, the participants found that all criteria are fulfilled with the exception of *intergenerationality*, as contact with people belonging to other generations is something to be implemented. The criterion of *universal accessibility* is fulfilled due to the ease of access to all services for all users. The criterion of *openness to life and emotional ties* is identified by the fact that the ACTIVA system "encourages physical activity" and is a tool for social inclusion in the digital world that helps "cohabitation units to strengthen and maintain emotional ties". Finally, the criterion of *technology* needs to be improved because "the technology is not mature" and "does not fully facilitate the daily activity of users and workers".

3.3   Collaborative Intelligence

On a quantitative level, the criteria *specialisation* and *focus and interdisciplinary* are fully developed. The criteria *human development* and *creative commons* are indicated 0.8 and 0.5, respectively, and can be further encouraged.

With regard to the qualitative level, the people interviewed find the criterion of *specialisation* a "greater degree of maturity so that in practice there is complete specialisation" and an increase in the quality of the service. In *human development*, the opinion is positive, with certain points regarding "efficiency in the strengthening of capacities", such as "improving the development of care teams". In the criterion *creative commons*, the need to reinforce commitments, mutual responsibility and collaboration between service professionals in order to generate an improved exchange for the care service is highlighted. Lastly, in the criterion *integrative and interdisciplinary approach* it is noted that the ACTIVA system is used to improve well-being and quality of life, but it would be advisable to "reinforce it through knowledge exchange" and "technological improvements in the system".

3.4   Well-Being

On a quantitative level, the criterion of *coordination and convergence of systems* has been evaluated with 1. The criterion of *prevention* has obtained an index of 0.95. Finally, the criteria of *proximity* and *adaptation of the environment and service to the person* are fulfilled, but could be reinforced.

In relation to the qualitative level, in the criterion *prevention*, the people interviewed evaluate it positively because "it facilitates the monitoring of the user" and "possible falls are detected and prevented" with the aim of promoting independent and autonomous living. In the criterion *proximity*, the people interviewed gave an almost positive evaluation due to the clarity of the communication channels, however, they pointed out that the senders of the communication do not have a specific action manual with indicated guidelines. The criterion of *coordination and convergence of systems* is evaluated positively due to "the coordination of tasks favoured by the ACTIVA system", but the need for "the data to reach the users" is pointed out. Finally, the criterion *adaptation of the environment and service to the person* highlights that the ACTIVA system "is not totally adapted to the life of the users" and that it is therefore necessary to promote the flexibility of the system by adjusting it to the needs of each person.

3.5   Co-responsibility

On a quantitative level, the criterion of *efficiency and continuous improvement* is fully met. The criteria of *creativity and innovation* and *diversity and equality* have been evaluated with indexes 0.8, both of them.

With regard to the qualitative level, in the criterion of *creativity and innovation*, it has a positive evaluation because it innovates in certain spaces and services such as residential homes and because it is a product in development and maturity, it is easier to adapt it to the target users. In the criterion of *diversity and equality*, the people interviewed think that the ACTIVA system is "a non-discriminatory technology" because it takes into account the diversity of the population and focuses on intervention with equality criteria. Finally, in the criterion of *efficiency and continuous improvement*, the evaluation was positive because the "monitoring of the ACTIVA system and its continuous evaluation" have helped to optimise the system and its possible continuity over time.

4 Conclusions and Future Works

This paper has presented a comprehensive social evaluation of the ACTIVA system using the DPA model. The results revealed that the ACTIVA system fulfills most of the criteria associated with the various principles of the DPA model with a score of 1/1 for both active ageing and ageing with dignity. Nevertheless, certain principles, including active participation, collaborative intelligence, and co-responsibility, require further development to enhance the system's effectiveness. The feedback from this evaluation indicated that the ACTIVA system is highly regarded as an effective and efficient tool for nursing homes. The valuable feedback obtained from these social evaluation will be utilized to inform and enhance the development of future activity recognition systems, ensuring they better carer to the needs and preferences of both older adults and their caregivers.

Therefore, we propose several future directions to address these findings. Firstly, the ACTIVA system should be applied to new residential or community environments to explore its adaptability and effectiveness in various contexts. Secondly, we encourage collaboration between the ACTIVA system and other compatible systems or services to further enhance its capabilities. Additionally, continuous evaluation and improvement of the system's functionalities, ethical considerations, and privacy measures are crucial for its sustained success.

References

1. Envejecimiento y salud. https://www.who.int/es/news-room/fact-sheets/detail/ageing-and-health. Accessed 7 May 2023
2. Albín-Rodríguez, A.P., De-La-Fuente-Robles, Y.M., López-Ruiz, J.L., Verdejo-Espinosa, Á., Espinilla Estévez, M.: UJAmI location: a fuzzy indoor location system for the elderly. Int. J. Environ. Res. Public Health **18**(16), 8326 (2021)
3. Albín-Rodríguez, A.P., Ricoy-Cano, A.J., de-la Fuente-Robles, Y.M., Espinilla-Estévez, M.: Fuzzy protoform for hyperactive behaviour detection based on commercial devices. Int. J. Environ. Res. Public Health **17**(18), 6752 (2020)
4. Cruz Lendínez, A.J., Rodríguez González, A.: La atención social y sanitaria a las personas en situación de dependencia: costes económicos, sanitarios y sociales del sistema. Index de Enfermería **26**(3), 205–209 (2017)
5. Espinilla, M., Verdejo, M., González, L., Nugent, C., Cruz, A.J., Medina, J.: Challenges of ethical evaluation models for intelligent assistive technologies. an initial ethical model based on linguistic decision analysis. Multi. Digit. Publishing Inst. Proc. **31**(1), 22 (2019)
6. López, J.L., Espinilla, M., Verdejo, Á.: Evaluation of the impact of the sustainable development goals on an activity recognition platform for healthcare systems. Sensors **23**(7), 3563 (2023)
7. Espinilla, M., Medina, J., Verdejo, M.A., Ruiz, J.L., Salguero, A.G.: ACTIVA (2021). https://www.safecreative.org/work/2111159810407

8. Ruiz, J.L.L., Espinosa, Á.V., Lendínez, A.M., Estévez, M.E.: OBLEA: a new methodology to optimise bluetooth low energy anchors in multi-occupancy location systems. J. Univ. Comput. Sci. **29**(6), 627 (2023)
9. Verdejo Espinosa, Á., Lopez, J.L., Mata Mata, F., Estevez, M.E.: Application of IoT in healthcare: keys to implementation of the sustainable development goals. Sensors **21**(7), 2330 (2021)

Human-Computer Interaction

Investigating Motivational Factors Influencing Users' Consumption of Video Streaming Services: A Human Factor Perspective

Sruti Subramanian[1]([⊠]) [iD], Katrien De Moor[1] [iD], and Kamil Koniuch[2] [iD]

[1] Department of Information Security and Communication Technology, Norwegian University of Science and Technology, Trondheim, Norway
{sruti.subramanian,katrien.demoor}@ntnu.no
[2] Institute of Communication Technologies, AGH University of Krakow, Kraków, Poland
koniuch@agh.edu.pl

Abstract. This paper presents comprehensive findings obtained from an online survey focused on understanding the motivational factors influencing users' consumption of video streaming services from a human factors perspective. The research uncovers a wide range of factors that motivate users to engage in video watching, encompassing both intrinsic and extrinsic motivations. Intrinsic factors include attributes such as *relaxation, inspiration, fun, happiness, enjoyment, good mood, laughter, learning.* On the other hand, extrinsic motivations are driven by *multitasking, recommended feeds, entertainment (time pass, interesting content and diverse content), music, social connection.* The results of the study have design implications, as they shed light on users' underlying needs, expectations, and preferences. Designers and developers can leverage these findings to create more tailored and engaging experiences that align with users' motivations, ultimately enhancing user satisfaction and engagement.

Keywords: Motivation · Human Factors · video streaming · intrinsic · extrinsic · influencing factors

1 Introduction

Video-on-Demand (VoD) streaming services, such as Netflix, YouTube, Amazon Prime, and Disney+ [17] have gained immense traction among users worldwide, offering the convenience of accessing an extensive library of digital content instantly. In this regard, the use of video streaming services has sparked large interest in diverse fields such as User Experience (UX) [2] and Quality of Experience (QoE) [16]. However, despite the exponential growth in streaming video production and consumption, there is a noticeable gap in understanding users' perspectives and the various human factors that influence users' engagement

[12]. Much attention has been given to Quality of Experience (QoE) and video delivery optimization concerning network conditions and application-level factors [1,8]. However, beyond this level, the impact of human factors on individual user experiences with video streaming services is still not fully understood.

Previous studies have started to investigated the influence of human factors, such as demographic and socio-economic background, physical and mental constitution, on the perceived quality and QoE of streaming services [24,27,30,31,33]. Additionally, subjective attributes like users' preferences, attitudes, values, and motivation have also been shown to impact users' experiences and QoE [18,23]. Motivation, in particular, is known to be a crucial factor in determining users' engagement and continued usage of an application. Although the influence of motivation on users' experiences with streaming service is acknowledged, this topic is still largely unexplored and characterised by limited available knowledge [14,16,23]. Understanding the driving forces and motivations for users can however provide valuable insights into user expectations, preferences, needs, and tolerance levels with streaming services. These insights can further inform the design and development of personalized and engaging streaming services that cater to the diverse needs of the user base. This understanding is particularly relevant as streaming services gain more prominence and are increasingly associated with excessive consumption patterns like binge-watching [9]. Comprehending the interplay between intrinsic and extrinsic motivational factors can further inform the design of effective incentives and rewards systems which can be strategically integrated into the platform to amplify users' intrinsic motivation. This knowledge can further aid in designing streaming services, (i.e., user interfaces, content recommendations, and personalized experiences) that align with users' motivations and their underlying needs, thereby enhancing user satisfaction and engagement. Furthermore, contributing to the overall understanding of how human factors influence the design and optimization of streaming platforms.

Therefore, the current study investigates the motivations influencing users' consumption of video streaming services from a human factors perspective. More specifically, in this paper we aim to address the following research question: "Which motivational factors do users draw on to watch video streaming content and what may this imply"? Through a comprehensive review of relevant literature and an online survey with 40 participants, we aim to contribute to the limited body of knowledge available. The remainder of the paper is structured as follows: relevant theory and related work is presented in Sect. 2, and the research methodology is briefly described in Sect. 3. Section 4 presents the results, and a discussion of the implications are further presented in Sect. 5. Finally, we provide a conclusion in Sect. 6.

2 Related Work

2.1 Theoretical Frameworks

Self-determination Theory: (SDT)[7], has been widely used to better understand human motivation. SDT identifies two different types of motivation. When one's motivation is intrinsic, the activity performed is perceived as inherently enjoyable (e.g., Engaging in a hobby or creative pursuit simply for the enjoyment of it). When driven by extrinsic motivation on the other hand, an activity is performed as it leads to a separable outcome (e.g., Studying for an exam in order to earn a good grade)[7]. SDT further proposes that experiences of the psychological needs of autonomy (personal agency), relatedness (social connectedness), and competence (sense of efficacy), enhance motivation and well-being [29]. For instance, the model presented in [19] builds upon SDT to design for motivation, engagement and well-being in digital experiences. SDT has been applied to active media contexts, such as video games and movement based video games [20, 22, 26]. Zhao et al. (2018) [32] used SDT to investigate different motivational forces (i.e., intrinsic and extrinsic) that drive live streamers' continuance broadcasting intentions. The authors indicate that both intrinsic and extrinsic motivation significantly influenced live streamers' performance expectancies, but not their perceived website attractiveness. Marika (2022)[13] further explored how audiences experience watching on-demand television, and validated and tested how a range of measures predict enjoyment. Results suggest that enjoyment is primarily explained by social significance, immersive viewing, lower levels of deliberate viewing, and positive perceptions of programmed paths.

Uses and Gratification Theory: Another relevant theory which was originally developed to understand mass communication/mass media from an "audience" perspective, is the Uses and Gratification Theory (UGT) [4]. UGT focuses on understanding why and how individuals actively choose and use media to satisfy their specific needs and desires. The theory suggests that specific needs and desires are what drive individual motivation to select and consume media. Rather than considering the audience as passive receivers of media, the theory highlights that individual users are active participants who proactively select and engage with media based on their personal motivations to fulfill their underlying needs. According to the theory, people seek out media content that fulfills their specific needs, such as information, entertainment, social interaction, personal identity, or escapism. It recognizes that individuals have different needs and preferences, and they actively select and consume media to gratify those needs, making media consumption a purposeful and goal-oriented process [4].

Various studies have analysed the usage motives of different types of media drawing one of the above or even other theoretical perspectives. A study on user motives of using social media identified the following seven unique uses and gratifications: social connection, shared identities, photographs, content, social investigation, social network surfing, and status updating [10, 11]. Similarly on a study conducted with young adolescents, Tanta et al. (2014) [28] mention that the use of Facebook primarily gratifies adolescents' need for integration, social interaction, information and understanding of their social environment. With respect to watching Television, McIlwraith (1998) [15] identified that "TV addiction" was an attempt by users to distract themselves from unpleasant thoughts, to regulate moods, and to fill time. Furthermore, with respect to understanding motivations to use online streaming services during the covid-19 pandemic using UGT and the Technology Acceptance Model (TAM), authors identified that individuals' perceived usefulness and ease of use of online streaming services were significant antecedents of their intentions to use the services. The authors further suggest that the participants sought emotional gratification from online streaming services, as it allowed to distract themselves into a better mood and to relax in their leisure time. They were also known to be using it to satisfy their needs for information and entertainment [6].

2.2   Human Factors

Human factors in computing refers to the aspect in which people think, perceive, remember, and act that influence the ways they interact with systems [3]. Furthermore, according to [23], a human influence factor (HIF) is any variant or invariant property or characteristic of a human user. The characteristic can describe the demographic and socio-economic background, the physical and mental constitution or the users' emotional state [23]. In this regard, it is widely recognized that users' interactions with streaming services are influenced not only by the attributes of the multimedia system but also by the surrounding environment and circumstances in which the experience occurs [5,16,23]. Multimedia experiences and their perceived quality may also vary significantly in terms of perception and endurance for different users, emphasizing their individuality and importance in shaping overall user satisfaction [21,34]. Wechsung et al. (2011) [31] investigated whether cognitive skills, mood, attitudes and personality traits influence quality perceptions, modality choice (speech vs. touch), and performance. It was identified that attitudes and mood are related to quality perceptions, while performance is linked to personality traits. Furthermore, it was stated that modality choice is influenced by attitudes and personality while cognitive abilities had no effect. In this regard some studies have explored the influence of specific human factors [16,24,31].

However, in most empirical studies related to multimedia services, human factors are only considered to a limited extent (e.g., HIFs gender, age, expertise level). Therefore, there is still limited understanding of how human factors influence quality of experience with multimedia services [16].

The current work aims to contribute to the gap in literature by investigating the motivational factors influencing users' consumption of video streaming services. By understanding what drives individuals to engage with streaming services, we gain valuable insights into their underlying needs, preferences, and behaviors. Investigating both intrinsic and extrinsic motivational factors further provides both a holistic understanding of users' engagement with streaming services and a more specific understanding of the implications of these two types of motivation for what users expect and need in order to have a good or worthwhile experience. Comprehending the interplay between intrinsic and extrinsic motivational factors can further inform the design of effective incentives and rewards systems which can be strategically integrated into the platform to amplify users' intrinsic motivation. This knowledge can further aid in designing streaming services, (i.e., user interfaces, content recommendations, and personalized experiences) that align with users' motivations and their underlying needs, thereby enhancing user satisfaction and engagement. Furthermore, it contributes to the overall understanding of how human factors influence the design and optimization of streaming platforms.

3   Research Methodology

An online questionnaire was developed to better understand users' experiences with and use of video steaming services. The questionnaire aimed towards gaining better insight into how, when, where and why video services are used, and what users care about when using video services in day-to-day life. The questionnaire comprised of a combination of open and closed questions and was structured such that these various aspects of users' experiences and consumption were captured. The questionnaire comprised of various parts. The results presented in this paper pertain to the analysed responses obtained from the specific section of the questionnaire which focused on influencing factors, specifically the motivational factors.

3.1   Sample Description, Data Collection and Analysis

A convenience sampling strategy was used to recruit participants. The online questionnaire was completed by a total of N=40 participants between 18 and 49 years of age (M = 30.5 and S.D. = 8.4). The sample further comprised of almost equal ratio of male (52.5 %) and female (47.5%) participants. The questionnaire was constructed using the online platform nettskjema[1] and circulated via different channels. The submitted questionnaires were further coded for themes using the Nvivo[2] software. An open coding approach was followed, which is an iterative data analysis technique where categories are added until induction thematic saturation [25] has been induced. It was an inductive process where the themes emerged from the data itself.

4   Results

The open coding process resulted in the identification of 13 categories, placed under the two overarching categories (i.e., intrinsic vs. extrinsic). In the following, we present unique selected quotes that illustrate the emerged categorization. The coding was further performed to keep the specific keywords intact as opposed to merging with other codes.

---

[1] https://nettskjema.no/.
[2] https://portal.mynvivo.com/.

4.1   Intrinsic Factors

The various categories that were intrinsic in nature (i.e., actions which are perceived inherently as interesting or enjoyable), were included within this categorisation. The concepts of *fun* (experience of amusement or playfulness), *enjoyment* (feeling of satisfaction and pleasure), *good mood* (state of positive emotions), and *happiness* (a state of well-being and contentment) are related but distinct emotional states. Therefore, they were grouped separately during the open coding process to allow for capturing nuanced user experiences. Separating these factors further allows for a more comprehensive understanding of how each emotional state influences users, leading to richer insights and targeted interventions.

Relaxation. Several of the participants indicated that they were motivated to watch videos for relaxation. Participant 14 (Male, 49 yrs) indicated to watch VOD content *"to get relaxed"*. Participant 31 (Male, 31 yrs) also indicated that *"I am usually relaxed after a movie"*. Other participants stated that watching videos allowed *"to disconnect"* (P18, Female, 22 yrs) and that they watch to *"[...] to take my mind off things"* (P25, Female, 26 yrs).

Inspiration. Participants indicated that they found watching videos to be inspiring and motivating with respect to different aspects. For example, to learn certain skills, such as playing the piano. As Participant 10 (Female, 20yrs) said: *"piano tutorials have motivated me"*. Similarly, P36 mentioned that book reviews inspire her to start a new book: *"book/series review on BookTube got me really excited to start reading a new book"* (P36, Female, 30 yrs). Another participant further indicates that such inspiration is dependent on content: *"Good content inspires"* (P38, Male, 37 yrs).

Fun. Several participants mentioned that they found watching videos to be a fun activity: *"It is fun"* (P11 Male, 22 yrs). It was further mentioned that it was an easy, not very demanding way to have fun. For instance, participant 07 (Female, 18 yrs) said that she watches videos *"when I want to have fun without effort"*.

Happiness. A number of participants mentioned that they feel happy watching videos, or as participant 32 (Male, 28 yrs) puts it, watching videos *"brings happiness"*. Participant 08 further says that watching funny content makes her happy: *"Often watch something funny so I feel happier"* (P08, Female, 21 yrs).

Enjoyment. Several participants mentioned enjoying watching videos. P37 also indicated that she enjoyed content related to a specific genre: *"I enjoy true crime series"* (P37, Female, 41 yrs).

Good Mood. Many participants mentioned that watching video content helps to get into a positive mood, e.g., Participant 05 (Female, 29 yrs) indicated to watch videos to *"get into a good mood"*.

Laughter. Numerous participants mentioned that they watch videos to get a good laugh: for instance, Participant 07 (Female, 18 yrs) said that when watching videos *"I laugh if I am alone"*. Similar to the other intrinsic factors, participants also indicated the importance of content with respect to laughter: *"Saw an* *episode of my favorite series that made me laugh out loud"* (P11, Male, 22 yrs). Or as participant 38 (Male, 37 yrs) phrased it: *"Watched a couple of comedy videos and had a good laugh".*

Learning. Learning was identified as another influencing factor as several participants mentioned that acquiring knowledge from documentaries, tutorials, lectures and other scientific content was a main motivation to watch videos. For instance, Participant 18 (Female, 22 yrs) referred to *"something professional to learn"* as a motivation to watch. Watching video content was also referred to as a way to get news and be up to date. Or as Participant 27 (Male, 38 yrs) put it, the motivation for watching is *"to get news and updates".*

4.2   Extrinsic Factors

The identified quotes which were extrinsic in nature (leading to a separable outcome) were classified within this category.

Multitasking. Participants indicated that they were motivated to watch videos as it was done in parallel to other activities. Numerous participants indicated watching videos while eating: *Videos are my company when I eat* (P05, Female, 29 yrs). Furthermore, it was also an activity done in combination with other tasks, e.g., in order *"not to be bored when doing menial tasks"*(P36, Female, 30 yrs).

Recommended Feed. The recommender systems providing videos was indicated to motivate participants to watch the suggested videos, for instance Participant 06 (Male, 26 yrs) was motivated to watch because *"I get recommended videos".*

Entertainment. Numerous participants further mentioned that being entertained was a main motivating factor: *"I watch for entertainment"*(P16, Female, 49 yrs).
Entertainment has further been categorized into the following three subcategories (i.e., Time pass, Interesting content, Diversity of content).

- **Time Pass.** Participants indicated that watching videos was a way to *"make time pass"* (P12, Male, 22 yrs). Participant 34 (Male, 25 yrs) hinted at a similar motivation : *"no benefits, just to get by the day".*
- **Interesting Content.** Watching interesting content was mentioned by several participants as motivating. For instance Participant 20 (Female, 42 yrs) referred to a *"particular interest in the video content"*, whereas Participant 13 (Male, 22 yrs) felt motivated by watching *"interesting video concepts".*
- **Diversity of Content.** Having diverse types of content to watch was further also mentioned as another motivating factor: *"I like that I can find more international movies or videos with more diversity of people"* (P29, Female, 33 yrs). Participant 33 (Male, 32 yrs) further mentioned that he watches different types of content : *"to explore my different areas of interest".*

Music. Listening to and watching music videos was mentioned as another motivation, for example by Participant 30 (Female, 25 yrs) who indicated that *"I like to watch music videos"*. It was also indicated that some participants only listen to music videos without watching the actual video.

Social Connection. Finally, watching videos with others as a social activity was mentioned by several participants. For instance participant 03 (Female, 29 yrs) describes her motivation as: *"Have a good time watching movies with friend(s). The social experience makes it pleasant for me"*.

5   Discussion

The findings above support the observation that different motivational factors may have different underlying needs, trigger different types of user behavior, and these various use cases have to be explored further to better understand user behavior pertaining to different contexts. However, from the identified motivations, we can infer the underlying needs and desires that these motivations are trying to fulfill.

- *Relaxation*: This factor may fulfill the need for stress reduction, mental rejuvenation, and a break from daily routines. People may seek relaxation while watching streaming videos to unwind, find peace of mind, and experience a sense of calmness.
- *Inspiration*: The need here might be to gain inspiration, by streaming videos that stimulate creativity, trigger new ideas, or showcase achievements, encouraging the users to aspire for personal growth and pursue their passions.
- *Learning*: The need for acquiring knowledge or learning may be characterized by accessing streaming videos that provide educational content, informative documentaries, or tutorials, allowing individuals to acquire new knowledge, develop skills, and expand their understanding of various subjects.
- *Fun, enjoyment, good mood, happiness, and laughter*: These are all factors that may fulfill the need for experiencing positive emotions. Streaming videos that are enjoyable, humorous, or uplifting may provide individuals with a sense of joy, amusement, and satisfaction (i.e., emotional gratification as mentioned in [6]). This may also be an attempt by users to distract themselves from unpleasant thoughts, and to regulate their moods [15].
- *Multitasking*: This may fulfill the need for productivity and efficiency. Users may watch streaming videos while engaging in other activities to optimize their time and achieve a sense of accomplishment in different areas.
- *Recommendations*: This may satisfy the desire for personalized suggestions and guidance. Users may seek recommendations to discover new and interesting content that aligns with their preferences, saving time and effort in finding engaging videos.
- *Entertainment* : The need for entertainment encompasses various aspects, including time pass (filling leisure time), interesting content, and diversity of content. The factors may fulfill the desire for engaging and diverse experiences

that cater to individual preferences and provide autonomy [7] by presenting a range of options to choose from. This may also be related to attaining emotional gratification [6] and a way of experiencing positive emotions.

– *Music*: Streaming videos that feature music may fulfill the need for auditory stimulation, enjoyment, and a connection to different emotions.
– *Social Connection*: It may fulfill the need for social engagement, community building, and connecting with like-minded individuals [28].

The results may further have the following implications. The coded intrinsic motivational factors (e.g., relaxation, inspiration, etc.,) influencing users to watch videos, seem to be largely dependent on the content. While which genre of videos is considered fun or relaxing or happy may be very subjective, it could be argued that the identified intrinsic motivational factors are dependent on content nonetheless. Therefore, content may be of more significance when the motivation is triggered by the identified intrinsic factors. More concretely, content may play an important mediating role to trigger intrinsic motivation. However, with respect to the extrinsic factors, the importance of content was reflected only with respect to entertainment, specifically in terms of having interesting and diverse content.

This raises the question whether video quality may be of more significance in the context of the intrinsic factors, since poor video quality may inhibit users from achieving the various intrinsic states. This is not to say that content and video quality are not important in the case of extrinsic factors. However, most of the extrinsic motivational factors place more significance on utilitarian aspects such as multitasking, time pass, listening to music, and social connection, which may in many cases not be too much dependent on content and quality as compared to all the intrinsic influencing factors which seem to be largely dependant on content and potentially its quality.

All these motivational factors may impact users' perception of the video, its quality and their overall satisfaction with the experience. Connected to this, the purposes for which the service are used may also largely influence users' experience. For example, if a user is watching a video for serious purposes such as learning/education, they may expect clear and accurate information that is presented coherently and in a way that is easy to understand. If the video fails to meet these expectations, the user's experience may be negatively impacted. On the other hand, if a user is consuming videos for entertainment purposes, they may overlook any minor technical issues if the video is engaging and enjoyable from their subjective experience. However, more lab and field studies explicitly addressing the link between viewing motivations, QoE and tolerance towards different types of impairments, are needed. Future studies can also focus on analysing the correlation between user needs, their motivations and user behavior. Such more fine-grained insights would allow for better tailoring of services to meet the specific needs and preferences of their users, leading to a higher level of satisfaction and better viewing experience.

6 Conclusion

The main aim of the study was to better understand motivational factors influencing users' consumption of video streaming services from a human factors perspective. The results indicate a combination of intrinsic and extrinsic motivational factors which users draw on, that might influence their overall experience with such services. The participants were intrinsically motivated by *relaxation, inspiration, fun, happiness, enjoyment, good mood, laughter, learning*, and extrinsically motivated by *multitasking, recommended feeds, entertainment (time pass, interesting content and diverse content), music and social connection*. By understanding what drives individuals to engage with streaming services, we gain valuable insights into their underlying needs, preferences, and behaviors. Investigating both intrinsic and extrinsic motivational factors further provides a holistic understanding of users' engagement with streaming services. The findings also suggest that when users are triggered by intrinsic factors, quality and content of the videos may be of more importance. However, when triggered by the extrinsic factors, utilitarian aspects such as multitasking, in addition to motivation factors such as time pass, listening to music, and social connection seem to be more important. The latter may not be too much dependent on content and quality of videos. Follow-up empirical studies are needed to validate these findings and to further understand the relation between intrinsic and extrinsic motivation for watching video content further.

Overall this knowledge can further aid in designing streaming services, that align with users' motivations and their underlying needs, thereby enhancing user satisfaction and engagement. Furthermore, it contributes to the overall understanding of how human factors influence the design and optimization of streaming platforms.

Acknowledgments. The research leading to these results has received funding from the Norwegian Financial Mechanism 2014-2021 under project 2019/34/H/ST6/00599.

References

1. Barman, N., Martini, M.G.: Qoe modeling for http adaptive video streaming-a survey and open challenges. Ieee Access **7**, 30831–30859 (2019)
2. Bentley, F., Silverman, M., Bica, M.: Exploring online video watching behaviors. In: Proceedings of the 2019 ACM International Conference on Interactive Experiences for TV and Online Video, TVX 2019, pp. 108-117. Association for Computing Machinery, New York (2019). https://doi.org/10.1145/3317697.3323355
3. Blandford, A.: Human Factors in Computing, pp. 791-794. John Wiley and Sons Ltd., GBR (2003)
4. Blumler, J.G., Katz, E.: The uses of mass communications: Current perspectives on gratifications research. Sage Annual Rev. Commun. Res. **iii** (1974)
5. Brunnström, K., et al.: Qualinet white paper on definitions of quality of experience (2013)

6. Camilleri, M.A., Falzon, L.: Understanding motivations to use online streaming services: integrating the technology acceptance model (tam) and the uses and gratifications theory (ugt). Spanish J. Marketing-ESIC **25**(2), 217–238 (2021)
7. Deci, E.L., Ryan, R.M.: Self-determination theory (2012)
8. Garcia, Marie-Neige., Argyropoulos, Savvas, Staelens, Nicolas, Naccari, Matteo, Rios-Quintero, Miguel, Raake, Alexander: Video Streaming. In: Möller, Sebastian, Raake, Alexander (eds.) Quality of Experience. TSTS, pp. 277–297. Springer, Cham (2014). https://doi.org/10.1007/978-3-319-02681-7_19
9. Granow, V.C., Reinecke, L., Ziegele, M.: Binge-watching and psychological wellbeing: media use between lack of control and perceived autonomy. Commun. Res. Rep. **35**(5), 392–401 (2018)
10. Joinson, A.N.: Looking at, looking up or keeping up with people? motives and use of facebook. In: Proceedings of the SIGCHI conference on Human Factors in Computing Systems, pp. 1027–1036 (2008)
11. Karapanos, E., Teixeira, P., Gouveia, R.: Need fulfillment and experiences on social media: a case on facebook and whatsapp. Comput. Hum. Behav. **55**, 888–897 (2016)
12. Lagger, C., Lux, M., Marques, O.: What makes people watch online videos: an exploratory study. Comput. Entertain. **15**(2) (2017). https://doi.org/10.1145/3034706
13. Lüders, M.: Self-determined or controlled, seeking pleasure, or meaning? identifying what makes viewers enjoy watching television on streaming services. Poetics **92**, 101639 (2022)
14. Malinovski, T., Vasileva, M., Vasileva-Stojanovska, T., Trajkovik, V.: Considering high school students' experience in asynchronous and synchronous distance learning environments: Qoe prediction model. Inter. Rev. Res. open Distributed Learn. **15**(4) (2014)
15. McIlwraith, R.D.: "i'm addicted to television": the personality, imagination, and tv watching patterns of self-identified tv addicts. J. Broadcasting Elect. Media **42**(3), 371–386 (1998)
16. Möller, Sebastian, Raake, Alexander (eds.): Quality of Experience. TSTS, Springer, Cham (2014). https://doi.org/10.1007/978-3-319-02681-7
17. Moro-Visconti, R., Moro-Visconti, R.: From netflix to youtube: over-the-top and video-on-demand platform valuation. Startup Valuation: From Strategic Business Planning to Digital Networking, pp. 309–339 (2021)
18. Msakni, H.G., Youssef, H.: Is qoe estimation based on qos parameters sufficient for video quality assessment? In: 2013 9th International Wireless Communications and Mobile Computing Conference (IWCMC), pp. 538–544. IEEE (2013)
19. Peters, D., Calvo, R.A., Ryan, R.M.: Designing for motivation, engagement and wellbeing in digital experience. Front. Psychol., 797 (2018)
20. Przybylski, A.K., Rigby, C.S., Ryan, R.M.: A motivational model of video game engagement. Rev. Gen. Psychol. **14**(2), 154–166 (2010)
21. Redi, J.A., Zhu, Y., De Ridder, H., Heynderickx, I.: How passive image viewers became active multimedia users: New trends and recent advances in subjective assessment of quality of experience. In: Visual Signal Quality Assessment: Quality of Experience (QoE), pp. 31–72 (2015)
22. Reinecke, L., Tamborini, R., Grizzard, M., Lewis, R., Eden, A., David Bowman, N.: Characterizing mood management as need satisfaction: the effects of intrinsic needs on selective exposure and mood repair. J. Commun. **62**(3), 437–453 (2012)
23. Reiter, U., et al.: Factors influencing quality of experience. In: Quality of Experience: Advanced Concepts, Applications and Methods, pp. 55–72 (2014)

24. Sackl, A., Masuch, K., Egger, S., Schatz, R.: Wireless vs. wireline shootout: Howuser expectations influence quality of experience. In: 2012 Fourth International Workshop on Quality of Multimedia Experience, pp. 148–149. IEEE (2012)
25. Saunders, B., Sim, J., Kingstone, T., Baker, S., Waterfield, J., Bartlam, B., Burroughs, H., Jinks, C.: Saturation in qualitative research: exploring its conceptualization and operationalization. Quality Quantity **52**, 1893–1907 (2018)
26. Subramanian, S., Dahl, Y., Skjæret Maroni, N., Vereijken, B., Svanæs, D.: Assessing motivational differences between young and older adults when playing an exergame. Games Health J. **9**(1), 24–30 (2020)
27. Suznjevic, M., Skorin-Kapov, L., Cerekovic, A., Matijasevic, M.: How to measure and model qoe for networked games? a case study of world of warcraft. Multimedia Syst. **25**, 395–420 (2019)
28. Tanta, I., Mihovilović, M., Sablić, Z.: Uses and gratification theory-why adolescents use facebook? Medijska istraživanja: znanstveno-stručni časopis za novinarstvo i medije **20**(2), 85–111 (2014)
29. Vansteenkiste, M., Ryan, R.M., Soenens, B.: Basic psychological need theory: Advancements, critical themes, and future directions. Motiv. Emot. **44**, 1–31 (2020)
30. Walton, T., Evans, M.: The role of human influence factors on overall listening experience. Qual. User Exper. **3**, 1–16 (2018)
31. Wechsung, I., Schulz, M., Engelbrecht, K.P., Niemann, J., Möller, S.: All users are (not) equal-the influence of user characteristics on perceived quality, modality choice and performance. In: Proceedings of the Paralinguistic information and its integration in spoken dialogue systems workshop, pp. 175–186. Springer (2011).https://doi.org/10.1007/978-1-4614-1335-6_19
32. Zhao, Q., Chen, C.D., Cheng, H.W., Wang, J.L.: Determinants of live streamers' continuance broadcasting intentions on twitch: a self-determination theory perspective. Telematics Inform. **35**(2), 406–420 (2018)
33. Zhu, Y., Guntuku, S.C., Lin, W., Ghinea, G., Redi, J.A.: Measuring individual video qoe: A survey, and proposal for future directions using social media. ACM Trans. Multimedia Comput. Commun. Appli. (TOMM) **14**(2s), 1–24 (2018)
34. Zhu, Y., Hanjalic, A., Redi, J.A.: Qoe prediction for enriched assessment of individual video viewing experience. In: Proceedings of the 24th ACM International Conference on Multimedia, pp. 801–810 (2016)

Translation and Validation of the AttrakDiff User Experience Questionnaire to Spanish

Gustavo López[ID], Ignacio Díaz-Oreiro[(✉) ID], Luis Quesada[ID],
and Kryscia Ramírez-Benavides[ID]

University of Costa Rica, San Jose 11501, Costa Rica
{gustavo.lopezherrera,ignacio.diazoreiro,luis.quesada,
kryscia.ramirez}@ucr.ac.cr

Abstract. The AttrakDiff questionnaire is a widely used instrument for measuring User Experience. However, a Spanish version of the questionnaire has yet to be validated. This represents a significant limitation, given the importance of the Spanish-speaking community. This study aims to translate and validate AttrakDiff to Spanish. Several techniques for translation were used, and the results were joined in a translation proposal. The translated version was evaluated in two scenarios. First, an evaluation with 200 + participants to assess the translation proposal. Second, an evaluation of three systems to perform a factorial analysis and determine the correlations between questions of the same dimension. The results of this study will contribute to the advancement of UX research and practice in the Spanish-speaking context and provide a valuable tool for practitioners and researchers who work with Spanish-speaking users.

Keywords: User experience · AttrakDiff questionnaire · Spanish · Translation · Psychometric evaluation · Cross-cultural adaptation

1 Introduction

The measurement of user experience (UX) has become increasingly important in the field of human-computer interaction (HCI) and product design [1]. One of the most widely used UX questionnaires is AttrakDiff, which assesses four dimensions of UX: pragmatic quality, hedonic quality, attractiveness, and stimulation [2].

The AttrakDiff questionnaire has since become a widely used tool in user experience research and has been adapted and validated for various product domains and user groups [3]. However, despite the popularity of AttrakDiff, it has yet to be validated in Spanish, which is a significant limitation given the importance of the Spanish-speaking community.

This study aims to translate the AttrakDiff questionnaire from English to Spanish and assess its psychometric properties with a Spanish-speaking sample. The validation of AttrakDiff in Spanish is expected to facilitate cross-cultural UX research and provide a valuable tool for practitioners and researchers who work with Spanish-speaking users.

© The Author(s), under exclusive license to Springer Nature Switzerland AG 2023
J. Bravo and G. Urzáiz (Eds.): UCAmI 2023, LNNS 835, pp. 243–254, 2023.
https://doi.org/10.1007/978-3-031-48306-6_25

The translation process followed established guidelines for cross-cultural adaptation of questionnaires, including forward and backward translation, expert panel review, and pilot testing. Furthermore, several translation techniques were applied to validate the translations. These techniques include crowdsourcing, expert translation, translation conducted by computer science experts, and automatic translation. The goal of this translation effort is to compare and validate the possible translations of the questionnaire and its possible interpretations.

The evaluation involved two executions. First, a general review with 200 + participants to evaluate the translated proposal. Second, the evaluation of three products by a group of participants with experience with digital products to assess the validity, reliability, and correlations between the questions.

Overall, the translation and validation of the AttrakDiff questionnaire in Spanish will contribute to the advancement of UX research and practice in the Spanish-speaking context and will provide a helpful instrument for evaluating digital products with this community.

2 Background

The field of user experience (UX) research has grown significantly in recent years, driven by the increasing importance of digital products and services in our daily lives. UX research is concerned with understanding how people interact with technology and how to design products that meet their needs and expectations. One widely used tool in UX research is the AttrakDiff questionnaire, which measures users' subjective experience of digital products regarding their attractiveness, usability, and hedonic qualities.

The AttrakDiff questionnaire consists of 28 items that use a 7-point semantic differential. The questionnaire is divided into two parts: hedonic quality and pragmatic quality. The hedonic quality measures the emotional and experiential aspects of the user's experience, such as the user's pleasure, arousal, and attractiveness of the product or service. Pragmatic quality measures the usability and functionality of the product or service, such as the user's efficiency, effectiveness, and control over the product or service [2–4].

While the AttrakDiff questionnaire has been validated and used extensively in English and German-speaking populations, there is a need for its translation and validation in other languages to support cross-cultural research and improve the usability and attractiveness of digital products for non-English speakers. Spanish is one of the most widely spoken languages in the world, and there is a growing demand for UX research and design in Spanish-speaking countries.

Translating a questionnaire involves more than just translating the words. It requires careful consideration of cultural and linguistic differences, as well as the context and purpose of the questionnaire.

It is worth noting that other UX questionnaires, such as the User Experience Questionnaire (UEQ) [5], have already been translated into multiple languages, including Spanish, to support cross-cultural research and design. However, the AttrakDiff questionnaire offers unique advantages, such as its ability to distinguish between pragmatic and hedonic aspects of user experience and its compatibility with other questionnaires in the UX field.

3   Related Work

This section describes research related to this study, mainly focusing on translations of user experience or usability instruments to Spanish. Some researchers have translated the AttrakDiff as part of the questionnaire applied in their research. The process is not described in detail in the papers, so the level of formality with which the translation was carried out is unknown. For instance, in 2008 [6], an Icelandic translation was made from the English version. It only explains that the translation was made by one of the authors and mentions some of the terms that presented problems due to language features. Furthermore, in 2011 [7], indicates that AttrakDiff was translated into Finnish by the paper's first author, with colleagues evaluating the result based on the English version.

In 2012, a version of the User Experience Questionnaire (UEQ) was proposed in Spanish. However, in this research, the methodology followed was a bilingual person's direct translation from German to Spanish. A retranslation process was followed until all translated words matched the originals. This process for translation is typical in research. However, it can introduce biases and often does not consider cultural differences [8].

Again in 2013, a follow-up paper described some validation of the Spanish version of UEQ proposed in [8]. In this paper, the authors used 94 students to evaluate amazon.es; in another study, 95 students evaluated Skype [9].

In 2015, the translation of AttrakDiff into French was presented. The questionnaire was translated from German into French by two HCI researchers, trilingual in French, German, and English, one of them a native German speaker. The English version of the questionnaire was used to refine the most difficult terms to translate. Then, a reverse translation process was carried out in parallel by two other researchers who were also trilingual. Of the 56 words (28 pairs) translated in reverse, 21 were consistent with the original German version. The other terms were reviewed and validated by a committee of the four researchers mentioned above, plus an external researcher specialized in UX and a regular user of the AttrakDiff in its English version [10].

In 2020, a version of the System Usability Scale was proposed in Spanish. Again in this research, the translation process included forward and backward translation. Forward translations were made by two native Spanish speakers who spoke English as their second language, and a backward translation was created by a native English speaker [11].

Furthermore, by 2023 an improvement proposal for the translation of UEQ was described in [12]. In this paper, the authors analyze changes in some items of the UEQ for use in the context of Costa Rican culture. The evaluation was conducted with 161 participants that examined both questionnaire versions.

4   Methodology

The objective of this study is to translate the AttrakDiff questionnaire to Spanish using different translation methods and compare their results. Five methods were selected: human translation by 38 English students, human translation by five experts in Human-Computer Interaction (HCI), human translation using crowdsourcing with 180 responses, automatic translation using Google Translate, Amazon, and Yandex, and automatic translation using ChatGPT.

Participants for the human translation methods were briefed on the study's objectives, the AttrakDiff questionnaire, and the translation process. They were provided with guidelines and instructions for translating the questionnaire, including information on the importance of maintaining the original questionnaire's meaning, structure, and format.

The 38 English students translated the questionnaire individually, using their skills and resources. They were given a specified time frame to complete the translation and were encouraged to consult resources such as dictionaries or grammar guides if necessary. However, they were instructed to avoid automatic translators, and they were also asked to leave translations blank if they needed clarification on a possible translation. The 38 responses were processed and analyzed based on frequencies to create one translated questionnaire version.

The five experts in HCI were selected based on their expertise in user experience research. The goal of this translation was to achieve a proposal created by people that usually use this type of questionnaire. Furthermore, they were asked to focus on maintaining the original questionnaire's meaning, structure, and format while also considering the specific terminology and language used in the field of Human-Computer Interaction.

For the human translation using crowdsourcing, online platforms, and social media was used to recruit participants. A survey was constructed using Google Forms in which pairs of words were presented individually. The research team later analyzed all responses for consistency, accuracy, and clarity.

For the automatic translation methods, the questionnaire would be translated using Google Translate, Amazon, and Yandex, which are popular machine translation services. The research team would then review the translations to assess their accuracy and relevance. In this section, the German version of AttrakDiff was also used to get six proposals, two from each platform. Additionally, the questionnaire was translated using ChatGPT, a state-of-the-art language model trained by OpenAI.

The final translation proposal included results from all the processes described above and a research team discussion to achieve consistency and ensure proper translation of all the questionnaire constructs.

5   Translation Insights

This section presents the results of each translation process and provides insights into each type of translation, focusing on the benefits and drawbacks of each method.

5.1   Modern Languages Faculty Students Translation

This translation process involved the participation of 38 English students. The students were requested to translate all pairs of words as part of a single exercise. Several translations yielded convergent results (see Table 1: 1L,1R, 4R, 6L, 7L, 7R, 8L, 12L,12R, 16R, 18R, 19L, 19R, 20L, 27R) with over 90% of convergence.

Conversely, words such as "alienating" demonstrated low convergence, with only 16% agreement. Additionally, words like "premium," "tacky," "undemanding," "unimaginative," "inventive," "isolating," and "unruly" exhibited less than 30% agreement. This translation exercise showed a considerable degree of divergence. This could be attributed to two factors: the number of participants involved and the varying interpretations of the semantic differential by each participant. Furthermore, it was observed that many translation proposals employed elaborate vocabulary, some of which still need to be officially recognized as valid variations of the word in the Spanish dictionary.

Another noteworthy observation in the translation process was the prevalence of negation in participants' proposals, as exemplified by translations such as "professional" and "not professional" or "stylish" and "without style."

5.2 Domain Experts Translation

The involvement of domain experts was pursued to further the translation process of the AttrakDiff questionnaire. These experts were selected based on their status as professors specializing in Computer Science, specifically focusing on Human-Computer Interaction, hailing from Latin America. Five experts were invited to participate in the translation process, all of whom provided their respective contributions. Each expert translated the questionnaire by leveraging their specialized knowledge in the field.

This approach showed the most congruent responses reflective of the original questionnaire. However, modifications were made to certain concepts to enhance the differentiation within the semantic differential or to utilize appropriate Spanish terminology. As a consequence of these adaptations, significant disparities emerged among the translations.

Specific terms garnered unanimous agreement across all translations (see Table 1: 1L, 3L, 4R, 5L, 5R, 6L, 7L, 8L, 9L, 9R, 12L, 12R, 14R, 19L, 16R, 18R, 19R, 20L, 23L). In contrast, specific terms such as "rejecting," "connective," "tacky," "premium," and "inviting" failed to find corresponding translations within any of the five experts' renderings.

While this methodology yielded more significant variability compared to other employed translation methods, it nonetheless proved valuable in identifying words that the experts believed should not be directly translated. Furthermore, specific direct translations caused a shift in meaning within the semantic differential, thereby posing a threat to the validity of the questionnaire.

5.3 Crowdsourcing Translation

The crowdsourcing approach yielded the highest number of responses, with 180 participants providing translations. However, participants were explicitly instructed only to translate word pairs in which they felt confident.

Within this translation context, words such as "good," "bad," "ugly," "attractive," "likable," "human," and "technical" generated a substantial number of convergent responses, each receiving more than 90 consistent translations. Conversely, words such as "cumbersome," "straightforward," "undemanding," "stylish," "alienating," and "tacky" elicited only ten similar translations.

Despite the convenience of crowdsourcing in gathering a large volume of responses, this particular translation method demonstrated the highest degree of dispersion. Moreover, many translations employed words not officially recognized in the Spanish Language Dictionary.

Another significant issue in this translation process was that many participants attempted to translate word pairs. Yet, they did not ensure that the proposed translations adhered to the expected antonymic relationship outlined in the questionnaire. This discrepancy could stem from a lack of comprehension regarding the nature and requirements of the questionnaire.

5.4 Machine Translation

Machine translation offers several benefits in the context of questionnaire translation. Firstly, it provides a quick and efficient method of translating, saving considerable time and effort compared to manual translation. Furthermore, some consistency is obtained because the rules or algorithms are identical.

However, one significant challenge associated with automatic translation is the inability to obtain the rationale behind the translation decisions made by the systems. While automated translation algorithms can produce linguistically correct translations, the underlying logic or reasoning behind those translations may need to be apparent.

Within this translation, several words had 100% convergence (see Table 1: 3Rm 7L, 9L, 12R, 15L, 15R, 16R, 19R, 23L, 25L, 25R). On the other hand, several words had low convergence (see Table 1: 6R, 10L, 10R, 13R, 14L, 14R, 17R, 22L). This convergence reached 3 out of the six translations, generally due to all translations from English or German being the same but different among languages.

Another area for improvement in the translation process was the inability to translate the questionnaire while preserving the semantic differential structure. This may have resulted in a loss of the nuanced meaning and relational context intended by the original questionnaire design.

The use of machine translation raised intriguing questions regarding the English version of the questionnaire, as the translations from English differed from those obtained through translation from German. This discrepancy prompted us to examine the possibility of improvement in the English version of the questionnaire. Furthermore, this discrepancy could indicate subtle differences in interpretation and understanding between the two language versions.

This observation underscores the importance of continuous refinement and validation of the questionnaire to ensure linguistic and conceptual equivalence across languages.

5.5 AI Translation

The final translation technique used was AI translation using ChatGPT. This translation had the same benefits as machine translation but addressed the translation issues while maintaining the semantic differential structure. Two approaches were conducted. First, each pair of words was translated separately, and the questionnaire was replicated. Second, the whole questionnaire was translated, stating to ChatGPT that it was a questionnaire with semantic differentials.

Interestingly, the main insight of this translation was the appearance of regionalisms of language used only in specific parts of Latin America. Furthermore, the tool proposed words such as: "manageable" and "captivador" that are not Spanish words.

5.6 Translation Proposal

The final translation proposal was formulated by consolidating insights derived from all translation techniques and through extensive discussions within the research team. Initially, a comprehensive analysis of the overall results obtained from each translation method was conducted, leading to the direct translation of 13 pairs of words. (See Table 1: 5, 9, 10, 12, 15, 16, 19, 20, 21, 22, 23, 26, and 27). These pairs demonstrated consistent translation outcomes in at least three employed translation techniques.

Following this initial set of words, a thorough discussion was held for pairs where one word was accurately translated while the second had multiple potential translations. During this discussion, not only the translation itself was considered, but also the contrast between the words was taken into account. Additionally, the Spanish Language Dictionary validated all proposed words from the first and second phases.

Furthermore, a decision was made to review all translations that were not direct adjectives or included negations (e.g., "professional," "not professional," "attractive," "not attractive"). In such cases, the research team delved deeper into the proposed translations to avoid negations.

In the second phase, 11 pair of words were translated and subsequently validated. (See Table 1: 1, 2, 3, 4, 7, 8, 11, 13, 14, 24, and 28). The final four pairs of words proved to be the most challenging to translate. The proposed translations for "professional - unprofessional," "rejecting-inviting," "unimaginative - creative," and "undemanding - challenging" were determined through in-depth discussions. In some instances, the original German version was also considered due to the lack of consensus among the various translation techniques. Furthermore, in cases where a direct adjective could not be found, the research team opted for either negative words or words rooted in other languages (i.e., "Frenchism") but present in the dictionary. Table 1 presents the proposed translation.

Table 1. Proposed translation for the AttrakDiff questionnaire

	English	Spanish	English	Spanish
1	human	humano	technical	técnico
2	isolating	aislante	connective	conectivo
3	pleasant	placentero	unpleasant	incómodo
4	inventive	inventivo	conventional	convencional
5	simple	simple	complicated	complicado
6	professional	profesional	unprofessional	amateur
7	ugly	feo	attractive	bello

(*continued*)

Table 1. (*continued*)

	English	Spanish	English	Spanish
8	practical	práctico	impractical	impráctico
9	likeable	agradable	disagreeable	desagradable
10	cumbersome	engorroso	straightforward	sencillo
11	stylish	con estilo	tacky	sin estilo
12	predictable	predecible	unpredictable	impredecible
13	cheap	barato	premium	valioso
14	alienating	excluyente	integrating	incluyente
15	brings me closer to people	me acerca a la gente	separates me from people	me separa de la gente
16	unpresentable	impresentable	presentable	presentable
17	rejecting	rechazable	inviting	atrayente
18	unimaginative	sin imaginación	creative	creativo
19	good	bueno	bad	malo
20	confusing	confuso	clearly structured	claramente estructurado
21	repelling	repulsivo	appealing	atractivo
22	bold	atrevido	cautious	cauteloso
23	innovative	innovador	conservative	conservador
24	dull	aburrido	captivating	cautivador
25	undemanding	poco exigente	challenging	desafiante
26	motivating	motivante	discouraging	desalentador
27	novel	novedoso	ordinary	ordinario
28	unruly	inmanejable	manageable	manejable

6   Evaluation

The evaluation of the translated version had two stages. First, a general review with 200 + participants to evaluate the translated proposal. Second, the evaluation of three products by a group of participants with experience with digital products to assess the validity, reliability, and correlations between the questions.

In the first evaluation, an exercise was conducted with the Spanish-translated version of the AttrakDiff questionnaire, involving a convenience sample of 208 individuals. Participants were asked to evaluate the instant messaging application WhatsApp. Of the 208 participants, 88 identified as female, 116 as male, and four preferred not to indicate their gender. The ages ranged from 14 to 80 years, with a mean of 27.44 years, a median of 23 years, a first quartile of 21 years, and a third quartile of 29 years. Regarding their relationship with technology, 138 participants indicated a high level (66.3%), 59 showed a moderate level (28.4%), and 11 indicated a low level (5.3%).

As a result of this evaluation, it was observed that the pair of concepts "barato" (cheap) and "costoso" (expensive) belonging to the Hedonic Quality of Identification category confused participants. Many attributed the cheap category to the product, assuming it meant "free" since no payment was required to use the application. Additionally, it was noted that there were high correlations between the concepts of Pragmatic Quality and Hedonic Quality of Identification, making it difficult to distinguish these two factors clearly.

With the insights of the first evaluation, a second evaluation was performed. In this effort, three products were evaluated: the subscription video-on-demand over-the-top streaming service Netflix, the online video sharing and social media platform Youtube, and the free and open-source learning management system Moodle implemented by a university. The participants in this second evaluation were all students linked to computer science ages ranging from 20 to 38. 90% of participants identified as male and 10% as female. All participants had a high relationship with technology.

The responses of 29 participants across the three products were used to calculate Cronbach's Alpha coefficients for the four scales of AttrakDiff. The results are as follows: Attractiveness, 0.92; Hedonic Identity, 0.71; Hedonic Stimulation, 0.71; Pragmatic Quality, 0.88. These coefficients are considered appropriate to demonstrate the consistency of the four scales.

It is worth noting that within the Hedonic Stimulation scale, the item "undemanding/challenging" showed negative correlations with five of the other six items on the scale. The AttrakDiff questionnaire considers "challenging" as the positive concept of the pair and "undemanding" as the negative concept, which may be interpreted differently by the participants, for whom an "undemanding" interaction would be perceived as positive. If this item is removed and the Cronbach's Alpha coefficient for the Hedonic Stimulation scale is recalculated, the value increases to 0.83.

Additionally, a principal component analysis with Varimax rotation was performed for the Identity, Stimulation, and Pragmatic Quality scales. The Attractiveness scale was not included in this analysis since hedonic and pragmatic qualities contribute to the rating of Attractiveness, according to the authors of AttrakDiff [2].

Bartlett's sphericity test was first conducted for the factorial analysis, yielding a significant result: Chi-square = 1309.24 (df = 210, n = 87), p-value < 0.01. Subsequently, the Kaiser-Meyer-Olkin (KMO) measure of sampling adequacy test was performed, resulting in a value of 0.9. Both results indicate that the correlation matrix is suitable for the factorial analysis.

The principal component analysis was conducted with the three factors corresponding to the Identity, Stimulation, and Pragmatic Quality scales. The results indicate that these three factors explain 59% of the variance. Table 2 presents the three identified factors and the corresponding loadings of each item on each factor.

The first factor corresponds to Pragmatic Quality and accounts for 25% of the variance, with loadings ranging from 0.53 to 0.82 for the seven items of the scale.

Table 2. Factorial Analysis with VariMax rotation for all three products (Netflix, Youtube, Moodle)

		Pragmatic	Stimulation	Identity
Identity	Isolating / Connective	0.31	0.32	**0.59**
	Cheap / Premium			**0.55**
	Tacky / Stylish		0.60	**0.44**
	Alienating / Integrating			**0.53**
	Unpresentable / Presentable	0.52		**0.49**
	Separates / Brings me closer to people			**0.37**
	Unprofessional / Professional			**0.33**
Pragmatic	Confusing / Clearly structured	**0.76**		0.38
	Cumbersome / Straightforward	**0.80**	0.36	
	Technical / Human	**0.48**	0.47	
	Unruly / Manageable	**0.70**		0.41
	Impractical / Practical	**0.69**		0.43
	Unpredictable / Predictable	**0.53**		0.32
	Complicated / Simple	**0.82**	0.32	
Stimulation	Dull / Captivating	0.39	**0.75**	
	Cautious / Bold		**0.64**	
	Conservative / Innovative	0.32	**0.77**	
	Conventional / Inventive		**0.64**	
	Ordinary / Novel		**0.82**	
	Undemanding / Challenging	−0.79	**−0.10**	−0.23
	Unimaginative / Creative	0.50	**0.64**	0.38

The second factor corresponds to Stimulation, with loadings between 0.64 and 0.82 for six out of the seven items. The only item that presents issues is "undemanding/challenging," which reports negative loadings across all three factors. This item had already been identified as problematic in the study of Cronbach's alpha coefficients.

The third identified factor corresponds to Identity, with loadings between 0.33 and 0.59 for the seven items of the scale. Table 2 also shows some items with significant correlations in scales different from their own.

Regarding the Attractiveness scale, a linear regression model was created with the three scales as predictor variables and Attractiveness as the response variable. The obtained F-statistic is 184.6, and the corresponding p-value is $< 2.2e\text{-}16$, for the three variables and 83 degrees of freedom, indicating that the model is statistically significant and explains 75.62% of the variance in Attractiveness ($R^2 = 0.7562$). This means that, as proposed by the original questionnaire, the Attractiveness scale is globally explained by the evaluations of the pragmatic and hedonic scales.

In conclusion, the results of the correlations in Cronbach's alpha coefficients, the factorial analysis, and the linear regression model support the validity of the translated version in Spanish.

7 Conclusions

The translation of user experience questionnaires is a critical step in cross-cultural research to ensure the validity and reliability of the instruments. In this study, we employed multiple translation methods to propose a translated version of the AttrakDiff questionnaire.

The human translation by students allowed for a comprehensive exploration of different interpretations and linguistic nuances. However, it also introduced variability and inconsistencies due to the students' varying expertise and understanding of the questionnaire. The translation by HCI experts brought valuable insights and domain-specific knowledge, enhancing the accuracy and relevance of the translated version. Crowdsourcing provided many responses, but it also resulted in a diverse range of translations and the inclusion of non-standard words not found in the Spanish Language Dictionary. Automatic translation, while efficient, could not consider the semantic differential and context, resulting in translations that needed further refinement.

Evaluating the translated version of the AttrakDiff questionnaire using a sample of 208 participants provided valuable insights into the consistency within categories and highlighted specific problems. Furthermore, a second evaluation with 29 participants evaluating three products was conducted. In this evaluation, several statistical techniques were applied to assess the validity of the translation proposal.

The results of the correlations in Cronbach's alpha coefficients, the factorial analysis, and the linear regression model support the validity of the translated version in Spanish.

Future research could focus on refining the translated version of the questionnaire based on the insights gained from this study. Further investigation is needed to enhance the clarity and distinction between the different dimensions of user experience. Moreover, it is worth exploring alternative methods and techniques for translation to broaden the scope of possibilities.

Acknowledgments. This research was partially funded by ECCI and CITIC at the University of Costa Rica, grant number 834C1013. In this work also contributed Hugo Villalta a student of the graduate program in computers and informatics at UCR.

References

1. Brito Maia, C.L., Sucupira Furtado, E.: A Systematic Review About User Experience Evaluation. In: Design, User Experience, and Usability: Design Thinking and Methods. DUXU 2016. Lecture Notes in Computer Science. pp. 445–455. Springer, Cham (2016). https://doi.org/10.1007/978-3-319-40409-7_42
2. Hassenzahl, M., Burmester, M., Koller, F.: AttrakDiff: Ein Fragebogen zur Messung wahrgenommener hedonischer und pragmatischer Qualität. In: Mensch & Computer 2003. Berichte des German Chapter of the ACM, vol 57. pp. 187–196. Vieweg+Teubner Verlag (2003). https://doi.org/10.1007/978-3-322-80058-9_19

3. Díaz-Oreiro, I., López, G., Quesada, L., Guerrero, L.A.: UX evaluation with standardized questionnaires in ubiquitous computing and ambient intelligence: a systematic literature review. Adv. Hum. Comput. Interact. **2021**, 1–22 (2021). https://doi.org/10.1155/2021/551 8722

4. Lallemand, C., Gronier, G.: Méthodes de design UX: 30 méthodes fondamentales pour concevoir et évaluer les systèmes interactifs. Librairie Eyrolles, Paris, France (2018)

5. Laugwitz, B., Held, T., Schrepp, M.: Construction and Evaluation of a User Experience Questionnaire. In: HCI and Usability for Education and Work, pp. 63–76 (2008). https://doi.org/10.1007/978-3-540-89350-9_6

6. Lárusdóttir, M.K.: Measuring the user experience of a task oriented software. In: VUUM 2008. , Reykjavik, Iceland (2008)

7. Raita, E., Oulasvirta, A.: Too good to be bad: Favorable product expectations boost subjective usability ratings. Interact. Comput. **23**, 363–371 (2011). https://doi.org/10.1016/j.intcom. 2011.04.002

8. Rauschenberger, M., Schrepp, M., Olschner, S., Thomaschewski, J., Perez Cota, M.: Measurement of user experience: a spanish language version of the user experience questionnaire (UEQ). In: Information Systems and Technologies (2012). https://doi.org/10.13140/2.1.1783. 9045

9. Rauschenberger, M., Schrepp, M., Perez-Cota, M., Olschner, S., Thomaschewski, J.: Efficient Measurement of the User Experience of Interactive Products. How to use the User Experience Questionnaire (UEQ).Example: Spanish Language Version. Int. J. Interact. Multimed. Artif. Intell. 2, 39 (2013). https://doi.org/10.9781/ijimai.2013.215

10. Lallemand, C., Koenig, V., Gronier, G., Martin, R.: Création et validation d'une version française du questionnaire AttrakDiff pour l'évaluation de l'expérience utilisateur des systèmes interactifs. Eur. Rev. Appl. Psychol. **65**, 239–252 (2015). https://doi.org/10.1016/j.erap. 2015.08.002

11. Sevilla-Gonzalez, M.D.R. et al.: Spanish Version of the System Usability Scale for the Assessment of Electronic Tools: Development and Validation. JMIR Hum. Factors. 7, e21161 (2020). https://doi.org/10.2196/21161

12. Hernández-Campos, M., Thomaschewski, J., Law, Y.C.: Results of a Study to Improve the Spanish Version of the User Experience Questionnaire (UEQ). Int. J. Interact. Multimed. Artif. Intell. InPress, 1 (2022). https://doi.org/10.9781/ijimai.2022.11.003

The Mind in Virtual Meetings: Comparing VR and Video Conferencing Environments Through Experiential Impact Assessment and EEG Analysis

Eric Kirchgessner[1]([✉]), Matías Sothers[1], Valentina Aravena[1], Nelson Baloian[1] [ID], and Gustavo Zurita[2] [ID]

[1] Department of Computer Science, FCFM, Universidad de Chile, Av. Beauchef 850, Santiago, Chile
{eric.kirchgessner,matias.sothers,
valentina.aravena.p}@ug.uchile.cl, nbaloian@dcc.uchile.cl
[2] Department of Information Systems and Management Control, Faculty of Economics and Business, Universidad de Chile, Diagonal Paraguay 257, Santiago, Chile
gzurita@fen.uchile.cl

Abstract. The advent of digital communication technologies has notably fostered remote collaboration. While platforms like Zoom are prevalent, emerging Virtual Reality (VR) technologies like Meta Quest 2 introduce new dimensions for virtual collaboration. This study investigates whether VR-based group meetings are more conducive to participant engagement, motivation, and non-hindrance of collaboration than traditional video conferencing platforms such as Zoom. This study is novel in employing Electroencephalogram (EEG) technology and questionnaires to assess the human factors of engagement, valence, arousal, motivation, flow, system usability, emotional state, and social presence. In this study, participants were engaged in a collaborative turn-based drawing activity in two distinct environments - a traditional video conferencing setting using Zoom and a VR-based setting using Horizon Workrooms. Both environments were configured to maintain equivalent functionality and settings. EEG data was collected using an EMOTIV EPOC + 14 channel wireless EEG headset. Findings reveal that the activity in VR was more intrinsically motivating than the activity in Zoom, independent of the participants' previous VR experience. Additionally, participants reported greater enjoyment (higher valence) and excitement (higher arousal) in the VR setting, with significant results from questionnaires. A strong correlation was found between EEG and questionnaire assessed arousal. Contrary to expectations, no significant differences were observed in usability, co-presence, focus, stress, and effort between the two platforms. These results provide insights into the potential of VR as a tool for fostering engagement and motivation in remote group activities and call for further investigation into the underlying mechanisms.

Keywords: Human-computer Interaction (HCI) · Human Factors · Virtual Reality (VR) · Video Conferences · Electroencephalography (EEG)

© The Author(s), under exclusive license to Springer Nature Switzerland AG 2023
J. Bravo and G. Urzáiz (Eds.): UCAmI 2023, LNNS 835, pp. 255–267, 2023.
https://doi.org/10.1007/978-3-031-48306-6_26

1 Introduction

The advent of remote collaboration technologies has revolutionized the way we communicate and collaborate, by virtue of facilitating group interactions irrespective of geographical distance. Since the start of the pandemic, traditional face-to-face meetings have been increasingly replaced by video conferencing platforms such as Zoom, Google Meets or Discord, which allow for remote collaboration. The emergence of Virtual Reality (VR) technologies has opened up new possibilities for education, social interaction, work and training [1], all situations where both in-person and virtual meetings are prevalent. VR offers an immersive and interactive experience that enhances participant engagement [2] and, immersion [3]. The use of VR for group meetings, as facilitated by platforms like Horizon Workrooms [4], may benefit from this effect. This is contrasted with the more conventional video conferencing platforms, which, despite their widespread use, may not provide the same level of engagement.

There is research comparing VR and Video Conferencing [5–7], and research using EEG to measure VR experiences [8–10], but there is limited research measuring VR and video conferencing experiences with EEGs. This study aims to fill this gap, providing empirical insights into the relative user experience of VR and video conferencing platforms. If a difference between the more common computer-based video conferencing and the newer VR technologies can be found, we can better design digital collaborative tools.

The primary objective of this study is not to quantify individual collaboration within a virtual reality environment. Instead, our focus is on identifying and examining the human factors that potentially impact the quality of such an environment. We aim to understand how these elements could enhance the environment's capacity to foster superior collaboration, particularly when comparing virtual reality with video conferencing for meetings.

The research hypothesis of this study is that, when used in a context similar to Zoom, VR usage for collaborative meetings over the Internet is more conducive to maintaining participant motivation, interest, and engagement without impeding collaboration due to stress, tedium, or difficulty focusing.

Virtual meetings, encompassing video conferencing, VR conferencing and beyond, have many applications, notably in education [11] and training [12]. Evidence suggests that higher engagement improves educational outcomes [13]. Low engagement in corporate video conferencing is considered a problem, the perception of which is affected by the technology used for the meetings [14]. In a professional setting, planning, problem solving and evaluation, activities that usually take place during meetings, are linked to the concept of Flow [15]. The sense of social presence plays a crucial role in instilling comfort within an environment, and can serve as a pillar in community building and trust establishment [16]. These are important factors to consider when evaluating a virtual meeting system.

2 Related Works

2.1 Virtual Reality Usage for Collaborative Meetings

Since its inception, VR has been the research focus for education, training, and therapeutic applications [17, 18]. The application of VR in group meetings has been receiving growing attention and could be considered part of the recent trend of "fourth wave of computing" also known as "Metaverse" [19, 20]. The capability of VR to enhance empathy and understanding among participants [21] and provide a sense of immersion, presence, and flow [22] are some of the potential advantages reported in the literature. Some preliminary comparisons between VR meetings and video conferencing have been made [5–7], with the main result being an improved sense of social presence [6] and immersion [7]. It is worth noting, the impact of VR on social-presence is still unclear [23]. Measuring human factors in VR meetings has mostly been carried out using questionnaires [5–7, 24]. VR has limitations halting adoption though, such as motion sickness [25].

2.2 Virtual Reality Usage with EEGs

Electroencephalogram (EEG) is a non-invasive method used to record electrical activity of the brain. It is widely used in cognitive and neuroscience research for understanding various cognitive processes, such as attention, memory, and emotion [26]. Consumer-grade EEG devices, such as those offered by EMOTIV, have made it easier and more affordable to record and analyze brain activity in real-world settings. These devices, while not offering the same precision as medical-grade equipment due to inferior signal quality, are still useful for non-critical and non-medical applications [27], such as this one.

The integration of EEG technology with VR can provide valuable insights into the cognitive and emotional impacts of VR usage. A few pioneering studies have begun to explore this combination, finding correlations between VR experiences and specific EEG patterns such as measuring emotional responses to imagery in virtual reality [8], estimating visual discomfort [10] and investigating spatial presence in VR [28]. However, the application of this approach in the context of collaborative meetings remains largely unexplored.

There are certain challenges when combining these two technologies, since they are both head mounted and movement can interfere with EEG measurement, as well as the fact that a VR HMD is a potential source of electromagnetic interference. These challenges and potential solutions were outlined by Tauscher et al. [29], and have been taken into consideration for the development of this experiment.

3 Method

3.1 Participants

Twelve engineering students (4 male, 5 female, and 3 non-binary or preferred not to answer) served as participants in the experiment, forming four groups. The age of the participants varied between 22 and 29, with the median being 24. The recruitment process

Fig. 1. Top left: A participant in virtual reality wearing the EEG. Top right: A participant in Zoom wearing the EEG. Bottom: The view from within Horizon Workrooms.

involved inviting students from the Faculty of Physical Science and Mathematics of the University of Chile to take part in the study. Prior experience with virtual reality was not a filter for recruitment; however, all participants had high proficiency levels with computers.

3.2 Activity

The study employed a collaborative turn-based drawing activity involving three partic-ipants, similar in spirit to the popular game "Exquisite Corpse". The goal of the game is to create a drawing centered around a specific pre-assigned theme, where on each participant's turn, they can only add a single line to the group's drawing. This activity was chosen deliberately to reduce the impact of the capabilities of the platform, allowing the subsequent analysis to concentrate on the influence of the digital environments and human factors instead. Unlike more complicated activities like brainstorming sessions or group presentations, each participant only had access to two tools: a pen and "undo". The assumption being that this enabled a more precise comparison between flatscreen meetings and VR meetings, without being overshadowed by the non-environmental characteristics and features of each platform.

Structuring the activity as a simple game with equitable participation guaranteed a more balanced and impartial evaluation, controlling for potential confounding variables such as varying skill levels, assertiveness, or personal inclinations among participants.

The study was executed sequentially in two distinct environments: Zoom (on a laptop) and Horizon Workrooms (VR). For both environments, the participants were sent to separate offices in the building. Both environments had microphones active. In Zoom, participants had their cameras active. In horizon workrooms, each participant had a different 3D avatar which mimicked their real head and hand movements. The avatar was not customizable. For the duration of the activity, participants remained seated, to reduce EEG interference and motion sickness, as well as in separate offices, such that they could not see or hear each other (See Fig. 1).

The rules of the game and the tools available to the participants remained consistent across both platforms. To prevent monotony and its conceivable impact on the results, a different theme was presented to the group in each iteration, the two options being "A Human Face" and "A Fantasy Creature". The sequence in which the groups engaged with the platforms was also varied to prevent any potential biases or learning curves associated with one environment from systematically influencing the results (See Table 1).

Table 1. Order of environments and themes for each group

Group	1st Environment	1st Theme	2nd Environment	2nd Theme
1	Virtual Reality	Human Face	Computer	Fantasy Creature
2	Computer	Fantasy Creature	Virtual Reality	Human Face
3	Virtual Reality	Fantasy Creature	Computer	Human Face
4	Computer	Human Face	Virtual Reality	Fantasy Creature

Participants were given an instruction sheet to read while the EEGs were set up and the environments were prepared. These instructions were later reiterated verbally. Participants were assigned a specific color pen, the color of which determined their position in the turn order. Participants were to draw a single, short, continuous line on their turn using a VR controller or a mouse on the laptop. Clear examples of permissible and impermissible contributions were demonstrated on the instruction sheet. Participants were limited to using only their assigned marker and had access to an 'undo' button for correcting errors.

After all participants confirmed their understanding of the rules, they were given 7 min on a countdown timer, and the game began. During this time, the research monitors did not interfere other than to answer questions about the rules, solve any technical issues, and let the participants know when they had 1 min left and when they had to stop. Beforehand, participants were encouraged to communicate and engage with each other during the game. Participants were informed that if at any moment they felt discomfort or did not wish to continue, they could halt the experiment.

Fig. 2. Collage of a selection of resulting drawings. Tagged by group number, platform, and concept. In order of creation from top to bottom for each group.

3.3 Apparatus

Multiple questionnaires were utilized to assess the factors under consideration and validate the assumptions made during the design of the activity. The final questionnaire given to participants consisted of 51 Likert-scale questions on paper. Each question was repeated twice, once for each environment being compared. Additionally, the questionnaire included two multiple-choice questions to gather specific information about participants' prior experience with both Zoom and VR.

Intrinsic Motivation Inventory. *(IMI)* [30]. A well-validated 5-point Likert-scale questionnaire for measuring intrinsic motivation. Of the subscales it offers, we included Interest/Enjoyment, Effort/Importance, and Pressure/Tension for 17 questions. Interest/Enjoyment is the most important subscale, corresponding to intrinsic motivation. A higher score means that the participant experienced more of the metric described by the name of the subscale.

Flow Short Scale (FSS). [31] A short 7-point Likert-scale questionnaire measuring flow state (focus) during an activity. The last three questions of this questionnaire were not analyzed, as they do not directly correspond to the flow scale, for a total of 10 questions. A higher score indicated that a higher state of focus was achieved.

System Usability Scale (SUS). [32, 33] A commonly used 10-question 5-point Likert-scale questionnaire for measuring the usability of a system. Unlike the other questionnaires, SUS scores go from 0 to 100, with 68 being considered "average usability", and 100 being the highest.

Self-assessment Manikin (SAM). [34] An image-based questionnaire with one question each for measuring valence, arousal, and dominance. Originally, these measures were captured on a 9-point scale. However, to enhance interpretability and practicality, we discretized this into a 5-point scale. Both Valence and Arousal have been inverted for easier interpretability with the EEG data. Valence is the positive/negative reaction to a stimulus; a higher score means a more pleasant experience and a lower score means a less pleasant one. Arousal is the intensity of emotion caused by a stimulus, where a higher score means more excitement and a lower score indicates calmness. Dominance is the control the stimulus has over the response.

Networked Minds Social Presence Inventory (SPI). [35] A questionnaire for assessing social presence. Of the three levels of presence that can be assessed, only the first, co-presence, was included in the questionnaire. For the subscale used, a higher score means the participant felt their co-participants were more present (perception of self), and they felt that their co-participants felt the same way (perception of others).

Electroencephalograms (EEG). During both activities, participants wore an EMOTIV EPOC + 14 channel wireless EEG headset. At the start of each setting, while participants read the instructions or idled waiting for other participants to be ready, a 3-min baseline measurement was taken. Then, another measurement was taken during the entire 7 min of each activity. Measurements were taken of the pre-processed EEG θ, α, β and γ frequency band powers provided by the EMOTIV software, with a polling rate of 8 Hz. The performance metrics offered by EMOTIV were not used, as they are a black box and designed for single-user use.

For its use with the VR HMD, the EEG was placed first, and then the HMD on top, we found this to be the best way to maximize comfort and not degrade contact quality. Despite the ability to stand up and walk to the whiteboard in Horizon Workrooms, in order to prevent bad data [29], participants were to stay seated and used the miniature whiteboard on their desks. Before each activity, the quality of the electrode contacts was verified. Three metrics were calculated from the band powers:

Engagement. This metric can be estimated by the ratio between the total beta band power and the sum of the total theta and alpha band powers [36].

Arousal. This metric can be estimated by the ratio of the sum of the beta band power at F3 and F4, and the sum of the alpha band power at F3 and F4 [37].

Valence. This metric can be estimated by the ratio of the alpha and beta band powers at F4, minus the ratio of the alpha and beta band powers at F3 [37].

3.4 Statistical Analysis

For all measurements, significance was considered at $p < 0.05$.

To Analyze the Questionnaires: Values were calculated according to the instructions defined by their original specifications (Sect. 3.3). All answers were aggregated to be compared between platforms. For this, we assumed dependency between the two trials. Because of the sample size, the Shapiro-Wilk normality test was applied to each questionnaire's subscale's difference per platform, to determine whether to use the paired sample t-test or the Wilcoxon signed-rank test to calculate the significance of the difference. For questionnaires that showed a significant difference, Pearson rank coefficient between the VR-related results and each participant's experience in VR was calculated.

To Analyze EEG Results: For the baseline and activity on each platform were compared for each participant. Descriptive statistics were then calculated from the ratios. Like the questionnaires, dependency was assumed between the two trials. The normality of the data was tested, and the significance of the results was calculated.

To Compare Questionnaire and EEG Results: Pearson rank correlation was calculated between EEG and questionnaire valence and arousal.

4   Results

All groups managed to successfully complete both tasks, producing a variety of drawings, in complexity, size and concept (See Fig. 2). Most participants seemed to have enjoyed the activity. Afterwards, two participants reported slight visual discomfort.

EEG Data could not be acquired from two of the twelve participants. For the rest, some non-essential electrode's signals had to be rejected due to poor signal.

Of the participants: 4 reported never having used VR, 5 reported having used VR a couple times before, 2 reported having used it 3–5 times before, and 1 reported using it weekly. As for virtual meetings, all reported having them more than once a month.

4.1   Intrinsic Motivation Inventory

The difference between VR and Zoom for Interest/Enjoyment is significant, for the rest it is not (Fig. 3). No significant correlation could be found with previous experience in VR for Interest/Enjoyment (coefficient $= 0.15$, p-value $= 0.64$).

4.2   Flow Short Scale

The mean FSS score for the activity in VR was 5.725 (sd $= 0.594$). The mean FSS score for the activity in Zoom was 5.600 (sd $= 1.124$). Normality is assumed (statistic $= 0.907$, p-value $= 0.198$). The null hypothesis can't be rejected (statistic $= 0.406$, p-value $= 0.692$).

4.3   System Usability Scale

The mean SUS score for Horizon Workrooms, when used for this specific activity was 69.17 and for Zoom it was 77.50. Both are above the threshold of 68 which is considered an average usability. Normality is assumed (statistic $= 0.946$, p-value $= 0.584$). The null hypothesis can't be rejected (statistic $= 1.09$, p-value $= 0.298$).

4.4   Self-assessment Manikin

The difference between VR and Zoom for valence and arousal is significant, and for dominance it is not (Fig. 4). No significant correlation with previous experience in VR could be found for arousal (coefficient $= 0$, p-value $= 1$) or valence (coefficient $= -0.03$, p-value $= 0.9$).

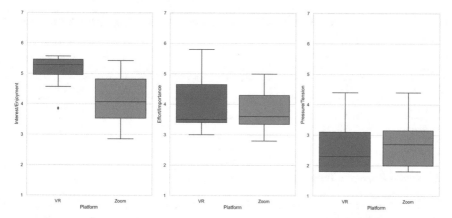

Fig. 3. Comparative boxplots of the different IMI subscales for each platform.

Table 2. Null hypothesis: there is no difference between the distributions of IMI scores for VR and Zoom.

Subscale	Shapiro-Wilk Test P value	Statistic	P Value
Interest/Enjoyment	0.414	3.99	**0.009**
Effort/Importance	0.491	1.81	0.096
Pressure/Tension	0.575	–0.48	0.640

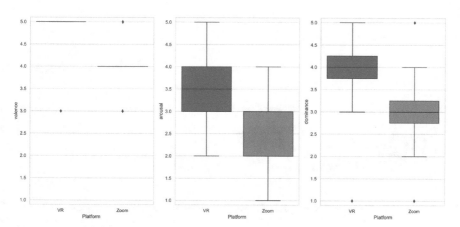

Fig. 4. Comparative boxplots of the different SAM subscales for each platform. *(Note: as mentioned in* Sect. 3.3, *valence and arousal have been inverted for more intuitive interpretation)*

Table 3. Null hypothesis: there is no difference between the distributions of SAM scores for VR and Zoom.

Subscale	Shapiro-Wilk Test P value	Statistic	P Value
Valence	0.003	0.0	**0.004**
Arousal	0.011	0.0	**0.003**
Dominance	0.024	−1.64	0.127

4.5 Social Presence Inventory

The mean co-presence for VR was 3.041 (sd = 0.25, perception of self: 3.104, perception of others: 2.979) and for Zoom it was 2.906 (sd = 0.39, perception of self: 3.031, perception of others: 2.916). Normality is assumed (statistic = 0.943, p-value = 0.543). The null hypothesis can't be rejected (statistic = 1.277, p-value = 0.227).

4.6 EEG Arousal, Engagement & Valence

All EEG metrics for all activities and participants were found to most likely not distribute normally (Fig. 5). Arousal difference p-value = 0.130, engagement difference p-value = 0.193, valence difference p-value = 0.566. None of the differences were found to be significant. With visual inspection, the EEG data is very noisy, and this noise is exacerbated by the usage of VR.

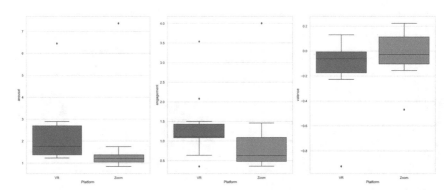

Fig. 5. Comparative boxplot of the difference in the EEG metrics per activity

However, a significant strong correlation was found between questionnaire and EEG arousal (coefficient = 0.508, p-value = 0.021). For questionnaire and EEG valence, the correlation was not significant (coefficient = -0.370, p-value = 0.107). For questionnaire Enjoyment/Interest and EEG engagement, the correlation was not significant (coefficient: 0.313, p-value: 0.177).

4.7 Summary

- The activity in VR was more motivating than the activity in Zoom (Table 2). No correlation was found with previous VR experience (Sect. 4.1).
- Participants enjoyed (higher valence) and were more excited (higher arousal) by the activity in VR, with significant results from the questionnaires and non-significant results with the EEG data (Table 3). No correlation was found with previous VR experience (Sect. 4.4).
- EEG and questionnaire assessed arousal was found to be strongly correlated (Sect. 4.6).
- No significant difference was found for usability, co-presence, focus (flow state), stress (pressure/tension), and effort (effort/importance) (Sects. 4.2, 4.3, 4.5).

5 Discussion

Our study evaluated user experience in VR and Zoom environments using various questionnaires and EEG data. The findings indicate that VR generated more intrinsic motivation and excitement compared to Zoom, according to the Intrinsic Motivation Inventory and the Self-assessment Manikin respectively. These emotional responses were not significantly influenced by prior VR experience, this suggests that VR's appeal extends beyond the novelty effect, where new experiences are inherently more exciting, pointing towards the inherent appeal of VR's immersive nature. The strong correlation between self-reported and EEG-measured arousal validated the combined use of subjective and objective measures.

Other human factors like focus, co-presence, and effort, as well as user experience factors like usability showed no significant difference between VR and Zoom. This implies that VR's immersive qualities may not necessarily enhance practical user experience aspects, however, they also do not detract from them, an important consideration when choosing a platform for meetings.

Our study faces several limitations. First, our small sample size and homogenous group of engineers limits the generalizability of our findings to broader, diverse populations, especially when it comes to levels of previous experience with Virtual Reality. This composition may particularly impact our usability outcomes, considering the possible higher technical proficiency within the group. Second, despite using advanced equipment, it is not medical grade, and the accuracy of EEG measures could have been compromised by the inherent noise [27], especially in VR conditions [29]. These limitations underscore the need for larger, more varied studies and refined EEG methodologies in future research.

Finally, our results highlight the role of the activity itself. In our study, all groups were able to complete their tasks in both environments and appeared to enjoy themselves, highlighting that the choice of task can greatly influence the user experience in both VR and non-VR settings.

In conclusion, this study provides empirical support (given equal conditions to traditional platforms) for the use of VR in scenarios where increased intrinsic motivation, enjoyment and arousal are desirable. Nevertheless, considering the non-significant differences in other key aspects and the shortcomings of this study, the decision to use VR should be context-dependent, considering not only its advantages but also its current limitations.

References

1. Hamad, A., Jia, B.: How virtual reality technology has changed our lives: an overview of the current and potential applications and limitations. Int. J. Environ. Res. Public Health **19**, 11278 (2022)
2. Allcoat, D., von Mühlenen, A.: Learning in virtual reality: effects on performance, emotion and engagement. Res. Learn. Technol. **26**, 2140 (2018)
3. Bowman, D.A., McMahan, R.P.: Virtual reality: how much immersion is enough? Comput. **40**, 36–43 (2007)
4. VR for business & business VR headsets I Meta for Work. https://forwork.meta.com/. Accessed 25 July 2023
5. Abramczuk, K., Bohdanowicz, Z., Muczyński, B., Skorupska, K.H., Cnotkowski, D.: Meet me in VR! can VR space help remote teams connect: a seven-week study with horizon work-rooms. Int. J. Hum. Comput. Stud. **179**, 103104 (2023)
6. Steinicke, F., Lehmann-Willenbrock, N., Meinecke, A.L.: A first pilot study to compare virtual group meetings using video conferences and (immersive) virtual reality. In: Proceedings of the 2020 ACM Symposium on Spatial User Interaction, pp. 1–2. Association for Computing Machinery, New York, NY, USA (2020)
7. Campbell, A.G., Holz, T., Cosgrove, J., Harlick, M., O'Sullivan, T.: Uses of virtual reality for communication in financial services: a case study on comparing different telepresence interfaces: virtual reality compared to video conferencing. In: Arai, K., Bhatia, R. (eds.) Advances in Information and Communication, pp. 463–481. Springer International Publishing, Cham (2020)
8. Horvat, M., Dobrinić, M., Novosel, M., Jerčić, P.: Assessing emotional responses induced in virtual reality using a consumer EEG headset: a preliminary report. In: 2018 41st International Convention on Information and Communication Technology, Electronics and Microelectronics (MIPRO), pp. 1006–1010 (2018)
9. Krogmeier, C., Mousas, C.: Exploring EEG-annotated affective animations in virtual reality: suggestions for improvement. Presented at the November 30 (2022)
10. Mai, C., Hassib, M., Königbauer, R.: Estimating visual discomfort in head-mounted displays using electroencephalography. In: Bernhaupt, R., Dalvi, G., Joshi, A., K. Balkrishan, D., O'Neill, J., and Winckler, M. (eds.) Human-Computer Interaction – INTERACT 2017, pp. 243–252. Springer International Publishing, Cham (2017) https://doi.org/10.1007/978-3-319-68059-0_15
11. Kavanagh, S., Luxton-Reilly, A., Wuensche, B., Plimmer, B.: A systematic review of Virtual Reality in education. Themes Sci. Technol. Educ. **10**, 85–119 (2017)
12. Xie, B., et al.: A review on Virtual Reality skill training applications. Front. Virtual Real. **2**, 645153 (2021)
13. Finn, J.D., Zimmer, K.S.: Student engagement: what is it? why does it matter? In: Christenson, S.L., Reschly, A.L., Wylie, C. (eds.) Handbook of Research on Student Engagement, pp. 97–131. Springer, US, Boston, MA (2012)
14. Kuzminykh, A., Rintel, S.: low engagement as a deliberate practice of remote participants in video meetings. In: Extended Abstracts of the 2020 CHI Conference on Human Factors in Computing Systems, pp. 1–9. Association for Computing Machinery, New York, NY, USA (2020)

15. Nielsen, K., Cleal, B.: Predicting flow at work: Investigating the activities and job characteristics that predict flow states at work. J. Occup. Health Psychol. **15**, 180–190 (2010)
16. Aragon, S.R.: Creating social presence in online environments. New Dir. Adult Contin. Educ. **2003**, 57–68 (2003)
17. Garcia-Palacios, A., Hoffman, H.G., Kwong See, S., Tsai, A., Botella, C.: Redefining therapeu-tic success with virtual reality exposure therapy. Cyberpsychol. Behav. **4**, 341–348 (2001)
18. Emmelkamp, P.M.G., Meyerbröker, K.: Virtual reality therapy in mental health. Annu. Rev. Clin. Psychol. **17**, 495–519 (2021)
19. Mystakidis, S.: Metaverse. Encyclopedia. **2**, 486–497 (2022)
20. Ning, H., et al.: A survey on the metaverse: the state-of-the-art, technologies, applications, and challenges. IEEE Internet Things J. **10**(16), 1–1 (2023). https://ieeexplore.ieee.org/doc ument/10130406
21. Shin, D.: Empathy and embodied experience in virtual environment: to what extent can virtual reality stimulate empathy and embodied experience? Comput. Hum. Behav. **78**, 64–73 (2018)
22. Mütterlein, J.: The Three Pillars of Virtual Reality? Investigating the Roles of Immersion, Pres-ence, and Interactivity. (2018)
23. Oh, C.S., Bailenson, J.N., Welch, G.F.: A systematic review of social presence: definition, antecedents, and implications. Front. Robot. AI. **5**, 409295 (2018)
24. Sadeghi, A.H., et al.: Remote multidisciplinary heart team meetings in immersive virtual reality: a first experience during the COVID-19 pan-demic. BMJ Innov. **7**, 2 (2021)
25. Chattha, U.A., Janjua, U.I., Anwar, F., Madni, T.M., Cheema, M.F., Janjua, S.I.: Motion sick-ness in Virtual Reality: an empirical evaluation. IEEE Access. **8**, 130486–130499 (2020)
26. Teplan, M.: FUNDAMENTALS OF EEG MEASUREMENT. Meas. Sci. Rev. 2, (2002)
27. Duvinage, M., Castermans, T., Petieau, M., Hoellinger, T., Cheron, G., Dutoit, T.: Performance of the Emotiv Epoc headset for P300-based applications. Biomed. Eng. OnLine. **12**, 56 (2013)
28. Baumgartner, T., Valko, L., Esslen, M., Jäncke, L.: Neural correlate of spatial presence in an arousing and noninteractive virtual reality: an EEG and psychophysiology study. Cyberpsy-chol. Behav. **9**, 30–45 (2006)
29. Tauscher, J.P., Schottky, F.W., Grogorick, S., Bittner, P.M., Mustafa, M., Magnor, M.: Im-mersive EEG: evaluating electroencephalography in Virtual Reality. In: 2019 IEEE Conference on Virtual Reality and 3D User Interfaces (VR), pp. 1794–1800 (2019)
30. Intrinsic Motivation Inventory (IMI) – selfdeterminationtheory.org. https://selfdeterminationt heory.org/intrinsic-motivation-inventory/. Accessed 22 May 2023
31. Engeser, S., Rheinberg, F.: Flow, performance and moderators of challenge-skill balance. Motiv. Emot. **32**, 158–172 (2008)
32. System Usability Scale (SUS) I Usability.gov. https://www.usability.gov/how-to-and-tools/ methods/system-usability-scale.html. Accessed 2023/07/24
33. Bangor, A., Kortum, P.T., Miller, J.T.: An empirical evaluation of the system usability scale. Int. J. Human-Computer Interact. **24**, 574–594 (2008)
34. Bradley, M.M., Lang, P.J.: Measuring emotion: The self-assessment manikin and the semantic differential. J. Behav. Ther. Exp. Psychiatry **25**, 49–59 (1994)
35. Biocca, P.F., Harms, P.C.: Guide to the Networked Minds Social Presence Inventory v. 1.2, https://web-archive.southampton.ac.uk/cogprints.org/6743/, last accessed 2023/06/05
36. Pope, A.T., Bogart, E.H., Bartolome, D.S.: Biocybernetic system evaluates indices of operator engagement in automated task. Biol. Psychol. **40**, 187–195 (1995)
37. McMahan, T., Parberry, I., Parsons, T.D.: Evaluating Player Task Engagement and Arousal Using Electroencephalography. Procedia Manuf. **3**, 2303–2310 (2015)

Human-Centered Navigation Systems: A Systematic Literature Review

Alejandro Najani León-Gómez$^{(\boxtimes)}$, Edgard Benítez-Guerrero, Carmen Mezura-Godoy, and Viviana Yarel Rosales-Morales

Faculty of Statistics and Informatics, Universidad Veracruzana, 91020 Xalapa, Mexico
zS22000342@estudiantes.uv.mx

Abstract. A navigation system (NS) helps a user to reach a destination from a given location. However, they provide a limited user experience as they are user-agnostic, i.e. they do not consider user characteristics, preferences, or conditions. This paper presents a systematic literature review (SLR) on human-centered NSs, which are NSs that personalize their interface and/or their contents based on a user model in order to improve the user experience. The aim of this SLR was to identify the most important elements of human-centered NS, their application domains, and the existing limitations and challenges. To perform this SLR, the PRISMA method was applied. As a result, we obtained 15 papers, which let us see that user models have been primarily described as ontologies, that this information is used in the selection of routes and their display, and that applications have been proposed for indoor and outdoor environments.

Keywords: Human Centered Computing · Navigation System · User Model

1 Introduction

Navigation encompasses all activities related to determining and reaching specific locations or destinations [1]. For this work, it refers to the techniques and technologies that assist users in their mobility across different environments, enabling them to move efficiently between two points. Nowadays, some of the most popular navigation systems known are Google Maps [2] and Apple Maps [3]. Google Maps is a constantly updated navigation system with the ability to display different layers in its interface that allow the user to view satellite images, the relief of the area's geography, use StreetView, display information about points of interest or choose a means of transportation. Apple Maps boasts a user-friendly and aesthetically pleasing interface, providing a distinctive user experience thanks to its integration with the brand's voice assistant, Siri [5], which enhances accessibility and usability. The above-mentioned systems are important, but they are not the only ones [4]. For instance, Active Bat, a system developed by AT&T, uses ultrasonic technology through a network or matrix of sensors installed in rooms. It applies a triangulation technique with the flight times of signals to calculate three-dimensional positions. Another system is Ubisense UWB [6], which utilizes radio frequency technology with a high penetration capacity in indoor spaces, reaching ranges

© The Author(s), under exclusive license to Springer Nature Switzerland AG 2023
J. Bravo and G. Urzáiz (Eds.): UCAmI 2023, LNNS 835, pp. 268–274, 2023.
https://doi.org/10.1007/978-3-031-48306-6_27

of up to 50 m. It can cover the area of a 20 x 20-meter building with four receiving points listening for UWB bursts emitted by devices within that coverage area.

On the other hand, considering user aspects to offer him/her personalized services is the main focus of Human-Centered Computing (HCC). HCC integrates human sciences, such as social, cognitive sciences, and computer science [7], to develop technologies that allow amplifying human abilities and improving their performance. Some of the capabilities considered under this approach may include the ability to perceive, decide, reason, perform cognitive work, and even maximize the user experience [8]. So, human-centered systems base their functioning on the retrieval and processing of personal characteristics and attributes of a user, generating information or content adapted to his/her abilities or conditions. The specific value for a describing attribute of each user can be obtained through various methods [9]. The definition of these attributes and their values is represented in user models that serve to identify behaviors or patterns, set preferences, and physical and mental abilities. With that is possible to achieve adaptation for each user depending on the system. The adaptation through these models and techniques is not limited to traditional preference aspects but can occur in different phases of each system and consider other aspects, i.e., a system can adapt its interface not only based on design preferences but also on the physical or cognitive capabilities of the user.

Preliminary literature research has allowed us to observe that no survey paper addresses human-centered navigation systems (HCNS), so our work is aimed to fill this gap. This paper reports a systematic literature review (SLR) that is aimed to understand the current state of human-centered navigation, identifying the main characteristics of related works and their application domains.

This paper is organized as follows. Section 2 presents the method used to conduct the literature review, the research questions to be answered, the search string, and the information sources consulted to retrieve literature based on the definition of inclusion and exclusion criteria. Section 3 provides the results of this literature review, the retrieved works, and their details according to the research questions. Finally, Sect. 4 concludes this paper.

2 Method

This research follows the PRISMA method designed to document the rationale, process, and findings transparently [10]. The method involves a guide for retrieving articles, using filters to select relevant works from scientific databases.

2.1 Research Questions

Two research questions were established to guide the investigation and cover relevant aspects of the research topic: (Q1) What are the user models found in the literature? and (Q2) What are the architectures of the systems found?.

2.2 Search String

The search string considered two fundamental concepts: navigation and user modeling. Based on these terms the search string was structured as follows: (((INDOOR OR

OUTDOOR) AND (NAVIGATION OR ROUTING)) AND (USER AND (PROFILE OR MODEL OR PREFERENCES))). Note that the terms INDOOR and OUTDOOR are important to define the context of navigation, excluding web navigation.

2.3   Information Sources and Inclusion/exclusion Criteria

The information sources for this research were Springer, ACM, IEEE, and ScienceDirect. These are sources of reliable content published on scientific topics related with this research and, in general, with Computing and Informatics.

Table 1. Inclusion and exclusion criteria.

Inclusion criteria	Exclusion criteria
There is institutional access to the publication	The publication is not written in English
The publication date is from 2002 to 2022	The publication is not from recognized academic research sources
From the abstract of the publication, one can check that it contributes to the research	Title or abstract of the publication is not related to user modeling in navigation systems

2.4   Extracted Works

The results of each source were filtered using the inclusion and exclusion criteria (see Table 1). Finally, the number of useful papers was 15. Table 2 shows the selected papers from each source.

Table 2. Selection of studies

Sources	Candidate	Eliminated	Included
Springer	70	63	7
ACM	46	45	1
IEEE	21	15	6
ScienceDirect	92	91	1

3   Results

The results obtained are explained next, answering the research questions Q1 and Q2.

3.1   User Models (Q1)

The main aspect considered in the user models found is physical capabilities, although some works take into account elements with more complexity as orientation ability, cognitive skills or behaviors [11] (see Table 3). Let us note that 5 works of these works operate in indoor environments, 2 in virtual mode, and 1 more in outdoors.

Table 3. Main characteristics of user models on works identified.

Ref	Humans aspects considered	Values retrieved using:	Use cases
[12]	Cognitive and spatial orientation ability	Exploration of virtual environment	Identify navigational behavior of user
[13]	Preferences	Pre-recorded movement trajectories of previous route	Generate recommendations for places with similar characteristics
[14]	Physical, cognitive or special conditions	Initial interaction between user and the system interface	Represent specific capabilities or limitations of the user
[15]	Habits, preferences and privileges	User data is entered through forms	Delivering a betterexperience between the environment and the user
[16]	Preferences	User data is entered through forms	Develop services for mobile users
[18]	Physical, cognitive, or sensory capacities	American Disability Act (ADA) standards	Proposes routes considering the (ADA) standards
[19]	Physical: visually impaired / motor disabilities	Listed user requirements	Develop a indoor navigation application for all users
[20]	Physical capabilities and preferences	Set of preferences according to the domain	Find the most natural path prioritizing the human factor

Let us note that OntoNav is at the base of User Profile Ontology (UPO) and other works, such as [18, 19]. In addition, C-NGINE's user modeling is based on UPO. [17].

3.2 System Architectures (Q2)

The architectures of the aforementioned systems have some elements in common (see Table 4). One can find modules to generate routes, to consider users characteristics, and to choose the best route according with user's conditions.

Anothe work is CoINS [20], and in its architecture the integration of the user model, the chosen route, and contextual information is used by the Preference.

Table 4. Main elements of systems architectures.

System	Module	Function
OntoNav [14]	NAV GEO SEM	Main Interface / Display routes Calculate the geometric route Choose the route that best suits the user's conditions
C-NGINE [21]	User Interface Knowledge base Services layer Reasoner Rule engine	Main Interface / Display routes Contains information in form of ontologies Identify navigational behavior of user Inferring logical events based on a set of affirmed facts Reasoning about the context through knowledge obtained in the form of rules

Assessment component, and the processes around these elements are controlled by a workflow engine, allowing the division of tasks.

4 Conclusion

This paper discusses the most notable works involving navigation systems under HCC approach. It was noticed that OntoNav is an important work that serve as base to others projects. The user model presented in OntoNav considers physical, mental and sensory abilities as well as preferences and user demographics. Similar aspects are considered in other systems depending on their functions. On the other hand, the most popular method to input values for the user model is by manual attribute setting through forms, being less common the use of methods that do not require explicit user interaction. it can also be noticed that most works focus on indoor environments. This is because in these environments it is easier to establish control parameters to evaluate the user and adaptive navigation. Besides, it is less expensive [22]. Another important aspect that can be noticed is that some works consider cognitive skills, but the way to obtain this information is manual. Considering cognitive skills is an important and complex topic into HCC and NS, but most of the works are focused on physical conditions, such as visual impairement or motor disabilites.

References

1. Karimi, H.A.: Me-Friends-Web (MFW): A model for navigation assistance through social navigation networks. In: 7th International Conference on Collaborative Computing: Networking, Applications and Worksharing (CollaborateCom) (2011)
2. https://www.google.com.mx/maps/ Google Maps Sitio Oficial de Google Maps (2023)
3. https://www.apple.com/mx/maps/ Apple Maps Sitio Oficial de Apple Maps (2023)
4. Cruz Alvarado, M.A., Sandí Delgado, J.C. Systems and technologies that facilitate the indoor positioning. Universidad de Costa Rica, Sede de Occidente: Pensamiento Actual, **17**(29) (2017)
5. https://www.apple.com/mx/siri/ SiriApple(MX) Sitio Oficial de Apple (2023)
6. Harle, R.: A survey of indoor inertial positioning systems for pedestrians. IEEE Commun. Surv. Tutorials. **15**(3), 1281–1293 (2013)
7. Jaimes, A., Sebe, N., Gatica-Perez, D.: Human-centered computing: a multimedia perspective. New York, NY, USA. In: Proceedings of the 14th ACM International Conference on Multimedia (2006)
8. Jaimes, A., Gatica-Perez, D., Sebe, N.: Human-centered computing: toward a human revolution. los alamitos, CA, USA. IEEE Comput. Soc. **40**(5), 30–34 (2007)
9. Eke, C.I., Norman, A.A., Shuib, L., Nweke, H.F.: A survey of user profiling: state-of-the-art, challenges, and solutions. IEEE Access, **7**, 144907–144924 (2019)
10. Page, M.J., McKenzie, J.E., Bossuyt, P.M., Boutron, I., Hoffmann, T.C., Mulrow, C.D., et al.: The PRISMA 2020 statement: an updated guideline for reporting systematic reviews. BMJ **372**, n71 (2021). https://doi.org/10.1136/bmj.n71
11. Koh, W.L., Zhou, S., Patkos, T., Antoniou, G., Plex-ousakis, D.: Modeling and simulation of pedestrian behaviors in crowded places. ACM Trans. Model. Comput. Simul, **21**(3), 1–23 (2011)
12. Sas, C.: User Model of Navigation. In: Masoodian, M., Jones, S., Rogers, B. (eds.) Computer Human Interaction. APCHI 2004. Lecture Notes in Computer Science, vol 3101. Springer, Berlin, Heidelberg (2004). https://doi.org/10.1007/978-3-540-27795-8_38
13. Takeuchi, Y., Sugimoto, M.: A user-adaptive city guide system with an unobtrusive navigation interface. Pers. Ubiquit. Comput. **13**, 119–132 (2009). https://doi.org/10.1007/s00779-007-0192-x
14. Anagnostopoulos, C., Tsetsos, V., Kikiras, P.: A human-centered semantic navigation system for indoor environments. Santorini, Greece: ICPS '05. In: Proc. International Conference on Pervasive Services (2005)
15. Salem, B., Rauterberg, M.: A human-centered semantic navigation sys tem for indoor environments. Berlin, Heidelberg: Springer Berlin Heidelberg (2004)
16. Kritsotakis, M. et al.: C-NGINE: A Contextual Navigation Guide for Indoor Environments. In: Aarts, E., et al. Ambient Intelligence. AmI 2008. Lecture Notes in Computer Science, vol 5355. Springer, Berlin, Heidelberg (2008). https://doi.org/10.1007/978-3-540-89617-3_17
17. Michou, M., Bikakis, A., Patkos, T., Antoniou, G., Plexousakis, D.: A semantics-based user model for the support of personalized, context-aware navigational services. Liverpool, UK: First International Workshop on Ontologies in Interactive Systems (2008)
18. Dudas, P.M., Ghafourian, M., Karimi, H.A.: ONALIN: on-tology and algorithm for indoor routing. In: 2009 Tenth International Conference on Mobile Data Management: Systems, Services and Middleware (2009)
19. Traubinger, V., Franzkowiak, L., Tauchmann, N., Costantino, M., Richter, J., Gaedke, M. The Right Data at the Right Moment for the Right Person — User Requirements and Their Implications for the Design of Indoor Navigation Systems. In: International Conference on Indoor Positioning and Indoor Navigation (IPIN) (2021)

20. Lyardet, F., Szeto, D.W., Aitenbichler, E.: Context-Aware Indoor Navigation. In: Aarts, E., et al. Ambient Intelligence. AmI 2008. Lecture Notes in Computer Science, vol 5355. Springer, Berlin, Heidelberg (2008). https://doi.org/10.1007/978-3-540-89617-3_19
21. Nikoloudakis, E., Kritsotakis, M., Bikakis, A., Patkos, T., Antoniou, G., Plexousakis, D.: Exploiting Se mantics for Indoor Navigation and User-Tracking. Symposia and Workshops on Ubiquitous, Autonomic and Trusted Computing (2009)
22. Tao, Y. Ganz, A.: Simulation framework for evaluation of indoor nav-igation systems IEEE Access **8**, 20028–20042 (2020)

Proactivity in Conversational Assistants: The mPLiCA Model Based on a Systematic Literature Review

Esperanza Johnson[1]([⊠]) [iD], Laura Villa[2] [iD], Tania Mondéjar[2] [iD], and Ramón Hervás[2] [iD]

[1] Spillskolen - Høgskolen i Innlandet INN, Hamar, Norway
esperanza.johnson@inn.no
[2] University of Castilla-La Mancha, UCLM, Ciudad Real, Spain
{Laura.Villa,Tania.Mondejar,Ramon.HLucas}@uclm.es

Abstract. In the last years, we have seen a rise in the use of assistants that are becoming more and more natural in their interactions with people. An emerging characteristic is the proactivity in the assistant interaction. The areas of use for these types of assistants range from health, education, to general assistance for tasks, among others., and the proactivity is usually a means to an end, usually to improve user engagement. Given the growing popularity, we have taken the opportunity in this paper to perform a systematic literature review which focuses on agents with a primary focus on them being proactive. During this, we have observed several interesting patterns, such as the main form of interaction for these agents is through verbal interaction, or the fact that they are usually robots. Many of these papers study user response and feelings to different levels of proactivity, with some defining a time-based proactive response, and other focusing on user involvement when defining proactivity levels. All these findings regarding proactivity make it possible to propose, a model based on the proactivity level and the agent's ability to learn from each interaction, which is what we are presenting in this paper.

Keywords: proactive · assistants · model · systematic review

1 Introduction

In the field of human-computer interaction, the use of conversational assistants, designed to help perform any kind of task has seen substantial growth. These assistants can take multiple forms, ranging from chatbots, virtual avatars, voice-based assistants, to embodied agents.

A key (but not very common) aspect in assistants is proactivity, which refers to their ability to anticipate the user's needs and offer help before it is requested. Through the analysis of user behavior patterns and other contextual data, proactive digital assistants can provide relevant reminders, suggestions, and alerts at the right time [35]. As example, Clippy, an early conversational assistant

J. Bravo and G. Urzáiz (Eds.): UCAmI 2023, LNNS 835, pp. 275–285, 2023.
https://doi.org/10.1007/978-3-031-48306-6_28

introduced by Microsoft in 1996, aimed to aid users in navigating Microsoft Office features. Represented as a paperclip with expressive features, Clippy was designed to proactively offer assistance based on user actions. Despite its innovative concept, it faced criticism for its intrusive behavior and imprecision in determining when help was needed, leading to its discontinuation in 2007[1].

In this paper, we have executed a systematic literature review, concentrating primarily on the aspect of proactivity within conversational agents. Drawing upon the findings about proactivity, we have devised a model that incorporates both the degree of proactivity and the assistant capability for learning from each interaction. This model is the primary subject of presentation within this paper.

2   Proactive Interaction: A Systematic Review

2.1   Systematic Review Design

As stated at the start of this paper, we aim to study the current state of proactive agents and their various characteristics. In order to do this, we carried out a systematic literature review following the 2020 PRISMA guidelines [22]. In order to begin, we first had to determine the search string to find as many related works as possible, which lead us to the following:

$$TITLE(proactiv * AND(assist * OR\,agent\,OR\,robot\,OR\,avatar)).$$

We settled on this string for a variety of reasons. Firstly, the keywords were all to be present in the title, in order to limit the scope of the papers encountered to those that involved proactivity and agents as the central theme. We then chose *proactiv** in order to ensure a wider range of word variations which would include proactive and proactivity, so as not to miss out on relevant contributions to the field. Finally, the last part of the string tried to cover as many types of agents as possible, including different types of assistants (or assistive agents). Through this, we found a total of 328 relevant works on Scopus, to which we applied a variety of criteria typically seen in systematic literature reviews, as well as one of our own.

Firstly, we selected those works in English, as well as those that were conference papers, articles, book chapters, reviews and short surveys, excluding those works that had been retracted. That left us with a total of 314 papers (-14) to screen more closely. We then performed a screening based on title and abstract and two different points, where we observed interesting patterns. Mainly, our inclusion of the word agents, which is also used in the area of Artificial Intelligence, yielded results focused on multi-agent approaches to problem solving. This differs from the more specific definition of agent used in the area of Human-Computer Interaction, which is that of a physical or virtual interactive system

---

[1] Sinclair J., 2001, "The Story of Clippy: Microsoft's Much-Hated Office Assitant", in https://edition.cnn.com/2001/TECH/ptech/11/26/clippy.obit.idg/index.html. Last accessed: 2023-05-20.

imbued with various capabilities to make it capable of interaction at different levels (as mentioned in the introduction). Another observation which explains the exclusion of many works during the title and abstract screen is that multiple papers spoke of proactive collaboration between humans and robots (or other types of systems), but focused exclusively on an industrial perspective, such as assembly lines, and thus, it was outside of the scope of our work and research area.

In total, we removed 229 different works when screening for title and abstract, leaving us with a total of 85 works. We then proceeded to apply the last inclusion/exclusion criteria. Firstly, we focused on works from the last 10 years (2012 to present day), observing that, while there are works published in this area much earlier, and many contributions since 2005, there was a steady increase in contributions starting in 2011. We also have a quality criteria of 2 citations per year the paper has been published, which is meant to measure impact of these papers. Papers published within this year or the last are not subject to this criteria, as they might not have had enough time to make an impact in their area. This left us with a total of 36 papers, from which we ended up removing 3 due to not being relevant due to the content not being align to the focus of this line of work, and 2 which were not accessible, giving us 31 papers in the end. Authors evenly distributed the work to both decide which papers were included/excluded, and any conflicts or borderline papers were decided among all authors.

Once we had all the papers we had selected, we then determined which data was of interest for us to study and which could help us reach conclusions regarding the state of proactive agents. Among the usual metrics, such as number of participants, average age of participants, we had area of application and type of agent used. We then added some which were more specific about the nature of the agents themselves, such as the cognitive services used, the type of interaction, if they were affective or not, and which level of affect they presented (according to Rosalind Picard's model of affect [25]). Finally, there were those aspects which are the focus of the paper and the mPLiCA model, which are the level of proactivity and the ability of the agent to learn.

All findings regarding these dimensions are explained in more detail in the following subsection.

2.2   General Findings

In this section we will discuss findings regarding more general aspects of the related work, such as average number and age of participants, area of application, and which technology has been used in order to implement the interactive capabilities of the agents. We will also comment on the affective capabilities of the proposals, and which level of affect they present.

For starters, we can observe that throughout all of the works, the average number of participants is of 60.29, although the mode is 15 users. This is because there are four different works that have over 100 participants, whereas the rest have anywhere between 0 and 60. Several of the works with no users as part of the evaluation also tend to be proof of concept, where the paper focuses on more technical aspects regarding validation of the system, rather than an evaluation of the system with users (be it an experiment in the wild, in a controlled environment, or the evaluation of user's opinions of the system). Because of this, the average age of participants is difficult to estimate, particularly since some papers did not mention a specific average age, and just mentioned a general bracket (i.e., university students or people over 55). Regardless, based on observation, most of the users that have participated in the various evaluation are adults, with only two papers having an average age of over 55. This lets us know that many of the applications for these proactive agents are focusing mostly on the general adult population.

When it comes to the technology involved in all of these works, most are supported through a variety of cognitive services that involve speech-to-text and text-to-speech technology, with most communication falling under it being verbal. This tracks with how technology in that area has been improving in the last years, allowing users to have a more natural interaction through speech, as has been seen in voice assistants such as Amazon's Alexa or the Google Home. In a couple of cases, some of that interaction is also supported to be through text, but it is more rare; whereas additional ways of interacting that include visual, tactile or gesture-based queues (eye-tracking, body-tracking, or tactile screens to interact with, for example) are more common.

The way the interaction is implemented is also likely strongly linked to the type of agents that are used in many of this works, with robots being the most used, in more than half of the cases (17), with avatars (4) and assistants (2) also adding to those numbers where verbal communication is the main for of interaction. There are some papers which use chatbots (3), and also some that use a simulation (5) (that is, no actual implementation of any type of agent).

As for overall findings regarding the area of application of these works, we 11 that serve to help in specific tasks (shopping, museum guide, household help), with 9 being applied in the area of health. Some of the contributions are still at the concept stage, and we found two reviews, as well as two works focused in the area of education. A detailed summary of these works can be found in the following section.

Finally, we wanted to know how many of these agents had any affective capabilities, and to what extent, finding that 26 out of the total of 31 works have agents with no affective capabilities. Of the other 5, one recognizes affect but does not portray it, one portrays affect but does not recognize it, and three can both portray and recognize affect. This subdivison of a system's ability to portray or recognize affect comes from Picard's paper on Affective Computing [25], where she classifies them into those four categories, with I being unable to recognize or portray affect, and IV being able to do both.

2.3   Selected Related Work

In Table 1 below, we summarize the main focus of each one of the works that we studied in more detail mentioning some key aspects such as their approach to a definition of proactivity, the main goal of the paper, or other interesting information.

Table 1. Table of works featuring Cognitive Computing, with a description of the services used

Author, year	Summary
(Patel, 2023)[23]	This paper studies how a variety of objects in a user's home change states throughout time, allowing the robot in place to infer and anticipate to the person's needs based on these routines and changes, by for example, noticing when a toothbrush is moved, inferring that the user is brushing their teeth, and then learning the behaviour and aiding the person to brush their teeth when the time comes
(Liao, 2023) [18]	This is a tutorial conference paper regarding a chatbot's ability to make proactive questions
(Villa, 2023) [31]	Proactive questions without context that are asked when user presence is detected
(Almada, 2023) [1]	Chatbot that acts according to student profile, takes feedback to enhance automatically, present multimodal (videos and others) content according to student learning style, suggests correlated controllable factors to students (proactively suggests to the student factors that may have a more significant impact on the student's results), redirects unsatisfied questions to the Lecturer/Assistant
(Li, 2022) [17]	SLR on Reactive vs Proactive robots, and people's trust in the service, revealing that proactive robots have a higher level of trust, but that said trust could be related to the company's reputation
(Xie, 2022) [33]	Starts interaction with customers (for example, offer help, approach to customer, initiate conversation, anticipate to customer needs) acting in advance when necessary
(Buyukgoz, 2022) [3]	Recognizes human intentions and acts to fulfill them (opening the door that they are about to cross), reasons about possible future threats or opportunities and acts to prevent or to foster them (predictions, recommending you to take an umbrella since the forecast predicts rain)
(Zargham, 2022) [34]	Evaluation of proactive conversations/interventions in several cases/scenarios when: Emergency is detected, health risk is detected, meeting reminders, cooking proposals/inspirations, technical support, nudging, fact checking, disagreement clarification, fact spoiler
(Kraus, 2022) [16]	Several levels: None (static, user explicitly request), Notification (static, Signal user "there is recommendations", but leaves the initiative to ask about them up to the user), Suggestion (static, directly propose the action), Intervention (static, executed a particular action in place of the user), Adaptative (dynamic, proactive dialogue with regard to user expectations)
(Shvo, 2022) [27]	Recognizes human cognition (RC), adapts to them (AC) and predicts them (PC), offers helps
(Faria, 2022) [6]	Proactivity based on recommending events based on user preferences and pre-defined medication reminders
(Chien, 2022) [4]	Compares the Pepper robot in commercial and product recommendation systems, combining two characteristics: proactive-reactive, and impassive-intimacy. Proactivity focuses on suggesting help and suggesting products
(Kraus, 2022) [15]	Scans the space for information and makes suggestions on household chores (tidying, cleaning, etc.), classifies proactivity by how quickly it reacts, and comments on the benefits on medium-level proactivity
(He, 2022) [9]	Household energy saving suggestions using AI
(Ko, 2021) [12]	Detects user proximity and makes an approach
(Kraus, 2021) [14]	Study on autonomy and trust, compares with a NAO 4 levels of proactivity: none, notification, suggestion, and decision
(Kraus, 2020) [13]	Monitors a performed task with a tool and intervenes in several situations (if it is the first time using a tool, or the person is idle for too long)
(Tan, 2020) [29]	Differences in the proactive behavior at different levels among anthropomorphic perception
(Iio, 2020) [10]	Behavior of different levels of proactivity among anthropomorphic robots for exhibits
(Zhu, 2020) [36]	Proactive interaction according to context with a greater or lesser explanation
(Ujjwal, 2019) [30]	Proactive agent for help in commercial space. It makes suggestions based on people's answers when it detects their presence
(Peng, 2019) [24]	High, medium and low levels of proactivity in a shop where the robot adapts to the user's need and contexts
(Bremmer, 2019) [2]	E-robot that acts to keep itself safe if (and only if) this does not conflict with obedience for human safety
(Sirithunge, 2019) [28]	Literature review about body language interaction with a robot
(Moulin-Frier, 2018) [21]	Defines Slow, Middle and Fast proactivity based on how quickly the robot (iCub) reacts to humans completing a series of tasks regarding movement of cubes between two areas of a table
(Falk, 2018) [5]	Narrating agent, which uses a Temporal Knowledge Map to parse the user input, and also uses Behaviour Trees to control agent behaviour, responding to two different types of user input. Will guide users through parts of stories they might have missed, with the proactivity being the ability to initiate the conversation based on engagement and user knowledge of the story.
(Liu, 2018) [20]	Study for a robot that proactively helps customers in a shop, where the proactivity was if the robot approached the customer first, or if it waited to help, ob serving that users responded well to a level of proactivity that involved its level of social appropriateness
(Garrel, 2017) [7]	Studies the reception of people to a proactive agent who initiates engagement, finding communication through gestures, verbal cues and motions, and were pleased it learned and recognized faces. Longer and more affective interactions were also positive.
(Liao, 2016) [19]	Authors study the impact of interruption by proactive interactions, which users used for 17 d. Users could personalize how social they wanted the agent to be, which impacted how human the communication would appear and the interactions they had. Interruptions were deemed unfavourable when users did not see the point, with authors suggesting context awareness, user-attribute awareness to mitigate this issue.
(Grosinger, 2016) [8]	This robot focuses on support for the elderly when it comes to ADL tasks, and anticipates to the useres needs by doing a temporal projection to predict future undesired states and identify actions that avoid them. They introduce the notion of opportunities to decide which goals to pursue and when. This opportunity detection helps in scheduling future tasks, or reduce the search space
(Ring, 2013) [26]	Authors of this paper aim to address loneliness in the elderly by means of an ECA that would proactively begin conversations with them, with a focus on increasing engagement when having the agent initiate conversations, and users feeling a sense of companionship towards the end of the evaluation.

3 Model of Proactivity Level in Conversational Assistants

3.1 mPLiCA: Proposed Model

Once interactive conversational agents have been thoroughly analyzed, a wide range of proactive behaviors can be identified. Some works also differentiate levels of proactivity, typically based on the autonomy and complexity of the agent's actions [16]. The highest level of proactivity is usually associated with interventions based on agent decisions without consulting the user. At a slightly lower but still medium-high level, there are decisions made by the agent that are consulted with the user before execution. Lower levels of proactivity are associated with basic dialogue actions or simple awareness (presence), without further actions in the agent or its environment. Additionally, as seen in Sect. 2, certain characteristics in these proactive agents recur more or less frequently, such as cognitive capabilities, affective capabilities, and the ability to learn from interactions with the user.

Based on all these elements identified during the systematic literature review, a model for characterizing proactive conversational agents is proposed, named mPLiCA, which is visually represented in Fig. 1.

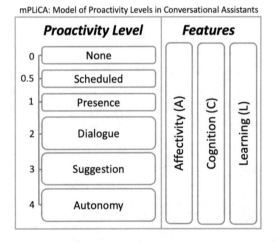

Fig. 1. mPLiCA, a model for characterizing the proactivity levels and features of conversational assistants

This model encompasses six types of proactivity, quantitatively rated from levels 0 to 4, with 4 being the highest level of proactivity in terms of autonomy and complexity:

- **Level 0. None:** The agent does not exhibit any proactive behavior.
- **Level 0.5. Scheduled:** This level is placed between 0 and 1 because when behavior is programmed for a specific time, such as reminders or alarms on

a calendar, the agent appears to display proactive behavior, but we consider it to be a very low level of proactivity, if at all.

- **Level 1. Presence:** The agent proactively manifests its presence to enhance interaction with it. This behavior typically includes visual signals [32] or voice messages [11], sometimes even personalized [12].
- **Level 2. Dialogue:** The agent proactively initiates a dialogue with the user. Unlike the previous level, intentions beyond mere presence are expressed, and an immediate response from the user is expected to continue the conversation.
- **Level 3. Suggestions:** The agent makes decisions regarding the user's needs, which typically involve actions to be carried out by the agent itself, the user, or within the environment. The agent consults with the user if they want the action to be performed.
- **Level 4. Autonomy:** The agent not only makes decisions but also executes them without consulting the user. The user may or may not be informed later.

The characterization of proactivity is complemented by other additional services or functionalities. These functionalities were seen during the literature review, where three specific characteristics stood out:

- **Cognition:** Refers to the use of cognitive services to facilitate conversational functions. This often involves using speech-to-text and vice versa services, as well as mechanisms to establish dialogues.
- **Affectivity:** Refers to the ability to perceive emotions in interactions with the user as well as displaying emotions by the agent.
- **Learning:** The ability of the agent to learn and adapt its behavior based on interactions and experiences with the user.

3.2   Coverage of the Model

The proposed model makes it possible to characterize all the papers selected in the systematic review. No important and significantly present characteristics have been detected in the papers that have been left out of the characterisation attributes. Neither are characteristics or levels of proactivity included that have a low incidence.

As general finding as the Fig. 2 shows, we see that cognitive functionality is very frequent (87%). The union of proactivity with affectivity is not common (17%), nor is it common to find agents that learn with time and experiences with the user (20%). As for the levels of proactivity, the most frequent levels present in agents are those who dialogue (level 2) and make suggestions (level 3).

Authors	Cognition	Affectivity	Learning	Scheduled	Presence	Dialogue	Suggestion	Autonomy
Patel 2023 [23]	1	0	1	0	0	0	0	1
Liao 2023 [18]	0	0	0	0	0	1	0	0
Villa 2023 [31]	1	1	0	1	1	1	1	0
Almada 2023 [1]	1	0	1	0	0	0	1	0
Xie 2022 [33]	0	1	0	0	1	1	1	0
Buyukgoz 2022 [3]	0	0	0	0	0	0	0	1
Zargham 2022 [34]	1	0	1	1	0	1	1	1
Kraus 2022 [16]	1	0	0	1	0	0	1	1
Shvo 2022 [27]	1	0	0	0	0	0	0	0
Faria 2022 [6]	0	0	0	0	0	0	1	0
Chien 2022 [4]	1	0	0	0	0	1	1	0
Kraus 2022 [15]	1	0	0	0	0	0	1	0
He 2022 [9]	1	0	0	0	0	0	1	0
Ko 2021 [12]	1	1	0	0	1	0	0	0
Kraus 2020 [14]	1	0	0	0	0	1	1	1
Kraus 2020 [15]	1	0	0	0	1	0	1	0
Tan 2020 [29]	1	0	0	0	0	0	1	1
Iio 2020 [10]	1	0	0	0	0	1	0	0
Zhu 2020 [36]	1	0	0	0	0	1	0	1
Ujjwal 2019 [30]	1	0	0	0	0	0	1	0
Peng 2019 [24]	1	0	0	0	0	1	1	0
Bremner 2019 [2]	1	0	0	0	0	0	1	1
Sirithunge 2019 [28]	1	0	0	0	0	0	0	0
Moulin-Frier 2018 [21]	1	0	1	0	0	0	0	0
Falk 2018 [5]	1	0	1	0	0	0	0	0
Liu 2018 [20]	1	0	0	0	1	1	0	0
Garrell 2017 [7]	1	1	0	0	0	1	0	0
Liao 2016 [8]	1	0	0	0	0	0	0	0
Grosinger 2016 [8]	1	0	1	0	0	0	1	1
Ring 2013 [26]	1	1	0	0	1	1	0	0

Fig. 2. Characteristics of the studies included in the systematic review and coverage of the model

4   Conclusion

In conclusion, and after reviewing multiple papers focused on proactivity through different agents, we can observe some interesting patterns. Firstly, many of the papers discuss the acceptance by users of proactive systems, where it was concluded that highly proactive systems do not count with as much acceptance as those with medium or lower proactivity levels. This is likely due to the user not being part of the decision making, therefore it seems of interest to have users informed of the actions and the decisions made by the system, which could potentially increase the acceptance of more proactive systems. Within those systems, it is also of interest to note that a remarkably low percentage of them had learning capabilities, i.e., the ability to decide on future decisions or actions based on past actions, behaviors and/or patterns they have detected while interacting with a variety of users. As another observation, it was interesting to note the small amount of works that featured affective computing, and it would be of interest to study if the agent's ability to engage with users emotionally improves user acceptance and perception. All in all, this study sheds light on the overall acceptance by users of proactive systems, as well as contribute a model to classify proactivity levels in conversational agents (mPLiCA) based on the information gathered from the literature reviewed. Future work will consist on implementing this model into a conversational agent and studying the acceptance and engagement of users with said agent.

Acknowledgements. This work was supported by SHARA3 project, funded by the Junta de Comunidades de Castilla - La Mancha (SBPLY/21/180501/000160).

References

1. Almada, A., Yu, Q., Patel, P.: Proactive chatbot framework based on the ps2clh model: An ai-deep learning chatbot assistant for students, vol. 542. LNNS, pp. 751–770 (2023). https://doi.org/10.1007/978-3-031-16072-1_54
2. Bremner, P., Dennis, L., Fisher, M., Winfield, A.: On proactive, transparent, and verifiable ethical reasoning for robots, vol. 107(3), pp. 541–561 (2019). https://doi.org/10.1109/JPROC.2019.2898267
3. Buyukgoz, S., Grosinger, J., Chetouani, M., Saffiotti, A.: Two ways to make your robot proactive: reasoning about human intentions or reasoning about possible futures, vol. 9 (2022). https://doi.org/10.3389/frobt.2022.929267
4. Chien, S.Y., Lin, Y.L., Chang, B.F.: The effects of intimacy and proactivity on trust in human-humanoid robot interaction (2022). https://doi.org/10.1007/s10796-022-10324-y
5. Falk, J., Poulakos, S., Kapadia, M., Sumner, R.: Pica: proactive intelligent conversational agent for interactive narratives, pp. 141–146 (2018). https://doi.org/10.1145/3267851.3267892
6. Faria, G., Silva, T., Abreu, J.: Active and healthy aging: The role of a proactive information assistant embedded on tv, vol. 1597. CCIS, pp. 70–84 (2022). https://doi.org/10.1007/978-3-031-22210-8_5
7. Garrell, A., Villamizar, M., Moreno-Noguer, F., Sanfeliu, A.: Teaching robot's proactive behavior using human assistance 9(2), 231–249 (2017). https://doi.org/10.1007/s12369-016-0389-0
8. Grosinger, J., Pecora, F., Saffiotti, A.: Making robots proactive through equilibrium maintenance. vol. 2016, pp. 3375–3381 (2016)
9. He, T., Jazizadeh, F., Arpan, L.: Ai-powered virtual assistants nudging occupants for energy saving: proactive smart speakers for hvac control 50(4), 394–409 (2022). https://doi.org/10.1080/09613218.2021.2012119
10. Iio, T., Satake, S., Kanda, T., Hayashi, K., Ferreri, F., Hagita, N.: Human-like guide robot that proactively explains exhibits 12(2), 549–566 (2020). https://doi.org/10.1007/s12369-019-00587-y
11. Johnson, E., González, I., Mondéjar, T., Cabañero-Gómez, L., Fontecha, J., Hervás, R.: An affective and cognitive toy to support mood disorders. In: Informatics, vol. 7, p. 48. MDPI (2020)
12. Ko, J.K., Koo, D., Kim, M.: A novel affinity enhancing method for human robot interaction-preliminary study with proactive docent avatar, vol. 2021, pp. 1007–1011 (2021). https://doi.org/10.23919/ICCAS52745.2021.9649834
13. Kraus, M., et al.: Was that successful? on integrating proactive meta-dialogue in a diy-assistant using multimodal cues, pp. 585–594 (2020). https://doi.org/10.1145/3382507.3418818
14. Kraus, M., Wagner, N., Callejas, Z., Minker, W.: The role of trust in proactive conversational assistants 9, 112821–112836 (2021). https://doi.org/10.1109/ACCESS.2021.3103893

15. Kraus, M., et al.: Kurt: a household assistance robot capable of proactive dialogue, vol. 2022, pp. 855–859 (2022). https://doi.org/10.1109/HRI53351.2022.9889357
16. Kraus, M., Wagner, N., Untereiner, N., Minker, W.: Including social expectations for trustworthy proactive human-robot dialogue, pp. 23–33 (2022). https://doi.org/10.1145/3503252.3531294
17. Li, D., Liu, C., Xie, L.: How do consumers engage with proactive service robots? the roles of interaction orientation and corporate reputation **34**(11), 3962–3981 (2022). https://doi.org/10.1108/IJCHM-10-2021-1284
18. Liao, L., Yang, G., Shah, C.: Proactive conversational agents, pp. 1244–1247 (2023). https://doi.org/10.1145/3539597.3572724
19. Liao, Q., Davis, M., Geyer, W., Muller, M., Shami, N.: What can you do? studying social-agent orientation and agent proactive interactions with an agent for employees, pp. 264–275 (2016). https://doi.org/10.1145/2901790.2901842
20. Liu, P., Glas, D., Kanda, T., Ishiguro, H.: Learning proactive behavior for interactive social robots **42**(5), 1067–1085 (2018). https://doi.org/10.1007/s10514-017-9671-8
21. Moulin-Frier, C., et al.: Dac-h3: a proactive robot cognitive architecture to acquire and express knowledge about the world and the self **10**(4), 1005–1022 (2018). https://doi.org/10.1109/TCDS.2017.2754143
22. Page, M.J., et al.: The prisma 2020 statement: an updated guideline for reporting systematic reviews. BMJ 372 (2021). https://doi.org/10.1136/bmj.n71, https://www.bmj.com/content/372/bmj.n71
23. Patel, M.: Longitudinal proactive robot assistance, pp. 775–777 (2023). https://doi.org/10.1145/3568294.3579982
24. Peng, Z., Kwon, Y., Lu, J., Wu, Z., Ma, X.: Design and evaluation of service robot's proactivity in decision-making support process (2019). https://doi.org/10.1145/3290605.3300328
25. Picard, R.W.: Affective computing-mit media laboratory perceptual computing section technical report no. 321. Cambridge, MA 2139, 92 (1995)
26. Ring, L., Barry, B., Totzke, K., Bickmore, T.: Addressing loneliness and isolation in older adults: Proactive affective agents provide better support, pp. 61–66 (2013). https://doi.org/10.1109/ACII.2013.17
27. Shvo, M., Hari, R., O'Reilly, Z., Abolore, S., Wang, S.Y., McIlraith, S.: Proactive robotic assistance via theory of mind. vol. 2022, pp. 9148–9155 (2022). https://doi.org/10.1109/IROS47612.2022.9981627
28. Sirithunge, C., Jayasekara, A., Chandima, D.: Proactive robots with the perception of nonverbal human behavior: A review **7**, 77308–77327 (2019). https://doi.org/10.1109/ACCESS.2019.2921986
29. Tan, H., et al.: Relationship between social robot proactive behavior and the human perception of anthropomorphic attributes **34**(20), 1324–1336 (2020). https://doi.org/10.1080/01691864.2020.1831699
30. Ujjwal, K., Chodorowski, J.: A case study of adding proactivity in indoor social robots using belief-desire-intention (bdi) model, vol. 4(4) (2019). https://doi.org/10.3390/biomimetics4040074
31. Villa, L., Hervás, R., Cruz-Sandoval, D., Favela, J.: Design and evaluation of proactive behavior in conversational assistants: approach with the eva companion robot, vol. 594. LNNS, pp. 234–245 (2023). https://doi.org/10.1007/978-3-031-21333-5_23
32. Villa, L., Hervás, R., Cruz-Sandoval, D., Favela, J.: Design and evaluation of proactive behavior in conversational assistants: Approach with the eva companion robot.

In: International Conference on Ubiquitous Computing and Ambient Intelligence, pp. 234–245. Springer (2022). https://doi.org/10.1007/978-3-031-21333-5_23

33. Xie, L., Liu, C., Li, D.: Proactivity or passivity? an investigation of the effect of service robots' proactive behaviour on customer co-creation intention **106** (2022). https://doi.org/10.1016/j.ijhm.2022.103271

34. Zargham, N., et al.: Understanding circumstances for desirable proactive behaviour of voice assistants: The proactivity dilemma (2022). https://doi.org/10.1145/3543829.3543834

35. Zhou, L., Gao, J., Li, D., Shum, H.Y.: The design and implementation of xiaoice, an empathetic social chatbot. Comput. Linguist. **46**(1), 53–93 (2020)

36. Zhu, L., Williams, T.: Effects of proactive explanations by robots on human-robot trust, vol. 12483. LNAI, pp. 85–95 (2020). https://doi.org/10.1007/978-3-030-62056-1_8

Conversational Agent Development Through Large Language Models: Approach with GPT

Laura Villa[(⊠)] [ID], David Carneros-Prado [ID], Adrián Sánchez-Miguel [ID], Cosmin C. Dobrescu [ID], and Ramón Hervás [ID]

University of Castilla-La Mancha, UCLM, Ciudad Real, Spain
{Laura.Villa,David.Carneros,Adrian.SOrtega,
Cosmin.Dobrescu,Ramon.HLucas}@uclm.es

Abstract. This study investigates the potential of a Large Language Model (LLM) as a Chatbot Development Platform (CDP) for designing dialog systems in the context of conversational agents. While traditional systems often combine rule-based and machine learning approaches via third-party platforms for dialog design, this research explored the utilization of GPT for this purpose, especially for intent and entity detection integral to CDPs. Through a fine-tuning process, two GPT models were adapted to enhance their performance in these tasks, resulting in time and resource efficiency. The resultant system offers a flexible and adaptable framework for chatbot design, with fine-tuning showing significant benefits in maintaining output consistency and saving GPT tokens. The intent classifier demonstrated high precision (99.056%). However, the system faced challenges with dynamic entities, such as dates, due to GPT's inability to access real-time data. Despite this, the study highlighted the immense potential of GPT and similar LLMs in developing conversational agents while drawing attention to the challenges of handling dynamic entities. These findings signal opportunities for innovative fine-tuning and system integration strategies, encouraging continued research in the dynamic field of conversational AI.

Keywords: Large Language Model · Dialog Model Design · Conversational Agents · Chatbot · Chatbot Development Platform

1 Introduction

In the current era of artificial intelligence, virtual assistants, chatbots, and conversational agents have revolutionized the way we interact with technology. They provide quick and personalized responses to user queries, thereby enhancing user experience. These systems are primarily classified into two types based on their design approach: rule-based and AI chatbots (also known as machine-learning chatbots) [5].

Third-party chatbots are some of the most well-known Chatbot Development Platforms (CDPs) for building conversational AI solutions. These systems often

© The Author(s), under exclusive license to Springer Nature Switzerland AG 2023
J. Bravo and G. Urzáiz (Eds.): UCAmI 2023, LNNS 835, pp. 286–297, 2023.
https://doi.org/10.1007/978-3-031-48306-6_29

use a combination of machine-learning techniques (such as *intent/entity classifiers*) and rule-based approaches for Natural Language Understanding (NLU) [9]. However, these platforms present some limitations, which motivates the development of other platforms with the same functionalities but more independent from third parties.

In this article, we will explore the use and efficacy of an LLM, (specifically GPT-3, given its versatility and power) as a CDP in the context of dialog flow design, taking advantage of their powerful NLP capabilities. In this way, the aim is to design a platform that is flexible, modifiable and adaptable to the specific language needs required. Thus, it would be a platform oriented to the expert developer, who would now have more control over the system configuration, moving away from the "black box" model common in third-party CDPs.

This paper is structured as follows: Sect. 2 discusses background and related research. Section 3 outlines our system's three main stages and the methodology. In Sect. 4, we detail our findings and their relevance to our objectives. Section 5 provides a broader interpretation of the results and their implications. Section 6 wraps up by summarizing the study's goals and contributions.

2   Related Work

2.1   Chatbot Development Platforms (CDPs)

Most well-known Chatbot Development Platforms for building conversational AI solutions are *third-party* systems. DialogFlow[1] (Google), IBM Watson Assistant[2], Microsoft Bot Framework[3], and Amazon Lex[4] are powerful NLU platforms that provide user-friendly interfaces for creating, designing, and deploying chatbots, conversational agents, and virtual assistants. These feature-rich platforms offer effective solutions for building interactive and effective virtual assistants for a wide range of applications, such as educational domains [7], healthcare [11,12] and early disease detection [4], and recommendations, support queries, and other services [3,10]. Their versatility and integration capabilities have made them popular choices for organizations seeking to develop conversational interfaces across multiple channels, industries, and domains.

However, third-party platforms have some limitations. They are tightly coupled to their intent recognition providers, and once the chatbot designer chooses a specific chatbot development platform, she ends up in a vendor lock-in scenario, especially with the NL engine coupled with the platform (and by extension, the training limitations of their NLP core). Similarly, current chatbot platforms lack proper abstraction mechanisms to easily integrate and communicate with other external platforms with which the company may need to interact [2]. Moreover, they are platforms that provide non-expert, user-oriented tools to deploy chatbots quickly, making them inflexible[5], non-adaptable frameworks that may be insufficient for expert developers.

---

[1] https://cloud.google.com/dialogflow.
[2] https://cloud.ibm.com/catalog/services/watson-assistant.
[3] https://www.botframework.com/.
[4] https://aws.amazon.com/es/lex/.
[5] https://www.chatbots.org/dialogflow. Last accessed: 15/09/2023.

2.2 Large Language Models

Pretrained Language Models (PLMs) (based on transformer architectures) have already demonstrated solid capabilities in addressing NLP tasks. In particular, groundbreaking advancements in the field of AI have emerged in recent years with the introduction of Large Language Models (LLMs). LLMs are large-sized PLMs that demonstrate powerful capabilities owing to their scale-up [13]. These systems are designed to understand and generate human language, and are able to process and generate text based on a massive amount of training data (a wide range of texts, such as books, articles, and websites, to learn patterns, structures, and semantic relationships within language). LLMs can generate coherent and contextually relevant text, making them powerful tools for a variety of NLP applications.

Developed by OpenAI, the Generative Pretrained Transformer (GPT) model [6] is one of the most remarkable LLM to date and is becoming very popular (especially owing to ChatGPT release). The use of GPT spans various domains, demonstrating its versatility and potential [1,8]. It has demonstrated impressive capabilities for text generation, translation, document summarization, and other NLP tasks, leading to its application in various areas.

Considering the NLP tasks that provide a CDP (intent detection, entity extraction, etc.), LLMs such as GPT can serve as powerful platforms for creating chatbot dialog flows, considering their powerful capabilities in information retrieval and NLP.

3   Materials and Methods

The three key elements in defining the dialog flow of a CDP are:

- **Intents**. These abstract concepts represent the purpose of a user in a query or text. This is the main way for the assistant to provide a response to the user, and their intent determines and guides the flow of the conversation.
- **Entities**. They are concepts, objects, or specific information that represent relevant concepts for the model. These were extracted from the user's input. Identified entities provide a more specific or customized service.
- **Dialog nodes**. They are usually represented as graphs or trees, and determine the conversational path, flow, or guidelines.

All of these elements are used to create *dialog flows*, which enable developers to design intricate conversational flows by mapping user inputs to appropriate responses. CDPs are used to identify user intents, extract entities, and traverse dialog nodes to generate suitable responses by leveraging Natural Language Processing (NLP) and machine learning techniques (Fig. 1). Thus, creating a conversational agent platform requires these three main processes: intent detection, entity extraction, and dialog node evaluation.

However, it is important to consider context-token limitations and costs when using the GPT models. Context tokens in GPT refer to the chunks of information

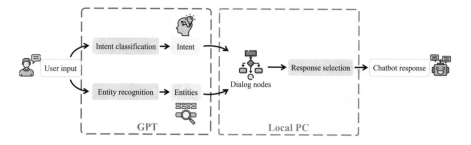

Fig. 1. Chatbot response generation process

that the model uses to generate relevant outputs, and if the input surpasses the model's token limit, it might truncate or ignore parts of it. Hence, the decision was made to dedicate GPT models for intent detection and entity extraction tasks, while the dialog node traversal was executed locally, reducing the context and token request count (Fig. 1).

Nevertheless, even with this division, the specific prompts needed for intent detection and entity extraction could still be in demand in terms of context tokens. To address this situation, we applied two techniques: fine-tuning and few-shot learning. Fine-tuning further trains pre-existing GPT models on specific tasks, enhancing their performance and effectively reducing the complexity of the required prompts. This mitigates the token-limit challenge. On the other hand, few-shot learning allows the model to adapt swiftly to new tasks using only a few examples without requiring extensive retraining.

These two techniques were employed simultaneously in our study, providing a robust solution to the context token problem while also offering flexibility in handling unseen or novel scenarios in dialog flows. Using this combination of techniques, we were able to craft efficient and tailored responses and cover a broad range of conversational contexts. For these tasks, we trained two GPT models: one for classifying intentions, and the other for recognizing entities.

3.1   Dialog Nodes

A sample dialog tree of a Spanish shop chatbot assistant is proposed (Fig. 2). Each node represents an intention that drives the flow of the conversation, returning specific predefined responses from the chatbot depending on the intents and entities detected. JSON format has been selected to define it due to its small file size and fast readability compared to other formats such as XML. The tree is a JSON file consisting of a list of dictionaries, where each one is a node of the dialog tree with several features: type (*standard* or *response_ condition*), title of the node, output (node responses, a list that can have many answers to have response variability), conditions (logic condition to be accomplished to run the node; condition words preceded by '#' mean identified intents, while those preceded by '@' mean extracted entities), dialog_node (unique node identifier), previous_sibling (previous sibling node identifier), parent (parent node identifier), and context (context variables in dictionary format).

Listing 1.1 shows an excerpt of a dialog node. As mentioned previously, the dialog tree contained in the JSON file is stored and traversed locally.

Listing 1.1. JSON dialog excerpt: parent and child node

```
1  {
2      "type": "standard",
3      "title": "What are your hours?",
4      "output": {},
5      "conditions": "#Know_Operation_Hours",
6      "dialog_node": "Hours of Operation",
7      "previous_sibling": "Opening"
8  },
9  {
10     "type": "response_condition",
11     "output": {
12         "text": ["We are closed on @holiday"]
13     },
14     "parent": "Hours of Operation",
15     "context": {},
16     "conditions": "@holiday:christmas or @holiday:new_year",
17     "dialog_node": "Hours of Operation - Closed"
18 }
```

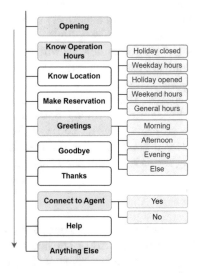

Fig. 2. Chatbot Assistant Dialog graph

In general, each node has a minimum condition and response, with the option of adding child nodes for extra information or specific responses depending on the condition (*response_condition* nodes). The assistant processes the dialog by navigating from the first to the last node, activating any node with a fulfilled condition. It then checks the user input against the conditions of the child nodes, progressing from first to last.

The assistant moves through the dialog tree in the following order: from the first node to the last, along each triggered node, and from the first child node to the last, along each triggered child node. This process continues until the assistant reaches the end of the current branch. If the assistant can not find the true condition among the child nodes, it returns to the base of the tree.

When building a dialog, the order of the nodes is important, as nodes are evaluated sequentially. The assistant uses the first root node whose condition evaluates to true without triggering subsequent nodes in the tree.

3.2 Intent Classifier

This model aligns with a natural-language classifier, where the classes are intentions that can be detected in the text. For the example of the flow of the dialog of the store assistant chatbot, the detectable intentions are *Yes, No, Thanks, Make_Appointment, Know_Operation_Hours, Know_Location, Help, Greetings, Goodbye*, and *Connect_to_agent*. Furthermore, we must add an additional generic class, *Irrelevant*, to encompass all requests that do not correspond to previous intentions.

Data Preparation. The first step in training our model is to create a training dataset. This dataset consists of a JSONL file with approximately 100 text examples for each class (Listing 1.2). The *prompt* column corresponds to the examples of sentences that express the intention presented in the *completion* column of the same row (class label). This dataset was created semi-automatically using the GPT-4 model, which it was given several base examples and asked to generate similar ones in that format. They were then reviewed by an expert to delete non-representative or incorrect examples from the dataset (training a previous model with a dataset generated by a later version is what we call *RetroTrain*).

Listing 1.2. JSONL intent dataset excerpt

```
1  {"prompt": "I prefer to talk to a real person", "completion": "
       Connect_to_agent"}
2  {"prompt": "see you later", "completion": "Goodbye"}
3  {"prompt": "hi!", "completion": "Greetings"}
4  {"prompt": "what kind of help can you provide?", "completion": "Help"}
5  {"prompt": "where are you located?", "completion": "Know_Location"}
6  {"prompt": "what time do you close on Sundays?", "completion": "
       Know_Operation_Hours"}
7  {"prompt": "can you book me an appointment for Monday at 5pm?", "completion
       ": "Make_Appointment"}
8  {"prompt": "not that", "completion": "No"}
9  {"prompt": "thank you for your help", "completion": "Thanks"}
10 {"prompt": "of course", "completion": "Yes"}
11 {"prompt": "I am a star wars fan", "completion": "Irrelevant"}
```

To optimize dataset processing and enhance training performance, label encoding was employed to assign each target label a numerical value (from 0 to $num\_classes - 1$).

When fine-tuning the OpenAI model, specific format characteristics were necessary in the training dataset. For instance, each prompt needed to conclude with a fixed separator to signal the model of the start of completion, and each completion had to begin with a whitespace because of our tokenization method. OpenAI offers a CLI data preparation tool that validates and reformats these datasets, thereby ensuring that they meet these requirements.

Fine-Tuning. The CLI data preparation tool was also used to split the dataset into training and testing partitions (80% and 20% of the original dataset, 848 and 212 entries, respectively). Based on OpenAI's recommendation for classification tasks, the *Ada* base model, which is a faster and cheaper method, was selected. This model is typically comparable in performance to slower and more expensive models for classification use cases.

3.3   Named Entity Recognition (NER)

This is a pattern-recognition model. In this case, the entities to be extracted are *sysdate* for dates, *systime* for times, and *holiday* for the names of festivities (christmas, halloween, christmas, new years, etc.).

Data Preparation. The training dataset consists of a JSONL file containing 105 text examples that can contain several combinations of extracted entities (Listing 1.3). The *prompt* column corresponds to examples of sentences with entities, and the *completion* column represents the extracted entities. This dataset was created in a similar way to the intention dataset: semi-automatically using GPT-4 and examples manually reviewed.

Listing 1.3. JSONL entities dataset excerpt

```
1  {"prompt": "we will meet at 3am on christmas", "completion": "holiday:['
       christmas']\nsystime:['03:00:00']"}
2  {"prompt": "today is 30/04!", "completion": "sysdate:['30/04/2023']"}
3  {"prompt": "tomorrow is Friday", "completion": "sysdate:['14/07/2023']"}
4  {"prompt": "good afternoon, how are you?", "completion": ""}
```

In this case, since many completions had multiple entities separated by '\n' (which is the character normally used at the end of these), it was necessary to perform a preprocessing step where each completion ends with a fixed 'END' stop sequence. This informs the model when completion ends.

The CLI data preparation tool was used to validate and reformat the dataset.

Fine-Tuning. Because of the complexity of the named entity extraction task, *Davinci* was used as the base model. This is the most capable GPT-3 model, able to perform any task that the other models can perform, often with higher quality.

4   Results

4.1   Intent Classifier

This model was evaluated using the test partition of the dataset (which contains 212 entries from the total of 1060). Following the fine-tuning process, we obtained a model with an accuracy of 99.056%. The confusion matrix is displayed in Fig. 3.

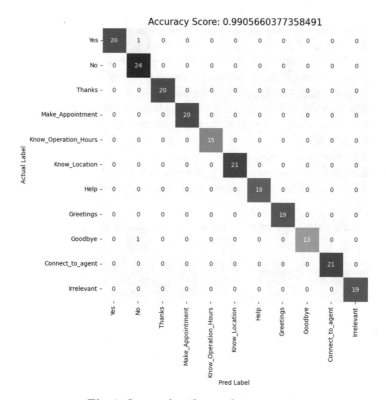

Fig. 3. Intent classifier confusion matrix

4.2 Named Entity Recognition (NER)

Given that this is not a classification model but rather a pattern recognition model, traditional techniques for evaluating classification models, such as direct model accuracy, cannot be employed. However, to assess this model, we used a dataset comprising 50 example entries with various combinations of the extracted entities. The metrics considered were as follows:

- **Row-Level Accuracy**: In this instance, a row is deemed "correct" if all the entities in that prompt row are accurately identified by the model. To compute this metric, we calculate the number of rows where the model's prediction aligns perfectly with the true data, and then divide this by the total number of entries.
- **Entity-Level Accuracy**: This metric evaluates the number of entities that the model has correctly identified from all appearances in the dataset. For each type of entity, we calculate the number of times the model correctly extracts that entity and then divide it by the total number of appearances of that entity in the dataset.

An interesting phenomenon was observed in the predictions of the dataset, particularly those related to the *sysdate* entity. While dates in the training dataset follow the 'dd/mm/yyyy' format, many prompts emphasize the day and month rather than the year, as they are often timeless in the yearly context (e.g., "July 20th is Friend's Day"). However, there are cases where the year is significant (e.g., "I want to make a reservation for July 27, 2025"), which is why the model was trained with full-date formats.

This discrepancy notably impacts the metrics, lowering the accuracy of the *sysdate* entity, as many prompts do not reference a specific year. The model defaults to 2023 if the context year is not specified. The metrics in Table 1 were obtained by strictly adhering to the full-date format ('dd/mm/yyyy'). However, when considering matches where only the day and month are referenced, regardless of the year, different results were obtained, as shown in Table 2.

Table 1. NER model accuracies

Table 2. NER model accuracies (year independent)

Row-Level Accuracy	
Correct Rows	**Accuracy**
31/50	0.620
Entity-Level Accuracy	
Entity	**Accuracy**
systime	0.889
sysdate	0.571
holiday	1.000
no_entities	0.800

Row-Level Accuracy	
Correct Rows	**Accuracy**
43/50	0.860
Entity-Level Accuracy	
Entity	**Accuracy**
systime	0.889
sysdate	0.905
holiday	1.000
no_entities	0.800

5   Discussion

The proposed system provides a flexible and adaptable framework for designing chatbots and conversational agents. The use of widely known formats such as JSON for creating custom dialog flows increases flexibility and simplifies the process, although a user interface application can further ease this task.

Fine-tuning is beneficial for maintaining a consistent output format and saving tokens. Moreover, the option to train custom models based on pretrained LLMs offers a significant opportunity to adapt these models to more specific requirements and applications.

5.1   Intent Classifier

The intent classification model showed impressive precision (99.056%), despite the use of the base *ada* GPT model, the least powerful variant. This indicates that fine-tuning GPT can effectively train natural language classification models with high performance.

The confusion matrix (Fig. 3) reveals specific error cases:

– The word 'bye', associated with a *Goodbye* intent, was misclassified as *No*. This was an intentional test to observe how a commonly used word from a different language (in this case, English) could be interpreted. Its classification as *No* is understandable.

– The phrase 'yes, there is no discussion', meant to represent a *Yes* intent, was classified as *No*. The phrase's definitive tone may have influenced its incorrect classification as *No*.

In the model construction, an essential feature is the *Irrelevant* class. This was deemed necessary because the classification model must always generate an output that must be one of the classes of the dataset. This class is very generic, but vital, encompassing a wide range of examples that diverge significantly from other specified intents.

5.2   Named Entity Recognition

The evaluation metrics of this model are listed in Tables 1 and 2.

Notably, handling dates and times (*sysdate* and *systime*, respectively) is the most challenging aspect of entity extraction. The main issue is that the model lacks awareness of its current temporal context and does not know the current date. Consequently, *sysdate* extraction is limited to the dates and times provided in the training examples. For example, if there is a reference to the current day in the training dataset, the model will explicitly associate it with the current date. In future inferences, the term 'today' will most likely be associated with the day it was trained, not the day on which information is being inferred.

Another challenge arises with dynamic entities, such as those when the model lacks the contextual cues that humans use when discussing dates. Any time reference based on the current time (e.g. 'next Monday') will be interpreted based on the training dataset time, since the model does not have the capability to know the current date. This is also related to the phenomenon mentioned in Sect. 4.2. If a date is mentioned without specifying the year, this could be because the year is not relevant in that context. However, the model defaults using the year from the training dataset.

Therefore, as a text-based AI, GPT cannot access real-time information such as the current date or time. In this particular experiment, for the *systime* entity, some examples could be inferred (e.g. prompts like 'see you at 15:00' are less time context dependent), hence the higher accuracy in the tables. However, other kind of time prompts are problematic (e.g. 'see you in 3 h'), as the model does not know the current time.

The extraction of the *holiday* entity yielded impressive results, with all instances being correctly identified. However, GPT's capacity for creativity can influence the entity values it extracts. It can accurately extract all *holiday* values proposed in the training dataset, but is also capable of extracting its own non-specified holidays. Depending on the system requirements, this can be either desirable or undesirable. This may be beneficial when not all holidays need to be enumerated, as the model learns the pattern of holidays and extracts others that have not been specified. Conversely, it could be detrimental if a tightly controlled model is required that only extracts those that have been explicitly specified.

Importantly, when the *holiday* value has not been specifically outlined, there may be instances where the same *holiday* is extracted but named differently because the model lacks a reference value (e.g. valentine's day was extracted as *dia_sant_valentine* and as *dia_sant_valentino* in other prompt). This inconsistency can present challenges in processing these instances within the system.

6   Conclusion

This research substantiates the capability of Generative Pretrained Transformers (GPT) as a foundation for Chatbot Development Platforms (CDPs) in designing dialog systems. The data highlight the advantages of GPT and similar LLMs in performing complex tasks such as intent detection and entity extraction with notable accuracy when fine-tuned.

The study revealed a high intent classification precision (99.056%), even with a lower-capacity GPT-3 model, the *ada* variant. However, GPT models have limitations, particularly with dynamic temporal entities such as dates and times, owing to their inability to access real-time data. This highlights the need to incorporate external context-providing services to improve the extraction accuracy of these entities.

Despite these challenges, the model demonstrated impressive proficiency in handling static entities(for example, *holiday* in the context of this experiment), correctly identifying all instances. However, its ability to generate novel instances based on a static entity (particular non-predefined holidays, in this case) suggests a latent potential for creative outputs, extending beyond the boundaries of the training data. This propensity could be advantageous or disadvantageous depending on the system requirements, emphasizing the importance of careful system design considerations.

This research suggests a practical approach to create models with predefined intents and entities, similar to third-party systems, reducing the need for specific custom models for specific dialog flows. This opens promising avenues for more flexible and scalable conversational AI systems.

In summary, GPT and similar LLM mark a significant advancement in conversational agent development, despite certain challenges with dynamic entities. Future research could explore other fine-tuning strategies, system integration approaches, comparison between approaches, or new applications to push the boundaries of this rapidly evolving field.

Acknowledgements. This work was supported by SHARA3 project, funded by Junta de Comunidades de Castilla - La Mancha (SBPLY/21/180501/000160); TAICare project, as part of the Proyectos Estratégicos Orientados a la Transición Ecológica y a la Transición Digital 2021, financed by Agencia Estatal de Investigación (TED2021-130296A-I00); SSITH project, under Proyectos Prueba de Concepto, funded by Ministerio de Ciencia e Innovación (PDC2022-133457-I00); and the 2022-PRED-20651 predoctoral contract by the UNIVERSITY OF CASTILLA-LA MANCHA.

References

1. Abdullah, M., Madain, A., Jararweh, Y.: Chatgpt: fundamentals, applications and social impacts. In: 2022 Ninth International Conference on Social Networks Analysis, Management and Security (SNAMS), pp. 1–8 (2022). 10.1109/SNAMS58071.2022.10062688
2. Daniel, G., Cabot, J., Deruelle, L., Derras, M.: Xatkit: a multimodal low-code chatbot development framework. IEEE Access **8**, 15332–15346 (2020). https://doi.org/10.1109/ACCESS.2020.2966919
3. Husak, V., Lozynska, O., Karpov, I., Peleshchak, I., Chyrun, S., Vysotskyi, A.: Information system for recommendation list formation of clothes style image selection according to user's needs based on nlp and chatbots. **2604**, 788–818 (2020)
4. MacEclo, P., Pereira, C., Mota, P., Silva, D., Frade, A., Madeira, R.N.: Conversational agent in mhealth to empower people managing Parkinson's disease. **160**, 402–408 (2019). https://doi.org/10.1016/j.procs.2019.11.074
5. Motger, Q., Franch, X., Marco, J.: Software-based dialogue systems: survey, taxonomy, and challenges. ACM Comput. Surv. **55**(5) (2022). https://doi.org/10.1145/3527450
6. Radford, A., Narasimhan, K., Salimans, T., Sutskever, I.: Improving language understanding by generative pre-training (2018)
7. Ralston, K., Chen, Y., Isah, H., Zulkernine, F.: A voice interactive multilingual student support system using ibmwatson, pp. 1924 –1929 (2019). https://doi.org/10.1109/ICMLA.2019.00309
8. Ray, P.: Chatgpt: a comprehensive review on background, applications, key challenges, bias, ethics, limitations and future scope. Internet of Things and Cyber-Physical Systems **3** (2023). https://doi.org/10.1016/j.iotcps.2023.04.003
9. Safi, Z., Abd-Alrazaq, A., Khalifa, M., Househ, M.: Technical aspects of developing chatbots for medical applications: scoping review. J. Med. Internet Res. **22**(12), e19127 (2020)
10. Samuel, I., Ogunkeye, F.A., Olajube, A., Awelewa, A.: Development of a voice chatbot for payment using amazon LEX service with eyowo as the payment platform, pp. 104–108 (2020). https://doi.org/10.1109/DASA51403.2020.9317214
11. Villa, L., Hervás, R., Cruz-Sandoval, D., Favela, J.: Design and evaluation of proactive behavior in conversational assistants: approach with the eva companion robot. In: Bravo, J., Ochoa, S., Favela, J. (eds.) Proceedings of the International Conference on Ubiquitous Computing & Ambient Intelligence (UCAmI 2022), pp. 234–245. Springer, Cham (2023). https://doi.org/10.1007/978-3-031-21333-5_23
12. Villa, L., Hervás, R., Dobrescu, C.C., Cruz-Sandoval, D., Favela, J.: Incorporating affective proactive behavior to a social companion robot for community dwelling older adults. In: Stephanidis, C., Antona, M., Ntoa, S., Salvendy, G. (eds.) HCI International 2022 - Late Breaking Posters, pp. 568–575. Springer Nature Switzerland, Cham (2022). https://doi.org/10.1007/978-3-031-19682-9_72
13. Zhao, W.X., et al.: A survey of large language models. arXiv preprint arXiv:2303.18223 (2023)

Enhancing Body Percussion Learning: An ICT Supplement for Home Practice Using Gesture Recognition

Paloma Bravo⬤, Andrea Arias, David Carneros-Prado⬤, Cosmin C. Dobrescu⬤, and José Bravo(✉) ⬤

MAmI Research Lab, Castilla-La Mancha University, Ciudad Real, Spain
{David.Carneros,Cosmin.Dobrescu,Jose.bravo}@uclm.es

Abstract. The first instrument played by primary school students was their own bodies, utilizing techniques such as hitting, clapping, and rubbing. By employing gesture recognition techniques, body percussion can be controlled by capturing the specific movements required for each percussion exercise. In this paper, an ICT supplement for the study of body percussions at home is proposed. By utilizing an app running on smartphones or tablets, students can observe and listen to exercises provided by their teacher. Furthermore, they can practice and receive feedback on their performance errors through the apps. These functionalities are enabled by the implementation of Artificial Intelligence techniques for gesture recognition. Additionally, this proposal aims to surpass mere execution of body percussion exercises, focusing instead on the promotion of students' psychomotor skill development.

Keywords: Body Percussion · Rhythm · Psychomotor Development · Gesture Recognition

1 Introduction

Musical pedagogy in the 20th century centered on active teaching and meaningful learning, emphasizing student participation and experiential exploration rather than relying solely on theoretical knowledge. In the context of music education, it is widely agreed upon that the body, along with the voice, should be the primary instrument of musical expression and communication due to their innate presence in children. This approach offers numerous benefits across various domains such as motor skills, emotional development, social interaction, and cognitive abilities [1].

Body percussion, an activity that involves creating sounds by striking different parts of the body and combining them to create rhythmic patterns, is considered to be the most effective way to incorporate the use of the body as an instrument in primary school music education. This is common practice in these methodologies.

Furthermore, these methodologies are based on the principle that learning should be performed. By introducing a playful approach to activities, students can achieve higher

© The Author(s), under exclusive license to Springer Nature Switzerland AG 2023
J. Bravo and G. Urzáiz (Eds.): UCAmI 2023, LNNS 835, pp. 298–303, 2023.
https://doi.org/10.1007/978-3-031-48306-6_30

levels of engagement and motivation. Consequently, new methodological trends have emerged, such as Game-Based Learning, which involves using games as a learning technique [2]. Additionally, gamification, a playful activity where students learn as if they are playing a game rather than specific games, has gained prominence [3]. Incorporating these types of methodological trends into the music education classroom is highly beneficial, as they capture students' attention, interest, and motivation.

In this paper, we propose the utilization of Information and Communication Technology (ICT) for learning body percussions through an app designed to assist students at home. The app provides visual representations and hand recognition capabilities, and offers feedback on students' percussion performance. This approach serves as an initial music learning experience while also enabling the observation of students' psychomotor development.

2 Related Works

Gesture recognition has emerged as a natural form of input in computers and intelligent environments. By analyzing visual images, it becomes possible to identify different body positions, which serves as the basis for certain languages. This capability enables the learning of sign languages specific to deaf-mute individuals through observation of a sequence of hand images displayed on a screen.

Hand-gesture recognition has been extensively studied for various computer interactions. In [4], a proposal for managing PowerPoint presentations was presented, whereas [5] suggested an approach to hand gesture recognition based on shape-based features. These studies utilized cameras to capture hand movements and employed algorithms for segmentation, orientation detection, feature extraction, and classification. Depth cameras have also been explored as segmentation techniques, isolating the hands by identifying key points [6–8]. Several authors have proposed hand-tracking solutions to determine the temporal and spatial information of hand positions within a sequence [9].

In the realm of body percussion, Howard Gardner developed the BAPNE method [9]. This method facilitates the development of multiple intelligences including biomechanics, anatomy, psychology, neuroscience, and ethnomusicology. The BAPNE brings significant benefits to primary education, particularly in the areas of visuospatial and kinesthetic intelligence. Body percussion, as defined in [10], involves creating rhythmic patterns by striking, clapping, and rubbing the human body, allowing for the creation of various rhythmic combinations using the body as an instrument. In addition to its application in music, body percussion has also been applied in physical education and sports [11].

3 A Proposal for Body Percussion Gesture Recognition

This section outlines the proposal, starting with the selection of an Artificial Intelligence technique for gesture recognition to provide feedback to students after each body percussion exercise. The development of an app that displays a score of pictograms and records gestures, dividing the video into corresponding parts, is crucial.

For gesture recognition using AI, MediaPipe [12] was initially selected for our project. MediaPipe is a machine-learning library that offers various solutions, including hand tracking, face detection, and pose estimation. The pose-detection module of MediaPipe provides 33 skeletal points for each frame of a given video. Real-time information from these skeletal points enables the identification of gesture patterns for the required movements. Simple gestures such as clicking fingers, clapping, or hitting the chest or leg with one or both hands were used as initial exercises for body percussion. The objective was to mimic the teacher's role in identifying errors in students. The confusion matrix illustrates a high degree of model accuracy with a majority of gestures correctly identified.

3.1 App for Body Percussion Homework

We completed the initial development of the app's first part, which aims to support the study of body percussion at home. When designing the user interface, we prioritized the selection of simple graphs that were easy to understand. This simplicity was a key consideration, based on feedback from music teachers at the primary school level.

During discussions with the teachers about the app's user interface, we focused on addressing the importance of gestures and tempo when children practice body percussion. The selection of specific gestures in the app was based on recommendations from music teachers aligned with established musical pedagogies [13]. By emphasizing the significance of gestures and tempo, we aimed to provide a comprehensive learning experience for students.

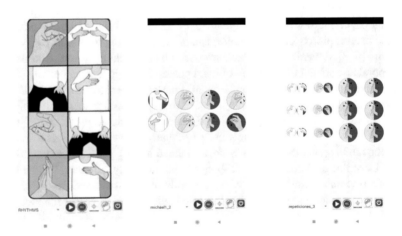

Fig. 1. App presentation and exercises screen

Figure 1 shows the app presentation, displaying pictures that provide visual guidance on how students should perform rhythmic exercises. Positioned at the bottom are several buttons, including options to play and stop the exercise visualization. The other button allows for video recording. In relation to the first part of the app, two additional buttons were incorporated to facilitate data preparation for the gesture recognition model. The first button captures the frequency of each gesture, while the second button converts the data into the necessary format for further processing. These features aim to streamline the data-preparation process for accurate gesture recognition.

For the model training, a dataset consisting of various gestures to be detected was assembled. This process entailed recording several volunteers performing gestures in sessions–1–2 min. The videos were subsequently segmented by gesture using the sound wave to pinpoint when the gesture occurred, and nine frames were captured (five prior, the gesture itself, and three following).

These videos were processed using a tool known as MediaPipe to extract the joint coordinates. Subsequently, a manual classification phase was implemented, during which videos where MediaPipe failed (owing to lighting issues, noise, etc.) or where the gesture was incorrectly performed, were discarded. This rigorous procedure resulted in a robust dataset of 2159 gestures.

Further processing was required for the spatial coordinates obtained from MediaPipe before they could be used as model inputs. Only the following joints were necessary to detect the targeted gestures:16-right_wrist, 14-right_elbow, 12-right_shoulder, 11-left_shoulder, 13-left_elbow, 15-left_wrist, 24-right_hip, and 23-left_hip.

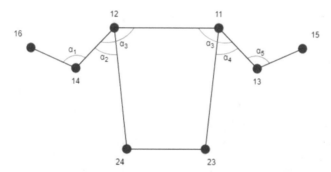

Fig. 2. Calculated Angles between Selected Joint Segments across Three Body Planes.

Data normalization involves calculating the angles between the segments formed by these specific joints. The calculated angles are shown in Fig. 2. To maximize the utility of this angle information, calculations were performed in three body planes: sagittal, frontal, and transverse. Consequently, the evolution of these angles across the nine frames was used as the input to the model.

A Support Vector Machine (SVM) was chosen as the machine learning algorithm for the gesture detection model because of its superior performance with the obtained dataset. The dataset was divided into two subsets for training:80% for training, and 20% for testing. This method achieved an accuracy of 94.44%. A deeper insight into the performance of the model on the test data is provided in Fig. 3, which shows the confusion matrix. The confusion matrix illustrates a high degree of model accuracy with a majority of gestures correctly identified.

After applying these Artificial Intelligence techniques, we intend to conduct this study at home. Trying to capture percussion gestures and tempos. Subsequently, a resume from each study session was sent to the teacher for the corresponding actuation.

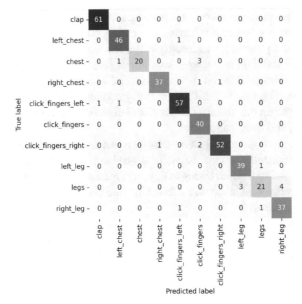

Fig. 3. Confusion Matrix of the SVM Model's Performance on the Test Data.

4   Conclusions and Future Works

In this study, we propose the first part of an ICT complement for body percussion learning by introducing one of the initial forms of music learning to students. By means of hand gesture recognition, it is possible to observe and correct incorrect movements and tempos from percussion exercises proposed by music teachers at primary schools.

This perspective opens a wide range of possibilities. On one hand, the study of percussion and, at the same time, it is possible to check students' psychomotor evolution. By using this kind of solution, teachers can detect anomalies and adequately solve this type of problem by means of remedial exercises.

Another important aspect is their capacity for attention and concentration. Children have little developed these capacities at this school age and, therefore, use short, simple, and playful activities to attract their attention and maintain their motivation. Despite this, the teacher facilitates of an increasingly way, long and complex exercises, presenting them as challenges to overcome to create in the student the desire for achievement and overcoming without realizing that they involve a high-er degree of concentration and attention on the same task. In this way, it is possible to develop these two fundamental aspects as well as to be able to diagnose again, objectively, the lack of them, and work to improve them.

Finally, we present a work in progress with the intention of evaluating this proposal when finished. The first part of this evaluation is presented in Sect. 3. The accuracy achieved for the gesture recognition was approximately 95%.

Furthermore, to develop other functionalities, such as gesture sequences by the teacher in a similar process when children practice at home or manage their children's mistakes to adequately send them to the teachers.

References

1. Willems, E.: The human value of Music Education. Editions Pro Música, Bienne (1975)
2. Correa Rodríguez, E.: Los beneficios de la música. Innovación y Experiencias Educativas. N° 26 (2016)
3. Simões, J., Díaz, R., Fernández, A.: A social gamification framework for a K-6 learning platform. Comput. Human Behav. **29**(2), 345–353 (2013)
4. Haria, A., Subramanian, A., Asokkumar, N., Poddar, S., Nayaka, J.S.: Hand gestures recognition for human computer interaction. In: 7th International Conference on Advances in Computing & Communications, ICACC-2017, Cochin, India, 22–24 August 2017 (2017)
5. Panwar, M., Mehra, P.S.: Hand gesture recognition for human computer interaction. In: Proceedings of the 2011 International Conference on Image Information Processing (ICIIP 2011) (2011)
6. Zhenyao, M., Neumann, U.: Real-time hand pose recognition using low-resolution depth images. In: Computer Vision and Pattern Recognition, pp. 1499–1505 (2006)
7. Yoo, B., et al.: 3D user interface combining gaze and hand gestures for large-scale display. In: Human Factors in Computing Systems, Atlanta, Georgia, USA, pp. 3709–3714 (2010)
8. Ren, Z., Yuan, J., Zhang, Z.: Robust hand gesture recognition based on finger-earth mover's distance with a commodity depth camera. In: ACM International Conference on Multimedia, pp. 1093–1096 (2011)
9. Chang, Y.J., Chen, S.F., Huang, J.D.: A Kinect- based system for physical rehabilitation: a pilot study for young adults with motor disabilities. Res. Dev. Disabil. **32**, 2566–2570 (2011)
10. Gardner, H.: Inteligencias múltiples. Editorial Paidós, Barcelona (1995)
11. Bevacqua, E., Desmeulles, G.: Real and virtual body percussionists' interaction. In: MOCO 2017, London, United Kingdom (2017)
12. Lugaresi, C., et al.: Mediapipe: a framework for building perception pipelines. arXiv preprint arXiv:1906.08172 (2019)
13. Romero-Naranjo, F.J.: BAPNE FIT: neuromotricity and body percussion in physical activity and sport sciences. Educ. Rev. USA **6**(2), 37–44 (2022)

User Modeling Through Physiological Signals: A Systematic Review

Heber Avalos-Viveros[✉], Carmen Mezura-Godoy,
and Edgard Benítez-Guerrero

Universidad Veracruzana, Xalapa, Mexico
zS22000341@estudiantes.uv.mx, {cmezura,edbenitez}@uv.mx

Abstract. Technology has acquired a fundamental role in our lives, enabling the integration of small and efficient sensors that can monitor and evaluate the physical and mental states of people during interaction with devices. This has led to significant advances in user modeling in the field of Human-Computer Interaction (HCI). This paper presents a review of studies that focus on user modeling through the analysis of physiological signals. While there have been substantial developments, challenges persist in implementing these elements in real-world applications. Future work recommends a broader and more diverse approach to the use of physiological signals and sensors, as well as further exploration of optimal techniques for data acquisition and processing in this field.

Keywords: User modeling · Physiological signals · Biometric sensors

1 Introduction

In recent years, the continuous growth of computer systems has led to emerging development trends that aim to meet user needs by creating adaptable systems that customize various aspects according to each user's preferences [1]. This customization is achieved through user models. However, it's crucial to consider various factors related to the user, such as cognitive abilities, previous experiences, emotional aspects, and others, all of which play a fundamental role in HCI [2]. As a result, new analytical tools have been adopted, leveraging biometric sensors to monitor users' physical and mental states through the acquisition of physiological signals [3]. Additionally, these tools have been adopted in a number of applications, such as [4]: a) usability tests for the adaptation and customization of software; b) user experience (UX) tests; c) monitoring the physical and mental states of people, among others. This article conducts a systematic literature review to analyze the methods, techniques, and algorithms used in user modeling through physiological signals. Additionally, it examines related works that include systematic literature reviews and concludes with an analysis of the results, followed by a discussion and conclusions based on the findings.

Supported by CONAHCYT.

J. Bravo and G. Urzáiz (Eds.): UCAmI 2023, LNNS 835, pp. 304–309, 2023.
https://doi.org/10.1007/978-3-031-48306-6_31

2 Related Work

Nowadays, extensive research has focused on analyzing user behavior within specific contexts by utilizing physiological signals as valuable data sources. Several notable studies have contributed to our understanding of this domain. As it points out, [5] systematically review methods for acquiring physiological signals and their application in assessing diverse emotional responses. They explore various methodological approaches, including machine learning-based classification and rule-based methods. Furthermore, [6] significantly contribute to affective computing by discussing a range of emotion models. They explore models rooted in emotional dimensions such as valence, arousal, and dominance, as well as those based on fundamental emotions like happiness, sadness, fear, and anger.

In an exhaustive investigation, [7] scrutinize prior research that harnesses sensors for emotion recognition through physiological signals. The sensors discussed encompass a variety of devices, including activity wristbands, smartwatches, and electrode-based sensors. Lastly, [8] meticulously analyze studies categorically assessing cognitive load in software developers, highlighting the importance of selecting suitable techniques for specific contexts.

3 Research Methodology

In this study, a systematic literature review was conducted following the guidelines outlined in the Preferred Reporting Items for Systematic Reviews and Meta-Analyses (PRISMA) Statement [9].

3.1 Research Questions

Initially, for the identification and selection of relevant studies, research questions are formulated to help define specific aspects of interest with respect to the research topic. In this context, the following research questions were defined:

- Q1. What physiological sensors allow us to effectively assess the physical and mental state of users during interactions with computer applications?
- Q2. What are the most relevant physiological signals currently relevant for user modeling, and how can they be optimally acquired and processed?
- Q3. Which physiological characteristics of the user have been interpreted during user modeling by using biometrics sensors?
- Q4. What are the possible applications of user modeling based on physiological signals?

3.2 Selected Database

In this systematic literature review, we have selected the Science Direct, Springer Link, Association for Computing Machinery (ACM) and IEEE Xplore databases for their broad coverage in the fields of computer science and engineering. In addition, they are providing accessibility and advanced search tools.

3.3   Search Strategy

During the initial phase of our search process, we employed keywords such as 'user modeling', 'user profile', 'user preferences', 'physiological signals', 'biosignals', and 'biomedical signals' combining these terms with Boolean operators 'AND' and 'OR' to refine our search. For our initial comprehensive search, we obtained a substantial amount of information, encompassing various types of publications, including scientific journals, conference papers, technical reports, books, and proceedings. As our research progressed, we further refined our search by critically analyzing the obtained results, evaluating their relevance, and assessing their similarity to our research topic. During this iterative process, we carefully considered terms such as 'user modeling', 'physiological signals', and 'biosignals' to optimize our search queries. Subsequently, we conducted a systematic search in September 2022 by combining terms that had demonstrated superior results, such as "User modeling" AND ("physiological signals" OR "biosignals"). This search strategy produced a total of 75 results in Science Direct, 479 in Springer Link, 476 in IEEE Xplore, and 107 in ACM. Before selecting articles, inclusion and exclusion criteria were defined as shown in Table 1.

Table 1. Inclusion and Exclusion Criteria.

Inclusion Criteria	Exclusion Criteria
– The work must have a context in computer science and engineering – Only articles written in English will be included – Only research articles that include the construction or evaluation of instruments related to the use of physiological signals in relation to user behavior will be considered – Articles published within the last 5 years will be included	– Studies limited to extended abstracts or literature reviews will be excluded – Articles lacking a structured process for processing physiological signals in their proposed solutions will not be considered – Works that do not mention any of the following physiological signals: electrocardiogram (ECG), electroencephalogram (EEG), electromyography (EMG), electrodermal activity (EDA), or galvanic skin response (GSR), will be excluded

3.4   Data Extraction

Following the specified criteria, we initially screened articles published between January 2017 and September 2022. These articles were categorized, with a primary focus on computer science and engineering. After removing duplicate publications, we identified a total of 105 articles across the selected databases. We then conducted a thorough review of the titles and abstracts, resulting in the exclusion of 76 articles that did not address our research questions.

Finally, we performed an in-depth selection process by carefully reviewing each paper in its entirety. This approach allowed us to identify the most pertinent, robust, and high-quality papers aligned with our research objectives. Consequently, we excluded articles that did not involve the development or evaluation of instruments related to physiological measures or that deviated from the context of user modeling. Additionally, literature review papers were excluded, as well as those lacking a structured signal processing approach and any mention of physiological signals such as ECG, EEG, EMG, EDA, and GSR in their proposed solutions. Upon completing the selection process, a total of 13 articles were identified as meeting our selected criteria.

3.5 Results

In this section, the results obtained in relation to the research questions posed will be presented. The relevant characteristics of each of these studies are summarized in Table 2.

– **Q1**: It was observed that various physiological sensors are used to interpret the physical and mental state of users in different contexts. The most common sensors include devices such as smart wristbands, brain-wearables, and

Table 2. Primary studies.

Ref	Year	Database	Q1 (Sensors)	Q2 (Signals)	Q3 (Features)	Q4 (Applications)
[10]	2017	Science Direct	ECG and GSR electrodes	ECG, GSR	Stress	Work stress
[11]	2017	Science Direct	–	GSR, EMG	Angry, happy, sad, surprised, etc.	Affective tutoring Systems
[12]	2017	Springer Link	Tobii EyeX, Kinect	Optical	Attention	Racing games
[13]	2018	ACM	EMOTIV insight headset	EEG	Calm, disgust and excited	Viewer experience design
[14]	2018	Springer Link	LaborCheck system, Fitbit	ECG	Stress	Stress recognition in desk jobs
[15]	2019	IEEE Xplore	Emotiv EPOC, Shimmer 2R5, electrodes GSR	ECG, EEG, GSR	Arousal and valence emotions	–
[16]	2019	Science Direct	Smart wristband	ECG, GSR	–	Behavior Monitoring in a real environment
[17]	2020	ACM	Forerunner, vivosmart 3, polar A370, empatica E4	ECG	Drowsiness	Driver drowsiness detection
[18]	2020	IEEE Xplore	Smart wristband	ECG	Excited, angry, sad, calm	Interactive gaming environments
[19]	2020	Science Direct	An instrumented glove was developed	ECG, GSR	Amusement, sadness, neutral	A machine learning model for emotion recognition
[20]	2021	Science Direct	Brain-wearable, Smart wristband	EEG, ECG	–	Emotion-aware mobile edge computing system
[21]	2022	IEEE Xplore	BioNomadix wireless modules	ECG, EMG, EDA	Stress	Virtual reality applications
[22]	2022	IEEE Xplore	Empatica E4, bioHarness 3, shimmer 3, y respiBAN	EEG, EDA, ECG, GSR	Arousal and valence emotions	Robust wearable emotion recognition

electrode-based sensors. These accessible and user-friendly devices promote the collection and analysis of physiological data. Approximately 92.31% of the studies employed these devices for acquiring physiological signals.

- **Q2**: The primary physiological signals used to characterize users have been identified, with ECG and GSR utilized in 77% and 46% of the studies, respectively. These signals play a pivotal role in capturing and comprehending the body's responses to a wide range of stimuli and situations. It is also important to mention that 60% of the works use more than one signal.
- **Q3**: Diverse user physiological characteristics have been analyzed utilizing biometric sensors. These characteristics encompass the identification of emotions such as stress, attention, mental workload, as well as broader emotional dimensions like arousal and valence.
- **Q4**: The potential applications of user models based on physiological signals are diverse, including the adaptation of tutoring systems and video games, stress detection systems in real environments, virtual reality applications, and user experience design.

4   Conclusions

A wide variety of techniques have been identified for modeling users using physiological signals, each with its own set of advantages and limitations. User models based on physiological signals hold significant potential within HCI. However, the implementation of these models in real-world environments poses technical, ethical, and methodological challenges that necessitate careful consideration and the development of appropriate solutions. For future research, it is advisable to focus on the development of efficient models, methodologies, and techniques.

Acknowledgements. This work was partially supported by the Universidad Veracruzana (UV), and by the National Council of Humanities, Sciences and Technologies (CONAHCYT) of Mexico.

References

1. DAdamo, M.H., Baum, A., Luna, D., Argibay, P.: Interacción ser humano-computadora: usabilidad y universalidad en la era de la información. Rev. Hosp. Ital. B. Aires **31**(4) (2011)
2. Arhippainen, L., Tähti, M.: Empirical evaluation of user experience in two adaptive mobile application prototypes. In: MUM 2003. Proceedings of the 2nd International Conference on Mobile and Ubiquitous Multimedia. Citeseer (2003)
3. Morales, E.: Medidas fisiologícas para evaluación de usabilidad: el caso del ritmo cardíaco. tesis, Facultad de Estadística e Informática (2018)
4. Ortega-Gijón, Y.N., Mezura-Godoy, C.: Usability evaluation process of brain computer interfaces: an experimental study. In: Proceedings of the IX Latin American Conference on Human Computer Interaction, pp. 1–8 (2019)
5. Tawsif, K., Aziz, N.A.A., Raja, J.E., Hossen, J., Jesmeen, M.: A systematic review on emotion recognition system using physiological signals: data acquisition and methodology. Emerg. Sci. J. **6**(5), 1167–1198 (2022)

6. Wang, Y., et al.: A systematic review on affective computing: emotion models, databases, and recent advances. Inf. Fusion **83–84**, 19–52 (2022)
7. Wijasena, H.Z., Ferdiana, R., Wibirama, S.: A survey of emotion recognition using physiological signal in wearable devices. In: 2021 International Conference on Artificial Intelligence and Mechatronics Systems (AIMS), pp. 1–6. IEEE (2021)
8. Gonçales, L., Farias, K., da Silva, B., Fessler, J.: Measuring the cognitive load of software developers: a systematic mapping study. In: 2019 IEEE/ACM 27th International Conference on Program Comprehension (ICPC). IEEE (2019)
9. Ciapponi, A.: La declaración prisma 2020: una guía actualizada para reportar revisiones sistemáticas, Evidencia, actualizacion en la práctica ambulatoria (2021)
10. Sriramprakash, S., Prasanna, V.D., Murthy, O.R.: Stress detection in working people. Procedia Comput. Sci. **115**, 359–366 (2017)
11. Petrovica, S., Anohina-Naumeca, A., Ekenel, H.K.: Emotion recognition in affective tutoring systems: collection of ground-truth data. Procedia Comput. Sci. **104**, 437–444 (2017)
12. Georgiou, T., Demiris, Y.: Adaptive user modelling in car racing games using behavioural and physiological data. User Model. User-Adap. Inter. **27**(2), 267–311 (2017). https://doi.org/10.1007/s11257-017-9192-3
13. Deja, J.A., Cabredo, R.: Using EEG emotion models in viewer experience design: an exploratory study. In: Proceedings of the 4th International Conference on Human-Computer Interaction and User Experience in Indonesia, CHIuXiD 2018, pp. 82–88. Association for Computing Machinery (2018)
14. Sanchez, W., Martinez, A., Hernandez, Y., Estrada, H., Gonzalez-Mendoza, M.: A predictive model for stress recognition in desk jobs. J. Ambient Intell. Humanized Comput. **14**(1), 17–29 (2018)
15. Santamaria-Granados, L., Munoz-Organero, M., Ramirez-Gonzalez, G., Abdulhay, E., Arunkumar, N.: Using deep convolutional neural network for emotion detection on a physiological signals dataset (AMIGOS). IEEE Access **7**, 57–67 (2018)
16. Kanjo, E., Younis, E.M., Ang, C.S.: Deep learning analysis of mobile physiological, environmental and location sensor data for emotion detection. Inf. Fusion **49**, 46–56 (2019)
17. Kundinger, T., Riener, A.: The potential of wrist-worn wearables for driver drowsiness detection: a feasibility analysis. In: Proceedings of the 28th ACM Conference on User Modeling, Adaptation and Personalization, pp. 117–125. Association for Computing Machinery (2020)
18. Du, G., Long, S., Yuan, H.: Non-contact emotion recognition combining heart rate and facial expression for interactive gaming environments. IEEE Access **8**, 11896–11906 (2020)
19. Domínguez-Jiménez, J.A., Campo-Landines, K.C., Martínez-Santos, J.C., Delahoz, E.J., Contreras-Ortiz, S.H.: A machine learning model for emotion recognition from physiological signals. Biomed. Signal Process. Control **55**, 101646 (2020)
20. Yu, Q., Xiao, W., Jiang, S., Alhamid, M.F., Muhammad, G., Hossain, M.S.: Emotion-aware mobile edge computing system: a case study. Comput. Electr. Eng. **92**, 107120 (2021)
21. Orozco-Mora, C., Oceguera-Cuevas, D., Fuentes-Aguilar, R.Q., Hernández-Melgarejo, G.: Stress level estimation based on physiological signals for virtual reality applications. IEEE Access **10**, 68 755–68 767 (2022)
22. Dissanayake, V., Seneviratne, S., Rana, R., Wen, E., Kaluarachchi, T., Nanayakkara, S.: SigRep: toward robust wearable emotion recognition with contrastive representation learning. IEEE Access **10**, 18 105–18 120 (2022)

And How Enjoyable? Converting a User Experience Evaluation Questionnaire into a Voice Conversation

Ignacio Díaz-Oreiro[(⊠)] [ID], Gustavo López[ID], and Luis A. Guerrero[ID]

University of Costa Rica, San Jose 11501, Costa Rica
{ignacio.diazoreiro,gustavo.lopezherrera,
luis.guerreroblanco}@ucr.ac.cr

Abstract. Voice interfaces have gained popularity, driven by the proliferation of intelligent assistants and the development of natural language processing. Given their progress, they could be used to implement self-reported instruments such as standardized user experience (UX) evaluation questionnaires, particularly in the response capture mechanism. This novel use of voice interfaces would make sense if the evaluations using the voice mechanism did not differ significantly from the evaluations carried out with the traditional written questionnaire. In addition, given different possible implementations of the voice mechanism, it must be analyzed whether the user experience is affected by the format of the instrument used. This paper compares two implementations of the User Experience Questionnaire in which the capture mechanism is implemented by voice. Results show that in both implementations the evaluations retain their validity, and that the use of conversational patterns improves the user experience and increases the quality of the answers, which present fewer inconsistencies.

Keywords: User experience · UEQ · Voice-based interfaces · Branching · Conversational patterns

1 Introduction

In recent years, voice interaction has become one of the fastest growing technology trends. Thanks to advances in artificial intelligence (AI) and natural language processing (NLP), voice assistants like Siri, Alexa, and Google Assistant have become more accurate and reliable, resulting in an increase in the adoption of voice-based technologies, particularly in the form of smart speakers and virtual assistants. With the ability to perform a variety of tasks, voice interaction is transforming the way we interact with technology and could become even more prevalent in the years to come.

User experience (UX) is a critical aspect of modern product design, as developers seek to create products and services that are intuitive, user-friendly, and enjoyable to use. To ensure that products meet these standards, UX designers often rely on standardized questionnaires to evaluate user experiences and gather feedback [1]. Standardized questionnaires provide a consistent framework for assessing user experiences, allowing

J. Bravo and G. Urzáiz (Eds.): UCAmI 2023, LNNS 835, pp. 310–321, 2023.
https://doi.org/10.1007/978-3-031-48306-6_32

designers to compare results across different products and user groups. These questionnaires are designed to measure pragmatic factors, being the functional and practical aspects of the user experience, and hedonic factors, or the emotional and affective aspects of the user experience [2]. By gathering this data, designers can identify areas for improvement and make data-driven decisions to enhance the user experience.

The aim is then to design a voice interaction that allows the capture of responses to a standardized UX questionnaire. This interaction must preserve the characteristics of the original written questionnaire in terms of the validity of the evaluation. Additionally, since several voice implementations are possible, two versions of this interaction are presented: one that translates the questionnaire items very directly, and another that uses item branching and includes conversational voice patterns seeking a more natural interaction. Both versions are compared in terms of the user experience and usability of using them, as well as in terms of the number of inconsistencies they generate and the precision of their results.

The concept of branching has existed for a long time in the context of telephone surveys [3, 4]. The term is taken from [5] where the branching approach is described. Regarding the evaluation of alternative formats of bipolar scales, [6] uses branching on a written scale and [7] analyzes whether branching reduces the number of unanswered questions. Moreover, the conversational patterns implemented follow the Natural Conversation Framework [8], which defines an interaction as natural if it meets three principles: design based on the recipient, minimization, and repair. These principles are based on the Conversation Analysis concepts of [9].

2 Related Work

Related works are presented in the following categories: conversational patterns, conversational Chatbots, and voice interfaces in survey contexts.

Regarding research that uses conversational patterns with the principles exposed by Moore and Arar [8], there are different studies, but no implementations of capture mechanisms for questionnaires. Among this research we can cite [10], in which a conversation system for applications in the Business Intelligence domain is presented. In [11] a Chatbot is used to search for information in a large company. In [12] the authors design a malicious Chatbot (*trollbot*) to investigate whether an automatic agent of this type could create conflicts with users without them realizing that the one causing the conflicts is not a person but a trollbot. Finally, in [13] it is discussed the topic of conversational agents within the aerospace industry, in which it is necessary to have quick and concise responses in complex situations.

Within the conversational interfaces we also find Chatbots. Although most Chatbots interact through writing, it is worth mentioning some studies that make comparisons with written questionnaires via the web, or that implement interaction not only in writing but also simultaneously by voice. In [14] researchers use the Wizard of Oz technique to create a Chatbot that provides help to UX evaluators reviewing recorded videos of participants interacting with a product. The evaluator, while reviewing a video, could ask questions of a trained artificial intelligence assistant who would provide supplementary information for analysis. The evaluator can write the questions or issue them orally. In

[15] it is investigated whether there are differences in the responses of a traditional web-based survey and a conversational survey based on a Chatbot, integrated into an instant messaging application. In [16] an experiment is carried out to test the effectiveness of teacher recruitment surveys through web forms and Chatbots, to later compare their effectiveness with telephone surveys. In [17] the perceptions of patients about Chatbots are investigated, comparing the experience of completing a health survey using a Chatbot versus an online form, both with visual display. In [18] they develop an application for a smartphone and carry out two case studies in which the questions are of the *yes/no* type or Likert scales that are answered by indicating the number associated with the answer. Responses can be delivered by voice or by selecting an option on the screen, but all the interaction is written on the screen.

Finally, regarding the use of voice interfaces in survey contexts, two articles from the health area were found. In [19] the authors study what types of questions are appropriate to be administered by smart speakers, in the collection of health data. For their study they present four questionnaires, where half of each questionnaire is answered by voice and the other half in writing. The questionnaires use yes/no questions and five- and seven-point Likert scales. For the oral version, the participants state the words *yes* or *no*, numbers for the Likert scale and also the phrases: *strongly disagree, disagree, neutral, agree,* and *strongly agree*. In all cases, the oral versions obtained similar results to their written counterparts. In [20] a *how often* type questionnaire is implemented in the form of a 6-point Likert scale. Participants respond in three modalities: in writing, by voice with *discrete* responses (only one number is uttered) and by voice with *open* responses (answers with open sentences that are later manually converted to numbers by three independent researchers). The results show higher correlations between written responses and discrete voice responses than open voice responses.

3   Development Process

Two voice–based versions of the User Experience Questionnaire (UEQ) [21] were built, each one around a different implementation of the semantic differential on which the UEQ is based. The first implementation, which we call Voice1–7, transforms the semantic differential into a single question, in which both the attitude direction and intensity are identified, as shown in Fig. 1.

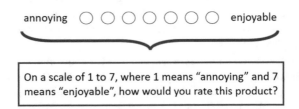

Fig. 1. Direct implementation of semantic differential

In the second implementation, as shown in Fig. 2, the semantic differential becomes two *branched* questions: a first question that indicates the direction of the attitude and a second question that unfolds based on the answer to the first question and determines the intensity of the attitude.

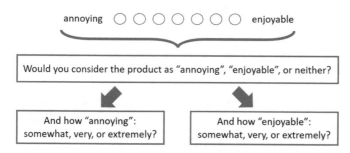

Fig. 2. Semantic differential implementation by means of branched questions.

Both voice implementations replicate the written UEQ questionnaire, using the terms of the official Spanish translation presented in [22] and were built using Voiceflow [23], a tool for developing interactive conversations that can be then incorporated into voice assistants or shared via a web interface. Voice1–7 implementation was explained in detail in [24], showing that an evaluation with 40 participants did not present significant differences compared to the same evaluation using the written questionnaire.

The second implementation, Voice2Q, includes the semantic differential version as branched questions and an additional set of conversational patterns presented in detail in [25]. These patterns fit the Natural Conversation Framework (NCF) proposed in [8], in which the conversation model is defined in three levels. At the inner level there are the conversational sequences, in this case, the implementation of the semantic differential as branched questions and their variants. Surrounding level 1 is level 2 in which sequences are administered, including patterns to modify a given response, request the meaning of concepts, or ask for a question to be repeated. Finally, at level 3 there are patterns at the level of the conversation, which includes the greetings at the beginning and end of the process, the possibility of asking about the skills of the program that captures the responses or at what point in the process the participant is.

4 Evaluation

The Voice2Q implementation was evaluated by 40 participants, computer science students, 32 males (80%), 6 females (15%) and 2 (5%) preferred not to disclose their gender. Ages were in the range of 20 to 33 years, with a median of 21 years.

Participants evaluated two products, alternating the voice-enabled version Voice2Q with the written version UEQ named UEQ-W. Table 1 shows the results for both implementations compared by a Wilcoxon rank test (UEQ analysis tool includes a t-Student test, but means showed no-normal distribution). No significant differences in the responses of the six scales were identified, as all p-values are higher than 0.05.

To further evaluate the UX of Voice2Q, focusing on voice interfaces we used UEQ + questionnaire (A modular Extension of the User Experience Questionnaire), which is an extension of UEQ presented in 2020 [26], where it is possible to modularly build a UEQ questionnaire according to what is to be evaluated. Currently, UEQ + has 20 scales to choose from, including the six traditional UEQ scales, plus other scales such as Trust, Novelty, Clarity, Haptic Sensation, among others.

Table 1. Means, Variances, and p-values of Wilcoxon rank test for independent samples for the six scales of UEQ-W and Voice2Q.

UEQ Scale	UEQ-W Mean (Variance)	Voice2Q Mean (Variance)	Wilcoxon test p-value
Attractiveness	1.38 (1.26)	1.38 (0.60)	0.7650
Perspicuity	1.31 (1.44)	1.24 (0.69)	0.5174
Efficiency	0.84 (1.12)	0.61 (0.59)	0.2174
Dependability	0.81 (1.20)	0.66 (0.57)	0.4959
Stimulation	1.32 (1.68)	1.04 (0.70)	0.1088
Novelty	1.71 (1.39)	1.59 (0.77)	0.3241

In [27] Klein et al. propose three new scales for systems with voice implementations that adapt to the UEQ + format: Comprehensibility (the voice assistant understands instructions and questions using natural language), Response behavior (the voice assistant behaves in a respectful, polite, and trustworthy manner), and Response quality (the responses of the voice assistant cover the user's information needs).

In addition to the three scales for voice implementation, two scales that were considered relevant to the study were added: Intuitive use (the product can be used immediately without any training or help), and Novelty (the design of the product is creative, captures the interest of users). The results for this evaluation are shown in Table 2. Both values of these scales and individual items vary between a minimum of -3 and a maximum of 3.

Participants also answered a set of Usability questions that are shown in Fig. 3. Of these questions, the most positive evaluations were obtained in the statements: *It is easy to know if the Voice Assistant understood the question I asked, It is easy to know if the Voice Assistant understood my answer, It is easy to ask the Voice Assistant for help, All concepts are easy to understand.* Also, in the last question in Fig. 3 participants indicate that it was not common for them to forget the question they were asked, which is a positive feature of the Voice2Q implementation.

On the other hand, the following questions had intermediate results: *The Assistant talks more than necessary, The way to interact with the Assistant is clear from the first question* and *It is easy to understand how many questions have been asked and how many are left to finish.*

Finally, the questions with negative evaluations were: *If I made a mistake while giving an answer, it is possible to correct it on the spot*, The length of the questionnaire is adequate, as well as the question that indicates that the questionnaire asks the same concepts more than once, which is considered as something negative in the process.

Table 2. Results of the voice interfaces UX evaluation of Voice2Q using UEQ +

UEQ + scale	Scale mean	Left item	Right item	Item mean
Comprehensibility	0.46	complicated	simple	0.23
		ambiguous	unambiguous	0.28
		inaccurate	accurate	0.18
		enigmatic	explainable	1.18
Response behavior	0.68	artificial	natural	0.20
		unpleasant	pleasant	1.28
		unlikeable	likable	1.30
		boring	entertaining	-0.08
Intuitive use	1.08	difficult	easy	1.10
		illogical	logical	1.35
		not plausible	plausible	0.75
		inconclusive	conclusive	1.13
Response quality	1.49	inappropriate	suitable	2.20
		useless	useful	1.73
		not helpful	helpful	0.93
		unintelligent	intelligent	1.13
Novelty	1.13	dull	creative	1.28
		conventional	inventing	0.88
		usual	leading edge	1.10
		conservative	innovative	1.25

Regarding the comparison of Voice1–7 and Voice2Q in the dimension of user experience for voice interfaces, Fig. 4 shows the means obtained in the five UEQ + scales analyzed. Voice1–7 scores better in Comprehensibility and Intuitive Use, while Voice2Q is better evaluated in Response Behavior, Response Quality and Novelty. However, Table 3 shows that only the Novelty scale presents significant differences (p-value < 0.05), as indicated a Wilcoxon rank test for unpaired samples conducted between Voice2Q and UEQ-W.

Figure 5 shows the detail of the comparison of the four items that make up each of the UEQ + scales used, including the three scales for systems with voice implementations, as well as the additional scales Intuitive Use and Novelty. As expected according to what is shown in Table 3, the most important differences are in the Novelty scale, particularly in the items: Conventional/Inventing, Usual/Leading Edge, and Conservative/Innovative, where Voice2Q reports better results.

Regarding the Usability questions, a comparison of Voice1–7 and Voice2Q is shown in Fig. 6. The values correspond to a transformation of the Likert scale, where 1 represents *strongly disagree*, 2 is *disagree*, 3 is *neither disagree nor agree*, 4 is *agree*, and 5 represents strongly *agree*.

For the four questions with a significant difference, marked with (*), the Voice2Q implementation achieves better results in three of these four.

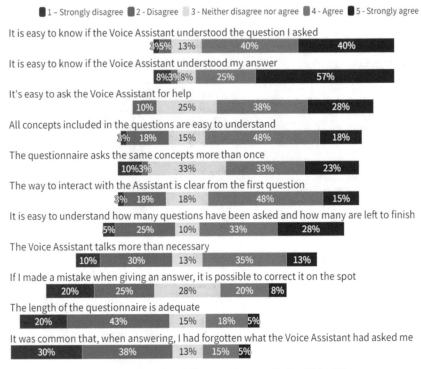

Fig. 3. Results of usability questions applied to Voice2Q.

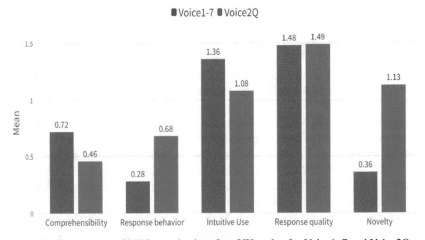

Fig. 4. Comparison of UEQ + voice interface UX scales for Voice1–7 and Voice2Q.

On the one hand, interaction with Voice1–7 appears more intuitive than Voice2Q, since the pattern is quickly recognized as a direct translation of the semantic differential, so the interaction is clear from the first question.

Table 3. Wilcoxon test for UEQ + evaluation of Voice1–7 and Voice2Q

UEQ Scale	Voice1–7 Mean	Voice2Q Mean	Wilcoxon test p-value
Comprehensibility	0.72	0.46	0.4070
Response behavior	0.28	0.68	0.1982
Intuitive Use	1.36	1.08	0.2716
Response quality	1.48	1.49	0.8128
Novelty	0.36	1.13	0.0176 (*)

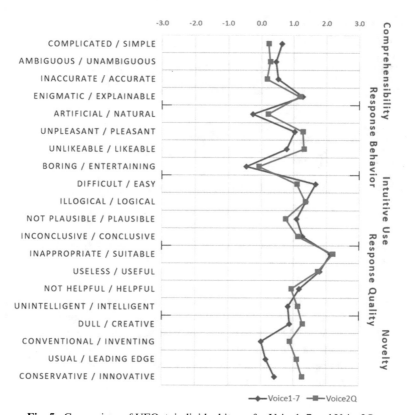

Fig. 5. Comparison of UEQ + individual items for Voice1–7 and Voice2Q.

On the other hand, in questions about the ease of knowing if the voice assistant understood the question, and asking the assistant for help, Voice2Q presents more favorable opinions than Voice1–7. This may be influenced by the fact that Voice1–7 only includes basic conversational patterns, not having the ability to answer how the interaction works.

Finally, on the question of whether it was common to forget what the assistant had asked me, Voice2Q is rated better than Voice1–7, since participants report that it was more common to forget the question in Voice1–7 than in Voice2Q, which is a positive feature since the conversion of the semantic differential into two questions seems to represent a lower mental load to participants than the direct version, in terms of remembering what was asked.

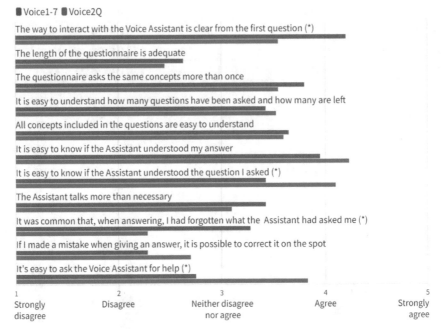

Fig. 6. Usability questions for Voice1–7 and Voice2Q. (*) indicates responses with significant difference identified.

In terms of inconsistencies, UEQ details when responses to items on a particular scale are inconsistent. These inconsistencies are reported by participant and by scale. Table 4 shows that Voice2Q responses presents significantly fewer inconsistencies than Voice1–7, both at the total level (inconsistencies in the six scales for the 40 participants) and at the level of responses considered critical (those responses from a participant with inconsistencies in three or more scales).

Table 4. Inconsistencies identified in Voice1–7 and Voice2Q.

Format	Inconsistencies	Critical answers
Voice1–7	60 / 240 (25.0%)	7 / 40 (17.5%)
Voice2Q	17 / 240 (7.1%)	0 / 40 (0.0%)

Another point worth mentioning is that the means of the responses from Voice2Q seem more precise than those obtained with Voice1–7: the interquartile ranges are smaller on four of the six scales and the extreme data are closer to the median on all six scales, as and as shown in Fig. 7.

Finally, in terms of the time it takes to fill out each voice-implemented questionnaire, it took participants an average of 7:46 min to complete Voice1-7, and 9:44 min to complete Voice2Q implementation.

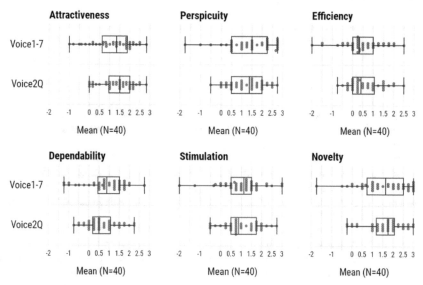

Fig. 7. Comparison of means for Voice1-7 and Voice2Q, by UEQ scale.

5 Discussion and Conclusions

Results show that a version of the UEQ questionnaire, whose capture mechanism is done by voice, provides results without significant differences from an evaluation of the same product using the traditional written UEQ.

When studying the two proposals for voice capture questionnaires, Voice1-7 and Voice2Q, the latter implementation, with the conversion of the semantic differential in branched questions plus the incorporation of conversational patterns, makes the user experience more favorable in factors such as Novelty and memory load associated with the process. In addition to this, the Voice2Q version presents significantly fewer inconsistencies than Voice1-7, making the response collection process more effective, despite the fact that the time to complete the questionnaire is longer.

Therefore, the results obtained with the Voice2Q implementation are promising, in the sense that it could be used as an alternative means to carry out user experience evaluations, with results equivalent to those obtained with the written version of UEQ. This would be an innovative use of conversational voice interfaces.

As future work, a comparative evaluation of the Voice2Q version with the written version UEQ-W will be carried-out in terms of the general user experience, number of inconsistencies found, scale precision and specific workload of both questionnaires. Additionally, the application of the format with conversational patterns to the standardized UX evaluation questionnaires AttrakDiff [28] and meCUE [29] will be explored, the first consisting of semantic differential scales, while the second poses additional challenges, since the meCUE questions are not presented as semantic differential scales but as Likert scales.

Acknowledgments. This research was partially funded by CITIC and ECCI at the University of Costa Rica, grant number 834-C1–013.

References

1. Lallemand, C., Gronier, G.:Méthodes de design UX: 30 méthodes fondamentales pour concevoir et évaluer les systèmes interactifs. 2eme édition; Éditions Eyrolles: Paris, France (2018)
2. Díaz-Oreiro, I., López, G., Quesada, L., Guerrero, L.A.: UX evaluation with standardized questionnaires in ubiquitous computing and ambient intelligence: a systematic literature review. Adv. Human-Computer Interact. **2021**, 1–22 (2021)
3. Miller, P.V.: Alternative question forms for attitude scale questions in telephone interviews. Public Opin. Q. **48**(4), 766–778 (1984)
4. Fabrigar, L.R., Krosnick, J.A.: Attitude measurement and questionnaire design, pp. 42–47. Blackwell Publishers, Blackwell encyclopedia of social psychology (1995)
5. Krosnick, J.A., Berent, M.K.: Comparisons of party identification and policy preferences: the impact of survey question format. Am. J. Polit. Sci. **37**(3), 941–964 (1993)
6. Yu, J.H., Albaum, G., Swenson, M.: Is a central tendency error inherent in the use of semantic differential scales in different cultures? Int. J. Mark. Res. **45**(2), 1–16 (2003)
7. de Rada, D., Igúzquiza, V.: ¿Influye el diseño de las preguntas en las respuestas de los entrevistados? Res. Revista Española de Sociología **31**(1), 9 (2022)
8. Moore, R.J., Arar, R.: Conversational UX Design: A Practitioner's Guide to The Natural Conversation Framework. Morgan & Claypool, New York, (2019)
9. Sacks, H., Schegloff, E.A., Jefferson, G.: A simplest systematics for the organization of turn taking for conversation. In: Studies in the Organization of Conversational Interaction, pp. 7–55. Academic Press (1978)
10. Quamar, A., Özcan, F., Miller, D., Moore, R., Niehus, R., Kreulen, J.: Conversational BI: an ontology-driven conversation system for business intelligence applications. Proc. VLDB Endow. **13**(12), 3369–3381 (2020)
11. Liao, Q.V., Geyer, W., Muller, M., Khazaen, Y.: Conversational Interfaces for Information Search. In: Fu, W.T., van Oostendorp, H. (eds.) Understanding and Improving Information Search. HIS, pp. 267–287. Springer, Cham (2020). https://doi.org/10.1007/978-3-030-38825-6_13
12. Vepsäläinen, H., Salovaara, A. Paakki, H.: What Would Be the Principles for Successful Trollbot Design? In: ACM CHI Conference workshop (2019)
13. Liu, Y., Arnold, A., Dupont, G., Kobus, C., Lancelot, F.: Evaluation of conversational agents for aerospace domain. In: Proceedings of the Joint Conference of the Information Retrieval Communities in Europe (CIRCLE 2020). July 6–9 Samatan, Gers, France (2020)
14. Kuang, E., Jahangirzadeh Soure, E., Fan, M., Zhao, J., Shinohara, K.: Collaboration with Conversational AI Assistants for UX Evaluation: Questions and How to Ask them (Voice vs. Text). In: Proceedings of the 2023 CHI Conference on Human Factors in Computing Systems, pp. 1–15 (2023)

15. Zarouali, B., Araujo, T., Ohme, J., de Vreese, C.: Comparing Chatbots and online surveys for (longitudinal) data collection: an investigation of response characteristics, data quality, and user evaluation. Commun. Methods Measures, 1–20 (2023)
16. Beam, E.A.: Social media as a recruitment and data collection tool: Experimental evidence on the relative effectiveness of web surveys and Chatbots. J. Dev. Econ. **162**, 103069 (2023)
17. Soni, H. et al.: Virtual conversational agents versus online forms: Patient experience and preferences for health data collection. Front. Digital Health, **4** 954069 (2022)
18. Sprengholz, P., Betsch, C.: Ok Google: using virtual assistants for data collection in psychological and behavioral research. Behav. Res. Methods, 1-13 (2021).https://doi.org/10.3758/s13428-021-01629-y
19. Wei, J., et al.: Understanding How to Administer Voice Surveys through Smart Speakers. Proc. ACM Hum. Comput. Interact. **6**(CSCW2), 1–32 (2022)
20. Maharjan, R., Rohani, D.A., Bækgaard, P., Bardram, J., Doherty, K.: Can we talk? design implications for the questionnaire-driven self-report of health and wellbeing via conversational agent. In: Proceedings of the 3rd Conference on Conversational User Interfaces, pp. 1–11 (2021)
21. Laugwitz, B., Held, T., Schrepp, M.: Construction and evaluation of a user experience questionnaire. In: Holzinger, A. (ed.) USAB 2008. LNCS, vol. 5298, pp. 63–76. Springer, Heidelberg (2008). https://doi.org/10.1007/978-3-540-89350-9_6
22. Rauschenberger, M., Olschner, S., Cota, M. P., Schrepp, M., Thomaschewski, J.: Measurement of user experience: a spanish language version of the user experience questionnaire (UEQ). In: The 7th Iberian Conference on Information Systems and Technologies (CISTI 2012) June 2012, pp. 1–6. IEEE (2012)
23. VoiceFlow homepage. https://www.voiceflow.com. Accessed 1 May 2023
24. Mata-Serrano, J.C., Díaz-Oreiro, I., López, G., Guerrero, L.A.: Comparing Written and Voice Captured Responses of the User Experience Questionnaire (UEQ). In: Rocha, Á., Ferrás, C., Méndez Porras, A., Jimenez Delgado, E. (eds.) ICITS 2022. LNNS, vol. 414, pp. 519–529. Springer, Cham (2022). https://doi.org/10.1007/978-3-030-96293-7_43
25. Díaz-Oreiro, I., López, G., Quesada, L., Guerrero, L.A.: Conversational design patterns for a UX evaluation instrument implemented by voice. In: Rocha, Á., Ferrás, C., Méndez Porras, A., Jimenez Delgado, E. (eds.) ICITS 2022. LNNS, vol. 414, pp. 530–540. Springer, Cham (2022). https://doi.org/10.1007/978-3-030-96293-7_44
26. Schrepp, M., Thomaschewski, J.: Design and validation of a framework for the creation of user experience questionnaires. Int. J. Interact. Multimedia Artif. Intell. **5**, 7–88 (2019)
27. Klein, A., Hinderks, A., Schrepp, M., Thomaschewski, J.: Construction of UEQ+ scales for voice quality (2020)
28. Hassenzahl, M., Burmester, M., Koller, F.: AttrakDiff: Ein Fragebogen zur Messung wahrgenommener hedonischer und pragmatischer Qualität. In: Mensch Computer. 2003: Interaktion in Bewegung, pp. 187–196 (2003)
29. Minge, M., Riedel, L. meCUE-Ein modularer Fragebogen zur Erfassung des Nutzungserlebens. In: Mensch & Computer, pp. 89–98 (2013)

A Collaborative Environment for Co-delivering Citizen Science Campaigns

Diego López-de-Ipiña[1]([✉]) [iD], Daniel Andrés Silva[1] [iD], Rubén Sánchez[1] [iD], Elena Not[2] [iD], Chiara Leonardi[2] [iD], Maite Puerta[1] [iD], Oihane Gómez-Carmona[1] [iD], Diego Casado-Mansilla[1] [iD], and Felipe Vergara[1] [iD]

[1] DeustoTech, University of Deusto, Bilbao, Spain
{dipina,d.silva,ruben.sanchez,mpuerta004,oihane.gomezc,dcasado,
felipe.vergara}@deusto.es
[2] Fondazione Bruno Kessler, Trento, Italy
{not,cleonardi}@fbk.eu

Abstract. Campaigns in Citizen Science (CS) experiments are co-production processes where a wide range of diverse stakeholders must be coordinated to achieve a common societal or environmental purpose, i.e., generate evidence for the validation of a hypothesis. This work describes a Collaborative Environment (CE) to aid the co-design and co-delivery of CS campaigns.

Keywords: Citizen Science · co-production · Collaborative Environment

1 Introduction

Citizen Science has risen as a potent method to involve and enable people of various origins in scientific exploration. By incorporating volunteers in gathering data, examining it, and drawing conclusions, Citizen Science initiatives tap into the combined strength of the community to produce meaningful findings and help tackle intricate environmental and societal issues. Nonetheless, overseeing such projects can present considerable organizational and coordination hurdles, necessitating creative strategies to boost teamwork and ensure successful campaign execution.

The EU's SOCIO-BEE project [1] seeks to develop, implement, and validate an advanced Citizen Science (CS) platform. This platform coordinates citizens equipped with wearable devices (refer to Fig. 6) for monitoring air quality (AQ) and simultaneously aids local policymakers and activist groups. The project's goal is to establish sustainable, scalable, and repeatable CS experiments that are collaboratively designed. As a result, SOCIO-BEE necessitates an environment for co-producing CS campaigns to facilitate their smooth execution. Within this framework, users can choose from various adaptable campaign templates, all of which utilize the tools from the SOCIO-BEE CS platform. These tools are designed to drive engagement initiatives and campaigns, empowering communities to actively participate in decisions aimed at reducing urban pollution.

J. Bravo and G. Urzáiz (Eds.): UCAmI 2023, LNNS 835, pp. 322–333, 2023.
https://doi.org/10.1007/978-3-031-48306-6_33

Co-production [2], also known as co-creation, is a joint effort between *service providers* (like government bodies or non-profit entities) and *service users* (such as citizens, patients, or clients) to shape, provide, and assess public services. This approach acknowledges the invaluable insights and expertise service users bring, enhancing the quality and efficiency of public services [3]. Within the co-production process, *co-design* and *co-delivery* stand as two separate stages, each with specific goals. Co-design involves collaboratively shaping the service, whereas co-delivery focuses on jointly implementing and offering the service that was co-designed [2].

The INTERLINK H2020 project [4] has explored ways to address the challenges faced by administrations when attempting to share and reuse services with private entities, including citizens. To this end, it introduced a tailored digital Collaborative Environment (CE) designed to facilitate the co-production of public services. This environment integrates a conceptual structure encompassing various action stages, as detailed in [5]. Consequently, INTERLINK has established a four-step co-production process for collaborators: a) the *co-design* phase, which is divided into two parts: *engagement* (forming a comprehensive stakeholder network) and *design* (shaping the service); and b) the *co-delivery* phase, which includes the *build* (implementation, evaluation, and joint delivery of the service) and *sustain* (long-term service management and utilization) stages.

Notably, this work explores the symbiosis of SOCIO-BEE's CS and INTERLINK's co-production approaches for better engagement and involvement of citizens in collaboration processes, mediated by INTERLINK's CE, targeted towards addressing, more effectively, societal, and environmental challenges, through the CS resources made available by SOCIO-BEE. Hence, this paper presents an examination of a CE adapted to support the co-delivery of CS campaigns. By elucidating its features, benefits, and implications, we seek to contribute to the broader understanding of the role of technology in advancing Citizen Science, fostering collaboration, and empowering communities to actively participate in scientific research. The structure of this paper is as follows. Section 2 deals with the related work. Section 3 describes INTERLINK's CE. Section 4 justifies why INTERLINK's CE suits perfectly the community orchestration and progress tracking needs of SOCIO-BEE. Section 5 concludes the paper and draws some further work reflections.

2 Related Work

Co-design and collaboration tools [6] are important enablers of co-production processes, as they help service providers and service users to work together in the design, delivery, and evaluation of public services, e.g., in the context of this work to collaboratively manage, execute and exploit CS campaigns. Besides, digital tools can facilitate new connections within the community, establishing relationships not possible before by overcoming problems of geographical dispersion of users, and empowering individuals by facilitating the sharing of sovereignty and responsibilities when it comes to service co-design [7].

Collaborative platforms enhance co-design and co-delivery by offering communication tools such as messaging systems, discussion boards, and video chats. These tools enable stakeholders to converse, develop a mutual understanding of issues, set objectives, and address disagreements. Studies indicate that successful digital support for collaborative endeavors hinges on factors like coordination, sustained engagement, open data accessibility, and information sharing [8]. Popular tools in this domain include Miro [9] – a digital whiteboard for joint brainstorming and diagram creation; Trello [10] – a web-based tool for overseeing and monitoring co-production project milestones; and Notion.so [11] – a platform for online collaboration, document sharing, and task coordination between service providers and users. However, while these tools offer adaptability and user autonomy, they often lack structured guidance. There is a gap in providing a comprehensive blueprint for the entire co-production journey, tailored to the specific nature of the service being co-delivered.

Combining co-production with the principles of a community of practice can significantly enhance the effectiveness of Citizen Science efforts [14]. A "community of practice" [13] represents a group of individuals or organizations united in their pursuit of positive change within their community or society, tackling initiatives like social justice, environmental sustainability, or economic development. The challenges in building and maintaining such communities, as seen in the context of Citizen Science, can be substantial. For instance, a community of practice centered on air quality may comprise experts from diverse fields like atmospheric science, public health, urban planning, and community organizing. Through collaborative efforts, these individuals can share their knowledge and insights, leading to a more comprehensive understanding of local air quality issues. Activities may involve data collection, analysis, identifying pollution sources, and devising strategies for addressing the problem.

However, overseeing communities of practice engaged in CS initiatives and experiments can be intricate, requiring digital tools to adeptly direct participants. Broadly speaking, digital collaborative spaces offer a foundation for the efficient organization, implementation, and utilization of outcomes from Citizen Science campaigns [12]. These spaces promote community involvement, data exchange, cross-disciplinary teamwork, collective discussions, reflections, and the broadening of scientific access. By tapping into these platforms, Citizen Science endeavors can amplify collective contributions, significantly impacting both scientific research and societal objectives. Therefore, this study aims to demonstrate how a standard co-production environment can be transformed into a potent instrument for managing change-driven communities in CS, offering structure, direction, traceability, and sharing functionalities to their tasks.

3 INTERLINK'S Collaborative Environment

The Collaborative Environments (CE) within INTERLINK is tailored to bolster the project's co-production approach, making it easier to adopt for the creation of new collaborative processes and services. This environment provides several key features: a) *management of co-producer teams and processes*; b) *a structured guide outlining the co-production steps*; c) *suggestions for knowledge and software tools* (termed INTER-LINKERs) that best fit the domain of the selected co-production activity; d) options to *select, initiate, assign, claim, and record the use of INTERLINKERs* (for instance, a CS campaign design template); and e) an *INTERLINKER catalog* that encourages the reuse of publicly accessible co-production tools and methodologies.

An assortment of co-production INTERLINKERs has been created to provide functionality useful in many co-production contexts, e.g.: a) *interlinker-googledrive* to deal with office documents, b) *interlinker-survey* to design and host answers for surveys; or c) *webpage augmenter* to annotate web ages. All those software enablers leverage on a common API to ease integration, previously reported in [5]. On the other hand, several knowledge INTERLINKERs have been defined, e.g., Stakeholder Mapping Canvas, Use Case Scenarios or Business Model Canvas templates, created declaratively by means of a JSON schema. Likewise, co-production schemas can be declared in JSON which are tuned to the specifics of a co-production process type, e.g., a Hackathon schema or a CS campaign execution schema.

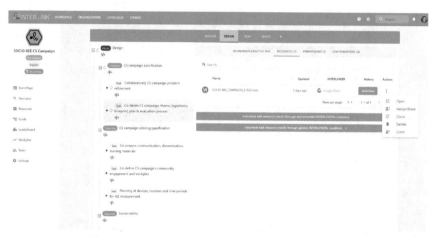

Fig. 1. Co-production process where DESIGN phase objectives and tasks are shown, together with resource instantiate for selected task. Notice actions that can be performed over a resource.

Observe in Fig. 1 how the generic BUILD sub-phase (at the top) provided by default by INTERLINK's co-production processes is replaced in the custom Citizen Science co-production tree/schema (below) by a RUN sub-phase, with different composing objectives and tasks. Importantly, the CE enables co-production processes to be customized (adding, modifying, and removing phases, objectives, and phases to a process) by clicking on the " +" sign present at the right-hand side of the last phase name. These features set our CE apart from other collaborative tools. Another noteworthy aspect is its capacity to be extended with additional tools, thereby enabling the incorporation of a given problem domain features, e.g., Citizen Science. By following a microservice approach and integrating through APIs, third-party open-source software components can be easily integrated into the platform. This flexibility empowers users to create new types of enablers and enhance the overall functionality. For instance, in our work a micro-volunteering engine [13] for crowdsourcing of data has been provided to aid in the CS campaigns co-execution.

Fig. 2. Success stories from which new processes can be cloned.

Fig. 3. Contributions pane of task view showing activities carried out by a user over the task and how that person can claim the work carried out by him.

3.1 Tackling Co-delivery: Replication and Sustainability

After reviewing the outcomes of INTERLINK's first iteration and post-piloting reflections, the importance of enhancing replicability became evident. This means there is a need to share successful outcomes and utilize them as a foundation for new processes. Additionally, the emphasis on bolstering sustainability was highlighted. For INTER-LINK to be sustainable, it is crucial to showcase advancements in co-production processes and recognize the efforts of team members; without this, sustaining long-term collaboration becomes challenging. To address these insights, several features were introduced to the Collaborative Environment to foster *replicability*:

- *Process cloning* – so that a new co-production process can be created based on an existing one. Only valid for own processes, it is useful for teams that want to reuse processes where they participated previously.
- *Cloning from success story* – the CE includes the ability to publish a co-production process in the form of a success story (see Fig. 2), allowing anyone to clone it. This is interesting to allow third parties to leverage from the co-production efforts of other teams for envisaged artefacts/results of similar nature.

To bolster *sustainability*, the CE has been combined with a Gamification Engine [14]. The objective is to track and recognize the contributions of each team member throughout the co-production process and its individual tasks by offering a points board. To achieve this integration, the subsequent features have been incorporated:

- *Activity timeline* – the Collaborative Environment has been improved to be able to review what the team has done over a process or what everyone has contributed with for the whole process or a single task (see Fig. 3's activities timeline popup).
- *Contribution Assignment, Claim and Validation* – team members can now be assigned work in resources and as result claim what they have done over the resources of a tasks (see Fig. 1's right hand side popup pane showing options "Assign/Share" and "Claim"). On the other hand, process admins may validate the team members contributions (see Fig. 3's Contribution pane's table showing contribution level of each participant in a task and "Award points" button). Thus, the CE calculates the contribution quota of a user in each task based on the configured complexity of the task and the level of contribution of the user in that task versus total contributions in task. Once a given task contributions are validated, it is closed, and point calculations realized. A central leader board showcases (see Fig. 4) the contributions, valued as points for the different team members.

4 A Collaborative Environment for Citizen Science

As previously indicated, holistic management and support of Citizen Science campaigns can be complex and present big challenges. There is a clear need for digital environments and a suitable set of accessible and usable tools and knowledge resources/enablers to streamline and democratize co-production processes.

4.1 SOCIO-BEE's Approach to Citizen Science

The SOCIO-BEE's approach to CS is based on the facilitation of structures to increase citizen engagement and awareness of climate change through experimentation and monitoring of the environment. Through an example taken from nature, SOCIO-BEE builds on the metaphor of bee colonies to develop effective behavioral and engagement strategies with a wide range of stakeholders, namely, Queen Bees, Drone Bees, Working Bees, and Bears. It co-creates through Citizen Hives long-lasting solutions against air pollution supported by technologies such as drones or wearables.

Figure 5 illustrates the roles various hive members, representing a community of practice, might assume within SOCIO-BEE. It also highlights the functionalities the

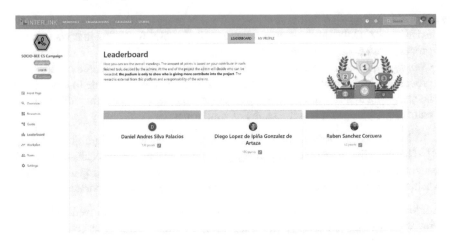

Fig. 4. Leaderboard showing scores of contributors to a SOCIO-BEE's coproduction process.

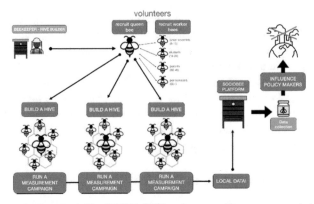

Fig. 5. Bee metaphor employed by SOCIO-BEE and co-creation process carried out by a Hive.

core SOCIO-BEE platform provides them for overseeing and participating in CS campaigns. It is important to note that the entire procedure undertaken by hive members is a collaborative co-production process involving multiple stakeholders, where:

- *Beekeeper*. Responsible to set up a new hive (team of bees) and the associated co-production process where they collaborate, i.e., CS campaign. S/he recruits members for that hive. S/he sets up and configures a new co-production process to manage a CS campaign, using the "SOCIO-BEE's CS co-production" schema, depicted before in Fig. 1, which comprises the phases, objectives and steps recommended for effective and sustainable execution of CS Campaigns.

- *Queen Bee*. S/he firstly configures CS campaigns (area, measurements' type, frequency and period, research questions to address and so on). For that it makes use of a core enabler part of the SOCIO-BEE toolkit for onboarding and engagement, named "CS Campaign template". Configuring a CS campaign entails not only to decide WHAT will be done, but also HOW, by WHOM, WHERE and WHEN. Besides for SOCIO-BEE, distribution of wearable devices (see Fig. 6) and installation of its mobile app are needed. To answer these questions, apart from the campaign template, other important enablers, namely "Campaign work program" or "community building" templates, may be co-edited by the Queen Bee in cooperation with some associate bees. Then, the Queen Bee is ready to launch, manage, and monitor the execution and evaluation of the campaign and measurements taken by Worker Bees, based on a set of indicators that were defined during the campaign co-design. During campaign execution, s/he also takes part in the interim publication of results for the hive's stakeholders by reflecting in workshops with other hive stakeholders, namely, bear, beekeeper, and hive members. Eventually, s/he closes the campaign and pushes results' wide dissemination and communication.

- *Worker Bees*: they are notified about new campaigns where they can partake through the SOCIO-BEE's mobile app or by email when they are made part of a campaign co-production process' team by the Queen Bee. Through SOCIO-BEE's mobile app, whenever they are available for pollination, i.e. air quality measurements' gathering, they get recommended possible cells to pollinize by means of SOCIO-BEE's and AQ sensing wearable and MVE [13]. They recurrently gather new measurements during a campaign's timespan. They regularly check how the pollination in the campaign is ongoing and access heat maps informing about air quality evolution in space and time. They are informed about end of campaigns and provide feedback of their experience through the mobile app. They take part in post-pilot reflection workshops.

- *Drone Bees*: they receive high-level information about the AQ CS campaigns arranged by a Beekeeper, which often may have a public or private organization (*Bear*) promoting and sponsoring it since it allows the Bear to gain information for better decision making. Drone Bees spread the results of campaigns published in easily graspable manner to enhance understanding of the effects of air pollution and remediation actions. They are invited to attend to reflection workshops to have a say regarding possible actions to be taken by the public administration.

4.2 Challenges Managing, Executing, and Exploiting CS Campaigns

The challenges associated to operating CS campaigns can be summarized in the following categories: organization, execution, and exploitation of campaigns. Regarding *Organization of CS Campaigns*, the most evident challenges are: a) *Campaign Design*: Designing a CS campaign involves defining clear objectives, selecting appropriate research questions, and determining the scope of the campaign; b) *Volunteer Recruitment*: attracting a diverse and motivated group of volunteers requires identifying target audiences, reaching out to them effectively, and addressing any barriers to participation; c) *Training and Support*: providing adequate training and support to volunteers is crucial for ensuring data quality and consistency; or d) *Ethics and Protocols*: establishing ethical guidelines and protocols for data collection.

Regarding *Execution of CS Campaigns*, the most evident challenges are: a) *Data Quality*: maintaining data quality implies ensuring that volunteers follow standardized protocols, validating data, and implementing quality control measures; b) *Participant Engagement*: keeping participants engaged and motivated requires providing regular updates and acknowledgements, fostering a sense of community, and offering opportunities for learning and skills development; c) *Data Management*: handling large volumes

Fig. 6. SOCIO-BEE's Air Quality (AQ) measuring wearables.

of data contributed by numerous participants, ensuring data accessibility and security, and implementing appropriate data storage and analysis methods; or d) *Communication and Collaboration*: facilitating effective communication between organizers, scientists, and volunteers is essential through clear communication channels, responsive feedback mechanisms, and collaborative platforms.

Regarding *Exploitation of CS Campaigns*, the most evident challenges are: a) *Data Analysis and Interpretation*: analyzing and interpreting the data collected from CS campaigns, dealing with large and diverse datasets requires robust analytical techniques and expertise; b) *Integration with Scientific Research*: integrating CS data with existing scientific research implies ensuring compatibility with established methodologies, and research practices to maximize the value of the collected data: c) *Policy and Decision-Making Impact*: translating CS findings into actionable policies and influencing decision-making processes, communicating results effectively to policymakers and stakeholders, and advocating for their consideration; or d) *Long-Term Sustainability*: sustaining Citizen Science initiatives beyond the project lifespan calls for securing long-term funding, developing strategies for project continuation, and building partnerships with relevant organizations and institutions.

4.3 Managing Citizen Science Campaigns through Collaborative Environment

In SOCIO-BEE, CS campaigns co-production, requires careful planning, collaboration between hive members and linked stakeholders, capacity building, effective communication strategies, and the use of the proposed app with its underlying MVE and wearable device. Overcoming these challenges can help ensure the success and impact of Citizen Science campaigns organized by SOCIO-BEE. Hence, to better govern and orchestrate the CS communities of change that will make use of its tools and resources, we decided to complement SOCIO-BEE's platform with INTERLINK's CE. Hence, a two-fold purpose has been met:

- Address lack of governance and support mechanisms to underlying co-production process which currently is not possible through SOCIO-BEE's own tools.
- Demonstrate that a Collaborative Environment can be used to co-design and co-deliver any artefact which may result of common interest and where cooperation among distinct stakeholders is mandatory, in this case a CS campaign.

Consequently, the adoption and adaptation of INTERLINK's CE to a Citizen Science (CS) domain has entailed the following steps:

1. *Generation of co-production process schema* which describes the main phases a CS campaign should go through to be fully realized. In Fig. 1, you can see the phases (ENGAGE, DESIGN, RUN, SHARE) that compose a default CS process in SOCIO-BEE, objectives for the "DESIGN" phase "CS campaign specification", "CS campaign piloting specification" and "Sustainability" and some of the tasks within those objectives. Besides, notice that for the selected task named "Co-ideate CS campaign: theme, hypothesis, blueprint, plan & evaluation process", the Queen Bee associated to the process, has leveraged a campaign template and co-edited for a campaign in Zaragoza, named "SOCIO-BEE_CAMPAIGN_2-ZGZ", in cooperation with fellow bees.

2. *Preparation of a co-production process representing a campaign.* Figure 1 represents an example of a CS campaign making use of the recommended co-production schema. Notice that pilots may customize such schema and adapt it to the specific needs of each pilot. Figure 7 shows some of the resources that have been instantiated thanks to the guidance and recommendation services offered by INTERLINK for a custom-built process for the Zaragoza pilot of SOCIO-BEE. Enablers to aid resolving the challenges managing CS campaigns have been added to the catalogue as SOCIO-BEE project progresses. For instance, apart from the mentioned Campaign specification template, a *Campaign blueprints* has been defined which helps choosing the strategy and data capture and analysis methods under which the campaign will be executed. Likewise, the pilot work programme and community building templates allow to clearly specify the timeline of a CS campaign together with the activities to be arranged.

3. *Execution of the co-production process by leveraging INTERLINKERs.* Once the preparations for a Hive to launch the campaign are completed, worker bees are mobilized by setting up measurement periods and regions resulting from the design of the campaign. In this process, worker bees use the wearable appointed to them and the SOCIO-BEE mobile app to perform the most suitable AQ measurements based on their time availability, current location and the level of measurements achieved in the cells of a given campaign's regions, as suggested by the MVE. SOCIO-BEE aims to achieve evenly distributed number of measurements per campaign cell part of the campaign's region and measurement slot comprising the campaign period. Face to face meetings is arranged regularly so that AQ and data analysis experts in cooperation with the Hive reflect about the measurements gathered and think on possible actions to reduce air pollution. The communication materials, datasets gathered, analytics performed, and the conclusions reached are registered as new resources within the tasks of the RUN phase shown in Fig. 1. Notably, a campaign does not conclude with the simple gathering of data and internal reflection by the Hive. For that, campaign participants take part in the "SHARE" phase's activities generating simple to grasp contents targeted to the public and curated reports targeted to public administration representatives.

Fig. 7. Fully customized co-production process for ZGZ pilot.

5   Conclusion and Future Work

This work has illustrated how a collaborative environment devised to support generic co-production processes has been applied to the Citizen Science domain. One of the biggest barriers for a more widespread adoption of CS practices is the complexity associated to handling the community and the work carried out by it, in the long run. Hence, a tool such as the CE supporting the community onboarding, guidance over the whole lifetime of a CS campaigns, acknowledging for collaborations, and tracking all its achievements (co-production resources) is very relevant.

A thorough assessment on how INTERLINK's CE tool can be useful for CS is currently being applied within SOCIO-BEE. New enablers to tackle different aspects and tasks of CS campaigns are being incorporated. Besides, the effectiveness of the in-place mechanisms for sharing, assigning, recognizing, and rewarding the outcomes in the form of resources contributed by co-producer teams or hives are also being analyzed in SOCIO-BEE's pilot iteration, concluding in September 2023. Third party tools to foster the joint dialogue and reflection over campaigns' findings among Hive's bees have also been introduced and analyzed, e.g. Loomio [15].

Acknowledgments. This work has been sponsored by INTERLINK H2020 project with Grant ID 959201. SOCIO-BEE H2020 project with Grant ID 101037648 and DEUSTEK5 (IT1582-22) Basque University system's A grade research team grant.

References

1. Wearables and droneS fOr CIty Socio-Environmental Observations and BEhavioral ChangE SOCIO-BEE Project Fact Sheet H2020 CORDIS European Commission. https://cordis.eur opa.eu/project/id/101037648/es. Accessed 6 March 2023
2. Brandsen, T., Honingh, M.: Definitions of Co-Production and Co-Creation. In: Co-Production and Co-Creation. Routledge, pp. 9–17 (2018)

3. Guarino, N.: Services as activities: towards a unified definition for (Public) services. In: 2017 IEEE 21st International Enterprise Distributed Object Computing Workshop (EDOCW), pp. 102–105 (2017). https://doi.org/10.1109/EDOCW.2017.25
4. interlink-project – Innovating goverNment and ciTizen co-dEliveRy for the digitaL sINgle marKet. https://interlink-project.eu/. Accessed 2023/03/10
5. López-De-Ipiña, D. et al.: A Collaborative Environment to Boost Co-Production of Sustainable Public Services. In: 2022 7th International Conference on Smart and Sustainable Technologies (SpliTech), pp. 1–6 (2022). https://doi.org/10.23919/SpliTech55088.2022.9854297
6. Cosgrave, C., et al.: From co-design to co-production: Approaches, enablers, and con-straints in developing a public health, capacity-building solution. Aust. J. Rural Health **30**, 738–746 (2022). https://doi.org/10.1111/ajr.12930
7. Paletti, A.: Co-production Through ICT in the Public Sector: When Citizens Reframe the Production of Public Services. In: Caporarello, L., Cesaroni, F., Giesecke, R., Missikoff, M. (eds.) Digitally Supported Innovation. LNISO, vol. 18, pp. 141–152. Springer, Cham (2016). https://doi.org/10.1007/978-3-319-40265-9_10
8. Gil-Garcia, J.R., Sayogo, D.S.: Government inter-organizational information sharing initiatives: understanding the main determinants of success. Gov. Inf. Q. **33**, 572–582 (2016). https://doi.org/10.1016/j.giq.2016.01.006
9. Miro Online Whiteboard for Visual Collaboration. https://miro.com/app/dashboard/. Accessed 14 March 2023
10. Home Trello. https://trello.com/. Accessed 14 March 2023
11. Notion – One workspace. Every team. https://www.notion.so. Accessed 31 March 2022
12. Luther, K., Counts, S., Stecher, K.B., Hoff, A., Johns, P.: Pathfinder: an online collaboration environment for citizen scientists. In: Proceedings of the SIGCHI Conference on Human Factors in Computing Systems, pp. 239–248. Association for Computing Machinery, New York, NY, USA (2009). https://doi.org/10.1145/1518701.1518741
13. Puerta Beldarrain, M.: A Micro-volunteering Engine to drive crowd-measuring of Air Quality in Citizen Science. Presented at the Splitech. In: 2023 8th International Conference on Smart and Sus-tainabile Technologies, Splitech, pp. 1–6, June 20 (2023)
14. Kazhamiakin, R. et al.: Using gamification to incentivize sustainable urban mobility. In: 2015 IEEE First International Smart Cities Conference (ISC2), pp. 1–6 (2015). https://doi.org/10.1109/ISC2.2015.7366196
15. Loomio: Involve people in decisions. https://www.loomio.com/. Accessed 24 July 2023

Author Index

J. Bravo and G. Urzáiz (Eds.): UCAmI 2023, LNNS 835, pp. 335–337, 2023.
https://doi.org/10.1007/978-3-031-48306-6

Printed in the United States
by Baker & Taylor Publisher Services